Springer Tracts in Modern Physics
Volume 188

Springer
Berlin
Heidelberg
New York
Hong Kong
London
Milan
Paris
Tokyo

Springer Tracts in Modern Physics

Springer Tracts in Modern Physics provides comprehensive and critical reviews of topics of current interest in physics. The following fields are emphasized: elementary particle physics, solid-state physics, complex systems, and fundamental astrophysics.
Suitable reviews of other fields can also be accepted. The editors encourage prospective authors to correspond with them in advance of submitting an article. For reviews of topics belonging to the above mentioned fields, they should address the responsible editor, otherwise the managing editor.
See also www.springer.de

Cornelia Denz   Michael Schwab
Carsten Weilnau

Transverse-Pattern Formation in Photorefractive Optics

With 143 Figures

Springer

Prof. Dr. Cornelia Denz
Dr. Michael Schwab
Dr. Carsten Weilnau

Westfälische Wilhelms-Universität Münster
Institut für Angewandte Physik (Institute for Applied Physics)
Corrensstr. 2–4
48149 Münster, Germany
E-mail: denz@uni-muenster.de

Library of Congress Cataloging-in-Publication Data

Denz, Cornelia.
Transverse-pattern formation in photorefractive optics/Cornelia Denz, Michael Schwab, Carsten Weilnau.
p.cm. - - (Springer tracts in modern physics, ISSN 0081-3869 ; v. 188)
Includes biblographical references and index.
ISBN 3-540-02109-4 (acid-free paper)
1. Nonlinear optics. 2. Photorefractive materials. 3. Pattern formation (Physical sciences) I. Schwab, Michael, 1972-
II. Weilau, Carsten, 1973- III. Title. IV. Springer tracts in modern physics ; 188.

QC1.S797 vol. 188
[QC46.2]
535'.2- -dc21 2003050354

Physics and Astronomy Classification Scheme (PACS): 42.65.-k: 42.65.Jx, 42.65.Sf, 42.65.Tg, 42.65.Wi, 42.65.Hw, 42.65.Pc

ISSN print edition: 0081-3869
ISSN electronic edition: 1615-0430
ISBN 3-540-02109-4 Springer-Verlag Berlin Heidelberg New York

Springer-Verlag Berlin Heidelberg New York
a member of BertelsmannSpringer Science+Business Media GmbH

www.springer.de

Typesetting: Camera-ready copy from the author using a Springer LaTeX macro package
Production: LE-TeX Jelonek, Schmidt & Vöckler GbR, Leipzig
Cover concept: eStudio Calamar Steinen
Cover production: *design & production* GmbH, Heidelberg

Printed on acid-free paper SPIN: 10867080 56/3141/YL 5 4 3 2 1 0

Preface

Systems under constant nonequilibrium external conditions may exhibit macroscopic spatial structures in steady state or, more generally, spatio-temporal structures – a phenomenon that is observed in a large number of different disciplines, such as biology, hydrodynamics, chemistry, solid-state physics and optics. These structures appear in systems as completely different as chemical reactions in shallow layers, thermal convection in fluid layers, periodically shaken layers of sand, structures on animal coats, growth of slime mold colonies, or in various materials interacting with laser light.

The wealth of different spatial structures is intriguing, and is motivating ongoing research aimed at gaining deeper insight into the stages of pattern formation or, in a more general sense, the creation of order out of a uniform, possibly chaotic state. Despite the richness of pattern creation, different systems display geometric patterns or structures that appear strikingly similar. It is therefore no surprise that universal behaviors of these complex systems have been determined and discussed. This fact has led to the wide interdisciplinarity of a field that is now referred to as nonlinear science or the science of complexity, and is also well known under the names of dissipative structures [1], synergetics [2], and self-organization [3]. Early research interests focused on questions of fundamental significance. Current activities involve the challenges of active application of these basic results, for control of turbulence in fluids or to understand the origin of cardiac arrhythmia and brain diseases, to name only a few examples.

What makes optics special compared to these disciplines? The general mechanism which gives rise to pattern formation is the combination of a nonlinearity with a coupling between different spatial entities. In other words, a nonlinear, enhancing effect and a redistributing effect such as diffusion compete. In the case of optical systems, the interplay between a nonlinear medium and light is sustained by propagation, which results in diffraction providing the transverse coupling mechanism that opposes the nonlinearity. Diffraction plays the same role as diffusion in other nonlinear systems, that of providing the competing mechanism to the nonlinearity. Therefore, it seems quite natural that transverse dynamic or static patterns form in nonlinear optics. Because optical systems may supply a variety of different nonlinearities, from Kerr-like to highly complex nonlinear dependences, various different kinds of patterns can be expected in optical systems. Moreover, the inherent parallelism of optics, the special feature of information processing without transport of mass, and the unique possibility of active control over the system parameters render nonlinear optical systems excellent objects of study for the investigation and application of nonlinear effects.

While in the early days of laser physics coherence and longitudinal effects attracted a great deal of attention, it is now clear that transverse effects are at least as important as the coherent nature of the light field. Though transverse effects had been known since the early studies on laser beam profiles, systematic examinations began rather late. This is mainly due to the fact that it was for a long time not "natural" to realize optical systems with a large transverse extent. Instead, in order to obtain a powerful and stable laser output in the simple Gaussian TEM_{00} transverse mode, which is desired for most applications, apertures or other means to reduce the transverse extent of the system were introduced into the laser cavity. As soon as these limitations in the direction transverse to the beam propagation direction were removed, the system generated patterns of varying degrees of complexity spontaneously. Because these phenomena were, in the case of lasers, mostly considered undesirable or difficult to understand or control, they were avoided even at the cost of lower energy output or a limited understanding of the physics of transverse effects in lasers.

Only in the last few years, a considerable interest has emerged in this field of pattern formation in the structure of the electromagnetic field in the plane orthogonal to the direction of propagation. This field is now often called *Transverse Nonlinear Optics* for short. Investigations of systems such as high-power lasers and surface-emitting lasers have favored the consideration of transverse nonlinear optics and have led to the discovery of a rich variety of optical pattern formation phenomena. The diversity of the contributions to this research field is as remarkable as the similarities in the experimental and theoretical results among the various interaction schemes and nonlinearities. It is now clear that spatial and spatio-temporal effects in the transverse plane are similar to phenomena of pattern formation and turbulence that are familiar in other fields, in particular hydrodynamics. The research devoted to hydrodynamic pattern-forming systems has provided the necessary language for the description and classification of the vast number of different systems and configurations. However, it is sloppy thinking to interpret optics as "dry hydrodynamics" [4].

Optics presents several outstanding features that are interesting and stimulating and can be expected to advance the field of pattern formation in general. First, optical systems are potentially very fast and have a large frequency bandwidth. Therefore, they lend themselves naturally to applications, for instance in telecommunications and information technology. One relevant example of the useful application of optical structures has already been provided by soliton transmission in optical fibers. Second, investigations in transverse nonlinear optics offer, in principle, the possibility of new approaches to parallel optical information processing, by encoding the information in the transverse structure of the electromagnetic field. The speed of processing and the ease of access to the relevant control parameters of a system are the most advantageous features of nonlinear optics for these applications. The third special feature is the tendency to display interesting quantum effects in transverse optical structures even at room temperature. Recent years have witnessed the start of investigations on quantum effects in transverse optical structures. This research is relevant from the viewpoints of fundamental quantum mechanisms, of quantum noise reduction (squeezing), and of applications, for example to noiseless transmission of images.

Although lasers are prototypes of pattern-forming optical systems, they are probably not ideal candidates for studies of pattern formation phenomena at present. In most practical lasers the design imposes severe limitations on the spatial patterns which can evolve, with the consequence that only a few linear modes of the laser are sufficient to describe the observations in most cases, i.e. the patterns are completely dominated by boundary effects. Moreover, the small space–bandwidth product and the low sensitivity of lasers do not allow one to access spatio-temporal phenomena in their complete variety. Finally, modeling of transverse effects in lasers is extremely difficult in the case of a high Fresnel number, which corresponds to the situation of a large aspect ratio in hydrodynamics, making a comparison with theoretical predictions quite difficult. Therefore, many effects of pattern formation, especially pattern dynamics due to competition and drift instabilities, as well as interaction effects between different patterns, have not been realized up to now in experiments or have not been analyzed and explained in a satisfactory way.

Although there are several nonlinear optical systems that can be made to exhibit spatial and spatio-temporal effects in a much simpler way than in the case of laser systems – such as liquid crystal cells, light valves and atomic-vapor systems – they mostly cover only a limited area of the possible pattern formation scenarios. In contrast, photorefractive nonlinear materials have some outstanding features that make them well suited as model materials for obtaining a complete overview of all possible spatial and spatio-temporal phenomena, beginning with the rather simple effects of soliton beam formation and reaching up to complex, uncorrelated spatio-temporal structures that may be interpreted as turbulent or chaotic states. The photorefractive nonlinearity is also advantageous for these investigations from an experimental point of view. First, the photorefractive nonlinearity is saturable. Therefore, the nonlinear effects do not depend on the intensity itself, but on the ratio of the beam intensity to that of a background illumination or a pumping beam, allowing one to realize all possible transverse-structure phenomena at very low laser intensities. Second, the saturation behavior includes the stage of a quasi-linear dependence of the refractive index on the intensity ($\Delta n \propto |E|^2$), which corresponds to the behavior of the ideal Kerr nonlinearity. Therefore, many simple theoretical models based on Kerr nonlinearities can be used to explain the experimental results obtained with a photorefractive nonlinearity.

Aim and Organization of this Book

The aim of the present book is to give an insight into the vast variety of transverse effects in nonlinear optical materials by using photorefractive materials as the nonlinearity of choice. Even a single laser beam entering such a nonlinear medium can experience complex transverse structuring due to modulation instabilities, which may lead to beam fanning or beam filamentation. In other cases, the interaction of two beams may lead to spatially regular patterns, as in the case of the single-feedback configuration, or to complex spatial patterns, as in the case of a nonlinear resonant system. Because the photorefractive effect may provide spatial structuring in the material itself due to the creation of refractive-index patterns, as well as energy exchange and therefore gain, it can be considered a slow equivalent to active materials that may exhibit laser-like behavior in oscillating systems.

These topics have led to the two main parts of this book, which deal with spatial photorefractive solitons and their interaction (Chaps. 4–7), and spontaneous pattern formation with feedback and photorefractive gain (Chaps. 8–11). We focus on the physical stages and conditions for the appearance of ordered patterned structures and describe the theoretical background of these effects. Our challenge is to go beyond the level of observation, classification and understanding of the nonlinear phenomena encountered.

Nonlinear optics allows the unique possibility of combining features of nonlinear effects with aspects of information technology. In this context, we try to understand the formation and interaction of spatial photorefractive solitons in order to exploit them for novel, adaptive concepts of waveguiding in telecommunications. The potential information capacity of a nonlinear optical system is determined by the number of different pattern states.

Photorefractive crystals provide the key to access a wealth of different self-organized patterns. We are especially interested in the key properties of the free-running nonlinear system, combined with different feedback mechanisms. Our special challenge is to go beyond the common stage of pure pattern observation, which is of course a basic prerequisite for further studies. We focus on the active addressing and selection of a desired pattern by suitably structuring the feedback signal. The large number of different spatial patterns or modes available in different systems is an excellent test bench for pattern control and selection methods. Our aim is to suppress unwanted patterns in favor of a desired structure, at the same time rendering the influence of the selection method on the system as small as possible.

In detail, this book is divided into eleven chapters that are organized in such a way as to retain continuity and to provide an insight into the whole area of nonlinear optical pattern formation. The book starts with the simplest configurations that may lead to instabilities and structure formation, and shows how pattern diversity is enhanced by adding feedback to the nonlinear system in various stages of complexity.

Chapter 1, on nonlinear waves and transverse patterns, is an introduction dedicated to a basic description of modulation instabilities in nonlinear systems. This chapter classifies the different situations that may arise in nonlinear systems and provide the physical background necessary to understand the following chapters.

Chapter 2, on light propagation in nonlinear optical media, describes the origin of self-focusing of beams in nonlinear optical media in general. It shows how these focusing effects can be exploited to create optical spatial solitons. Starting with one-dimensional transverse soliton phenomena, we show the general features of soliton formation which lead to a description by the Nonlinear Schrödinger equation. On the basis of these introductory investigations, the existence conditions for two-dimensional spatial optical solitons are considered, which lead to novel solitary structures with exciting interaction features.

The photorefractive nonlinearity is described in Chapter 3, beginning with the origins of the space charge field arising in these materials from their interaction with light. The band transport model equations are used to formulate the basic equations for the photorefractive nonlinearity. The complexity is gradually increased, starting with one beam interacting with the nonlinear material and then taking a second beam into account to show the beam-coupling properties of photorefractive crystals. An important feature of a diffusion-dominated photorefractive material is the possibility of performing an energy exchange between interacting beams in a wave-mixing process. The two-beam coupling properties are thoroughly discussed for the special case of a counterpropagating geometry, which is the configuration of choice for the investigations of pattern formation effects described in chapters 8 to 10. Special care has been taken to introduce the main variables and equations that represent the starting point for subsequent considerations.

In Chapter 4, spatial photorefractive solitons are discussed. This chapter is dedicated to the physics of soliton formation in a photorefractive material, and shows that the nonlinear formation of a space charge field due to the incident light beam can be balanced by screening an externally applied electric field, thus leading to a self-guided, stable beam. From a mathematical as well as a physical point of view, this solitary wave is very similar to spatial solitons in saturable media. The anisotropy of the photorefractive effect due to the application of the external electric field leads to an asymmetry of the refractive index created by the beams that strongly influences the formation of spatial photorefractive solitons. The most peculiar features are elliptically shaped solitary solutions, beam bending and diameter oscillations during soliton formation.

In Chapter 5, the interaction of spatial solitons in saturable photorefractive media, which may lead to novel, surprising scenarios, is treated. We distinguish between coherent and incoherent soliton interaction. For coherent interaction, we show that the relative phase of the interacting solitons is essential to the outcome of the interaction, leading to exciting new effects such as attraction and repulsion, soliton birth, annihilation and phase-dependent energy exchange. For the case of incoherent soliton interaction, the anisotropy of the photorefractive effect leads to a refractive index that displays regions of both attraction and repulsion – a previously unknown and surprising feature of the interaction between incoherent spatial solitons. In the direction of the applied electric field, solitons may attract or repel, depending on their relative separation, whereas they always attract when oriented along the Y-direction. This anomalous interaction gives rise to a wealth of new phenoma that were not expected from the general theory of spatial solitons in saturable media. Among these phenomena, rotation, oscillation and complex spiraling behavior show the complexity of photorefractive soliton interaction.

Chapter 6, waveguiding in photorefractive solitons for soliton-driven photonics uses the investigations of the preceding chapters to exploit the refractive-index modulation that is built up during soliton formation for waveguiding purposes. Owing to the strong wavelength selectivity of the photorefractive effect, it is possible to guide light of different wavelength regions. Thereby, several different scenarios of the interaction of coherent and incoherent solitons can be demonstrated, leading to waveguide couplers, dividers and splitters. Furthermore, have we been able to use the inherent parallelism of optical beam propagation to create two-dimensional lattices of photorefractive solitons that allow simultaneous waveguiding of a large number of information channels. By combining the interaction of photorefractive solitons with this feature, highly parallel and adaptive solitonic crystal structures can be formed, representing the basis of soliton-driven photonic applications.

When a multimode waveguide is induced not by a single, but several optical beams, vector solitons in a photorefractive medium can be constructed, these are the topic of Chapter 7. In this case the participating modes are not guided in a passive way, but interact with the medium and contribute equally to the formation of the required refractive-index profile. This is a substantially different approach from that described in the preceding chapters, since two or more beams copropagate in a joint refractive-index profile generated by themselves. Vector solitons can therefore be considered as molecules of light, and their stability is one of the most crucial questions in complex spatial-soliton optics. We show that combination of higher-order modes with the fundamental Gaussian mode can lead to stabilization of multipole vector solitons, whereas vortex beams, as a general rule, decay into multipole vector solitons. Moreover, we demonstrate the stabilization of multicomponent solitons and soliton clusters.

In Chapter 8, on self-organized pattern formation in single-feedback systems, the complexity of the material–light interaction is enhanced by allowing the system to feed the beam back into the nonlinear material. A single-mirror feedback scheme with a photorefractive nonlinearity constitutes an excellent model system for studying pattern formation effects. An extensive description of the main properties of the free-running system is the topic of this chapter. "Free-running" here refers to the single-mirror feedback experiment with unstructured input and feedback.

In Chapter 9, multiple patterns and complex pattern competition, we present experimental results which reveal parameter regions of simultaneously stable patterns and of pattern competition. The most surprising feature of a photorefractive single-feedback system is the existence of multistability, which provides a wealth of new patterns far beyond the threshold. In order to understand the complex dynamics of pattern interaction, detailed investigations of the uncontrolled system are necessary, and provide an appropriate and essential starting point for the implementation of manipulation and control schemes for active pattern selection.

In Chapter 10, which covers manipulation and control of self-organized patterns by spatio-temporal techniques, we approach the desire of researchers in this field to adjust and stabilize defined pattern states instead of passively observing spontaneous pattern formation. An additional system parameter, a frequency detuning of the pump beams, is introduced and allows us to choose a particular pattern state when two or more patterns are accessible for one given set of parameters. The unique real-time access to the spatial frequency domain that is available in optics is exploited for the realization of a spatial control scheme based on the manipulation of wave numbers in the Fourier plane. Direct filtering in the feedback arm is shown to be an excellent tool for forcing the system to a desired solution. The feedback system and, thereby, its solutions are changed by any intervention in the system, which motivates us to consider an extension of this invasive scheme. We have realized minimally invasive control units, which are based on the condition that the original pattern-forming system remains unchanged. Manipulation and control are provided by an additional bypass in order to encourage rather than force the system to adopt a desired solution.

Chapter 11 describes the outlook for still more complex configurations, in which interaction can occur; this leads to consideration of transverse patterns in active photorefractive oscillators. For small Fresnel numbers, boundary effects become important, and the system behavior is dominated by pure transverse resonator modes and their interaction. We demonstrate that the main difference from conventional laser modes is based on the strong frequency pulling in an unidirectional photorefractive oscillator due to the interaction of the pump beam with the resonator signal. For larger Fresnel numbers, the influence of the boundaries vanishes and material-dominated effects become significant. Here, fine-grained structures appear, with a correlation length that decreases to the scale of the smallest structure in the transverse plane. These completely uncorrelated states can be associated with turbulence.

The work described here is mainly experimental with a strong emphasis on the realization of predicted pattern formation scenarios and on configurations designed to analyze and exploit pattern formation in the direction of possible applications. However, in order to set the results into their proper relation to existing work, important theoretical derivations have been included. Most of these theoretical investigations were performed in collaboration with theoreticians in order to explain surprising experimental results or to verify numerical calculations. Several derivations that are well known but are indispensable for understanding our investigations are also briefly reviewed to provide consistency.

In this broad field, the physical description involves a number of different variables, that have been named in a historical fashion in the different fields. Therefore, in some parts of this book, different variables can obtain the same name, as e.g. the propagation constant β in soliton formation and the photorefractive thermal excitation constant β. In order to account for these, historically grown names, we left this ambiguity in our descriptions. However, within the frame of a chapter the definition of the variables is clear so that no confusion in the meaning of the variables should occur.

Since the beginning of our work on transverse pattern formation, the field has grown considerably, attracting a lot of interest from different fields. Many of our investigations started in the framework of projects in the Sonderforschungsbereich 185 “Nichtlineare Dynamik” of the Deutsche Forschungsgemeinschaft. This growth of the field has led to the formation of various different fields of investigations such as soliton formation and interaction, spontaneous pattern formation and active feedback systems. We are pursuing these topics in various projects funded by the Deutsche Forschungsgemeinschaft and by several binational exchange projects, thereby changing the focus slightly from basic analysis to application-oriented investigations. We therefore acknowledge suppport from the Graduiertenkolleg “Nichtlineare Kontinuierliche Systeme” at Westfälische Wilhelms-Universität Münster, from the Deutscher Akademischer Austauschdienst (DAAD) and from the generous Fazit-Stiftung, Frankfurt.

Many of these research directions and projects indicate that the original aim of this field of study – analysis of pattern formation in a model nonlinear system – is gradually being left behind, leading to opportunities for exploitation of phenomena far beyond threshold and for applications in optical information processing. In fact, our knowledge about the formation of different patterns, their interaction and the ways to stabilize, track and manipulate them is far enough developed to allow us to change the field in this direction now. Throughout this book, numerous hints can be found as to how the results can be exploited to pursue this direction – topics that are far beyond the scope of this book.

With the Help of ...

A project such as the investigation of transverse pattern formation in its various manifestations involves the collaboration of a number of people, and the work described in this book has been no exception. We are greatly indebted to those people who have helped us through their support, comments and criticism to transform the initial idea into reality. We appreciate their assistance and effort, especially all the support from our coworkers and collegues at the Institute of Applied Physics, Westfälische Wilhelms-Universität Münster (AP-WWU) and of the Institute for Applied Physics, Darmstadt University of Technology (AP-DUT). In particular, we are indebted to Prof. T. Tschudi for the ideal scientific framework at AP-DUT, where these investigations started and to the members of our research group, who have contributed freely to discussions and thus influenced many parts of this book.

The achievements of our research group that are presented in the following chapters owe much to the work of Markus Ahles and Jürgen Petter at AP-DUT, who realized soliton interaction, and of Denis Träger and Jochen Schröder at AP-WWU, who realized the first spatial-soliton lattices. Markus Sedlatschek and Deniz Tamer from AP-DUT investigated pattern formation and control of patterns in the single-feedback system. That work was considerably extended by Philip Jander and Oliver Kamps at AP-WWU, and directed strongly by those people into the control and understanding of patterns beyond the first bifurcation. Gerhard Balzer, Oliver Knaup and, especially, Ralf Nicolaus (AP-DUT) constructed and adjusted the unidirectional ring configuration, with a lot of effort to clearly distinguish nonlinear spatial effects from external noise sources. Ralf Nicolaus also implemented the phase-sensitive pattern control method – a challenging task that he performed successfully thanks to his outstanding experimental skills.

Many of the theoretical investigations that strengthened our experimental results were done in the group of Prof. Dr. F. Kaiser, Institute of Applied Physics, Darmstadt University of Technology, especially by Andreas Stepken and Kristian Motzek in the case of soliton interaction and by Oliver Sandfuchs in the case of spontaneous pattern formation in the single-feedback configuration. The perfect graphical representation of many of the images was performed by Barbara Hackel (AP-DUT) and Jochen Schröder (AP-WWU).

We are also grateful to several guest scientists who shared our interest in the field and contributed to several topics. Dr. Tokuyuki Honda, formerly at the National Research Laboratory of Metrology, Tsukuba, Japan, was involved in our first steps in spontaneous pattern formation, whereas Dr. Wieslaw Krolikowski, now at the Laser Physics Centre, Australian National University, Canberra, shared our excitement about photorefractive spatial solitons and their interaction. Sussie Juul Jensen, formerly at Risø National Laboratory, Roskilde, Denmark, supported the hard experimental work on stabilization of our unidirectional ring oscillator and on the investigation of pattern manipulation and control in the single-feedback system. We acknowledge also many helpful discussions with and numerical support from Prof. Mark Saffman, formerly at Risø National Laboratory, Roskilde, Denmark, and now at the Faculty of Physics, University of Wisconsin at Madison, Wisconsin, in the case of our work on solitary beam formation and on pattern formation. This cooperation has also led to several fruitful scientific exchanges, during which one of us, MS, appreciated the outstanding working conditions and excellent scientific supervision.

The theoretical support for the work on complex soliton structures was provided by Prof. Dr. Yu.S. Kivshar and his Nonlinear Physics Group at the Australian National University in Canberra, especially his coworker Anton Desyatnikov. Without their continuous active help, many of the experimentally observed configurations could not have been explained in such a clear way. Moreover, our collaboration led to several exchanges with the Australian groups. Among us, CW profited from the kind Australian hospitality of both Krolikowski's and Kivshar's groups, an excellent working environment and an inspiring and motivating supervision during his stays at the Australian National University.

CDs personal thanks go primarily to Wilfried Denz, who has tolerated preoccupations, shared disappointments, and still found time to provide encouragement and support. CD also thanks Tobias and Silas for their patience, and Irmgard and Horst Denz for their encouragement. MS expresses his sincerest gratitude to Brigitte and Bardo Schwab for their enduring support and encouragement. CW is personally indebted to Brunhilde and Horst Weilnau, and also to Martina Schadeck, Stefanie Güttel, Melanie Behringer, Erik Benkler and the Blechbläserzentrale Gallus for endless patience and motivation.

Certainly, the quality of the book has been improved because of them all – if there remain weak parts, the blame lies on us.

Münster, *Cornelia Denz, Michael Schwab, Carsten Weilnau*
June 2003

References

1. G. Nicolis and I. Prigogine. *Self-Organization in Nonequilibrium Systems: from Dissipative Structures to Order through Fluctuations.* Wiley, New York, 1977.
2. H. Haken. *Synergetics. An Introduction. Nonequilibrium Phase Transitions and Self-Organization in Physics, Chemistry and Biology.* Springer, Berlin, Heidelberg, 1978.
3. V.I. Krinsky. *Self-Organization, Autowaves and Structures Far from Equilibrium.* Springer, Berlin, Heidelberg, 1984.
4. C.O. Weiss, *Spatio-temporal structures part II,* Phys. Rep. **219**, 311-338 (1992).

Table of Contents

1 Introduction – Nonlinear Waves and Transverse Patterns

The study of solitary waves, transverse optical structures and their bifurcations to more complicated pattern states in systems far from thermal equilibrium is one of the most fascinating topics in science and has attracted considerable fundamental and practical interest. When a system is excited beyond a certain threshold, general principles of self-organization lead to spontaneous temporal and/or spatial instabilities, which are manifestations of a macroscopic cooperative behavior of the constituents of the underlying physical, chemical or biological system.

Extended, open nonlinear optical systems represent the classical objects of studies of structural nonlinear effects in the transverse plane. Though transverse effects have been known since the early studies of laser beam profiles, systematic investigation began rather late. The research on classical hydrodynamic pattern formation systems provided the necessary language for the description and classification of the vast number of different manifestations of pattern formation. Nevertheless, nonlinear optics is far more than yet another nonlinear system that exhibits pattern formation effects. We shall see later on that optics possesses some prominent and unique properties which render optical systems ideal objects of study for all kinds of nonlinear structure formation effects, with the possibility of externally controlling the system parameters and monitoring the power spectrum using a single lens.

We shall deal throughout this book with light–matter interactions which affect the structure of a smooth laser beam profile. Light–matter interaction is, of course, only possible if there is a nonlinearity in the description of the material response. This means that the simplest case to be treated is the passage of a smooth laser beam through a nonlinear medium. Even this simple situation exhibits a large number of different effects, among them beam fanning, phase conjugation, and the creation of localized solitary structures where the shape of the light distribution remains unchanged even though the inevitable effect of diffraction always tends to spread a beam of finite size. Nonlinearities are a key in explaining structural changes of an extended physical system, for example a laser beam with its enormous (but not infinite) number of degrees of freedom represented by the different spatial frequencies of the spatio-temporal distribution. One can imagine that the degree of complexity will rise considerably when we deal with the interaction of two (or more) beams inside the nonlinear medium. The fascination comes from the fact that nonlinearity enables different beams to communicate via the nonlinear interaction. Even here, a broad spectrum of different scenarios is possible under certain circumstances, starting from simple energy exchange between the beams, and extending to creation of an exact time-reversed replica of the incoming beam (phase conjugation), repulsion, attraction of beams and even the annihilation or birth of beams. These examples obviously show the potential applications of nonlinear materials in modern communication systems, where they may serve as coupling or waveguide-dividing elements.

Extensive research on the structural changes of a coherent light beam when it passes through a nonlinear medium started at the beginning of the 1960s and still represents a young, lively and growing field of research. The interpretation and prediction of effects on the spatial structure is one of the key elements in the field of *Nonlinear Photonics*. While in the early days of laser physics coherence and longitudinal effects attracted a great deal of attention, it is now clear that transverse effects are at least as important as the coherent nature of the light field. Transverse perturbations (nowadays referred to as *"Transverse Effects"*) have been known since the very early days of laser physics. Then, the aim was to prevent those unwanted effects. In view of this background, it is no surprise that the reported "optical damage" of a nonlinear crystal is today interpreted as the first report of a photorefractive effect in the history of optics. However, the key to the description of spatial instabilities or transverse confinement of a laser beam lies in the description of the nonlinearity. As it turns out, transverse effects are much more universal than the mathematical formulations of different nonlinearities suggest – systems that seem very different can exhibit the same macroscopic behavior. The universality of macroscopic behavior even connects hydrodynamics with optics, even though the physics and the mathematical equations are – at first sight – completely different.

A number of different mechanisms exist that can lead to a spatio-temporal instability of a physical system. One class of mechanisms arises from the existence of constraints and conservation laws – a case that often occurs in hydrodynamic systems, such as in the Rayleigh–Bénard convection, where a fluid is placed between horizontal plates. Here, the buoyancy force attempts to lift the whole mass of the fluid – representing the conservation of mass – and the top plate provides the constraint against this motion. The conducting uniform state is represented by a linear temperature profile between the lower and the upper plate, and a fluid which remains at rest. Above a threshold for the temperature gradient ΔT between the two plates, a cooperative convective motion sets in, which results in macroscopic patterns with a certain wave number k. The resulting convective roll pattern forms the basic element of more complicated patterns, which are based on a superposition of underlying roll patterns. A generic pattern is the hexagonal one, which is, owing to its inherent excellent symmetry properties, predominant in a number of different systems. Other examples involve square patterns or secondary instabilities of roll patterns that lead to zigzag patterns, to name only two examples. A detailed description of the fundamental phenomenology of pattern formation in hydrodynamic systems can be found in [1], we will adapt this derivation in the following to the case of optical pattern formation.

Another mechanism for finite-wave-vector instabilities is provided by competition between elementary units. This mechanism is wellknown in magnetic materials, where it leads to antiferromagnetism. It is also ubiquitous in chemical, biological and neural systems, where it is known as "local excitation in combination with lateral inhibition", mainly due to dispersion effects caused by diffusion.

In optics, the coupling between different spatial units is provided by diffraction during propagation of a light beam in a nonlinear material, which plays the same role as diffusion in nonlinear chemical reactions and biology. Nonlinear optics provides all the necessary ingredients for pattern formation, including strong local nonlinear dynamics. In contrast to diffusion and convection, the spatial coupling process of diffraction represents a massless mechanism, which implies considerable advantages. The inherent parallelism of optics, the special feature of information transport without transport of mass and, as a consequence, the unique possibility of active control over the system parameters render nonlinear optical systems excellent objects of study for the investigation and active control of pattern formation effects. The natural access to the Fourier spectrum in the transverse plane is unique to optical systems and motivates the use of control techniques that work in the spatial-frequency domain. Consequently, transverse effects, in particular the spontaneous formation of patterns from a smooth laser beam profile and the creation of localized states hsa attracted considerable interest over the last decade. *Transverse Nonlinear Optics* constitutes a lively field of research that contributes with new materials, new model equations and novel phenomena to the general field of research devoted to nonlinear spatio-temporal dynamics in optics.

One common phenomenon in pattern forming systems is the existence of a finite critical wave number k_c which grows exponentially above the instability threshold. The value of this wave number depends on the inverse of a characteristic length scale of the physical problem. In the case of Rayleigh–Bénard convection discussed above, this length scale is represented by the distance between the two opposing plates.

In the most general case, the appearance of a spatial frequency k_c is accompanied by a temporal frequency δ_c which manifests itself in oscillations of the pattern. Although a detailed understanding of the instability mechanism and its dependence on the system parameters requires a complete characterization of the system by macroscopic equations, the estimation of the characteristic wave vector k_c and the characteristic frequency δ_c is the crucial challenge understanding a nonlinear dynamical system. The idea is to simplify the problem and to transfer the description of the system behavior to the action of one mode k_c with a temporal frequency δ_c, which slaves all other modes (see [2, 3, 4] for the principles of slaving and self-organization). This simplification in the theoretical description is the direct cause of the macroscopic experimental behavior of a complex physical system beyond the instability threshold: a simple cooperative behavior occurs, which can be described by just two parameters: the wave number k_c and the corresponding frequency δ_c.

Driving a spatially uniform system in a situation near equilibrium by increasing a control parameter γ takes the system further from equilibrium, and results in a qualitatively similar behavior in all disciplines. In the most general case, a system becomes unstable to infinitesimal perturbations with a wave number k_c and a frequency δ_c above a defined threshold value $\gamma = \gamma_c$. The transverse mode with the critical wave number k_c grows exponentially with a growth rate proportional to the distance from the instability threshold $\gamma - \gamma_c$. Saturation of this explosive instability is usually provided by saturating terms of the order of $(\gamma - \gamma_c)^{-n}$, where $n \geq 2$.

A stationary pattern occurs for $\delta_c = 0$, whereas for $\delta_c \neq 0$ an additional frequency appears, leading to an oscillating pattern. Spatial pattern formation usually arises from the existence of a nonzero k_c, which allows one to identify generic mechanisms for finite-wave-vector instabilities. This situation is depicted schematically in Fig. 1.1**a**.

A linear stability analysis of the steady-state solutions of the corresponding equations of motion leads to a dispersion relation which separates the stable from the unstable wave vectors for a given control parameter γ. The global minimum of this threshold curve indicates the minimum threshold control parameter γ_c for which a modulation instability occurs. The system becomes unstable to perturbations with a corresponding wave number k_c, and a transverse pattern with a band of wave numbers centered around k_c appears above this instability threshold. Fig. 1.1**b** shows the corresponding picture in the Fourier plane: only a small band of wave numbers is active and participates in the pattern forming process. The shape of the threshold curve, as depicted in Fig. 1.1**a** is a quite general feature and can be found in a large number of different pattern forming systems. These represent the classical cases of hexagons, squares or roll patterns with wave numbers k_c. Nevertheless, the absolute minimum of a threshold curve such as that shown in Fig. 1.1**a** might equally well correspond to the mode $k = 0$, which corresponds to a plane-wave instability, or to modes with large k, i.e. $k \to \infty$ (high-k instability).

In such a general investigation, three types of instabilities can be distinguished in all structure and pattern forming systems:

- If $k_c \neq 0$, $\delta_c = 0$, the instabilities are stationary and periodic in space. Because of the richness of periodic structures in two- and three-dimensional space, many different ideal patterns may appear above threshold, the sim-

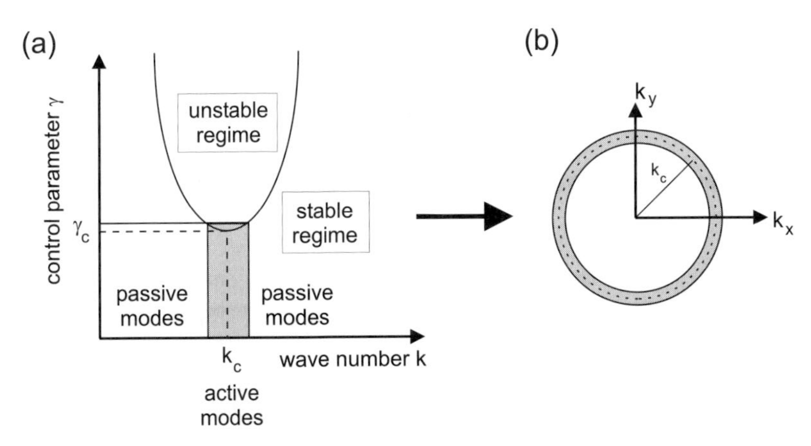

Fig. 1.1. (**a**) Representation of a physical system in the phase plane created by the control parameter γ as a function of a perturbation wave number k. The instability threshold curve separates the stable from the unstable regime with respect to the linear stability behavior of the solutions of the system. Above a threshold value γ_c of the control parameter, perturbations with wave numbers centered around k_c grow exponentially. (**b**) For a control parameter γ slightly above the instability threshold, only a small band of wave numbers (modes) is active (i.e. have a positive growth rate)

plest case being the roll state, characterized by a single wave vector with a certain absolute value of k_c and an arbitrary orientation in space. Other regular patterns, such as hexagons or squares in two dimensions, can be formed by superposition of the elementary rolls.

- If $k_c = 0$, $\delta_c \neq 0$, the instabilities are uniform in space and oscillatory in time. The ideal state does not display any spatial pattern, unlike the previous case. Uniform periodic oscillations with a frequency δ appear at threshold: the frequency of these oscillations depends on the distance $\gamma - \gamma_c$ from the instability threshold. We shall focus in this book on spatial modulation instabilities with $k_c \neq 0$. Therefore, the case $k_c = 0$ will play only a minor role in the following chapters.
- $k_c \neq 0$, $\delta_c \neq 0$ describes the most general case, of instabilities that are periodic in space and oscillatory in time. The simplest ideal pattern in this case involves a traveling wave, with a dispersion relation $\delta(k)$. Such nonlinear waves are sustained by a competition between the drive $\gamma - \gamma_c$, the dispersion and the nonlinearity in the medium. A subclass of such waves, that occurs generally in nondissipative or Hamiltonian systems is the *soliton*, or more exactly, in the case of nonconservative sytems, the *solitary wave*. In nonlinear physics, waves can be generated that do not disperse or diffract – they retain their size and transverse shape during propagation. In this case, the nonlinearity balances dispersion in the temporal domain or

diffraction in the spatial domain, and the wave propagates in the absence of a drive. If the delicate balance between dispersion (or diffraction) and nonlinearity is lost, the soliton solution vanishes immediately. In nonlinear systems, gain and loss effects that may provide a stabilizing influence can come into play, and novel solitary solutions – called *dissipative solitons* – can be created.

Solitons have become one of the most attractive and interesting paradigms of nonlinear physics; in many respects, this is due to their experimental discovery in transverse nonlinear optics. A soliton wave is more than just light or energy restricted to a limited region of space or time. Solitons can be either static or dynamic, they can be set in motion by an external force thereby obeying Newton's first and second laws of motion. When two solitons collide, they end up unchanged, as if no collision had occurred. Physical units with these special properties are usually called particles. In analogy to this particle–light dualism of scalar solitons, multicomponent solitons (often referred to as *vector solitons*) can be interpreted as molecules of light. Scalar and vector solitons will be considered in detail in Chaps. 4 and 5. Besides the creation of bright solitons, it is also possible to confine a distribution of zero intensity on a background illumination and create a *dark soliton*. Increasing the complexity of the nonlinear system, e.g. by placing a nonlinear material inside a resonator, requires not only a balance between diffraction and nonlinearity, but also a delicate balance between the pumping and dissipation of energy. Localized states with properties that satisfy these two distinct conditions are called *cavity solitons*.

In a more general case, the effects of driving and dissipation are just as important as dispersion due to diffusion. Nonlinear waves of this general kind are often called *auto waves* or *dissipative solitons* and have many distinctive properties. When two wave trains of such a type collide, they generally do not interfere, unlike the linear case. They form a domain boundary, or shock, which may remain stationary or may consume one wave or the other. In two and more dimensions, all the richness of the patterns described above, such as the existence of different structures or stability balloons, is added to the basic dynamics of nonlinear waves, leading to complex interaction effects.

Boundaries, *defects*, and *inhomogeneities* are natural limitations that perturb the ideal pattern, which has an infinite spatial structure. A *boundary* discretizes the previously continuous band of wave vectors inside the stability curve. Two-dimensional extended systems are not only affected by boundaries in terms of the wave number of the instability, but also the specific boundary situation may give rise to a preferred spatial pattern which is not the selected pattern for the ideal infinite system. Further restrictions on the allowed wave number band are possible, as well as a preferred orientation of the spatial pattern that arises. This is especially important in experimental two-dimensional pattern forming systems, where often one spatial pattern with a preferred orientation is found, even though the equations of motion are completely degenerate concerning the orientation on the instability circle. *Defects* are deviations from the ideal pattern, but the most useful limit to consider is a localized structure, embedded in an otherwise ideal pattern. Defects may be stationary or may move, and their structure often reflects topological characteristics of the ideal patterns in which they are immersed. They play an essential role in the dynamics of real patterns, either in selecting a particular regular pattern or in the transition of a stable pattern into dynamic fluctuations. The transition is, in that case, often accompanied by the appearance of defects in the otherwise ideal pattern. One prominent example of a defect is a penta-hepta-defect in a hexagonal pattern. Often, in the early stages of a pattern formation process, these defects are annihilated during the buildup of an extended geometrical pattern. Here, the defects are often pushed to the boundaries, leaving an ideal pattern behind.

Another reason for deviations from the ideal system is *inhomogeneities* in the material response. Owing to incomplete control of the growth process, a nonlinear crystal is never entirely homogeneous. For exapmle, photorefractive crystals often show variations of the photorefractive gain within their volume; it is even possible that growth domains are responsible for a completely different nonlinear behavior. These inhomogeneities affect the nonlinear response of the medium, which is also a function of spatial position inside the nonlinear material. This modifies the underlying systems of equations of motion and therefore leads to results that differ from the calculated result for an ideal homogeneous medium.

In real experimental systems, all three ingredients, namely boundaries, defects and inhomogeneities are present adding up to a spatial distribution of boundary values which have to be satisfied by the extended nonlinear system. It is clear that only a discrete set of solutions is likely to appear beyond the linear instability threshold. In systems that spontaneously form symmetric patterns, the origin of pattern formation and the selection of a particular pattern with a preferred geometry are two equally challenging questions. Under given external conditions, there often exist several stable solutions of the equations of motion, for example all states in the wave vector band defined by the stability balloon at a given γ. From a mathematical point of view, it is clear that for autonomous deterministic equations, the state found at long times depends uniquely on the initial conditions. In experiments, the state observed is often insensitive to the specific preparation conditions, thus masking the nonlinear dynamics of the system. It is one of the most challenging tasks of such experimental investigations to separate these effects and extract the nonlinear dynamics from the superposition with external noise sources and boundary constraints. Dynamic mechanisms may arise either from external noise or from the dynamics of the creation of the pattern itself. For example, Cross and Hohenberg [1] found that if a roll state grows via a front advancing into an unstable uniform state, then it has a unique wave vector and velocity independent of the initial conditions, rather than a continuous band. Generally speaking, a primary mechanism for pattern selection is the motion and interaction of defects, since they provide a way for a region of space with an "unfavorable" pattern to give way to a more favorable one.

Defects and perturbations such as inhomogeneities play an equally important role in the appearance of spatial solitons. Achieving a balance between diffraction and self-focusing is a quite delicate experimental challenge; the nonideal character of the physical system often renders the adjustment of this balance extremely difficult. The result is often an oscillation of waves around the soliton solution. Depending on the nonlinearity, the interplay may lead to complex effects such as soliton oscillations, energy dissipation between solitons and the formation of filamentation as the solitary solution breaks up. Other complex phenomena are likely to emerge in this region at the border of balance, giving rise to effects of overcompensation or of dominance of diffusion or diffraction. If the balance between the competing effects of diffraction and self-focusing is lost, the generated soliton may become unstable and cease to exist – here also, a linear stability analysis is an important means to decide whether small perturbations from the stationary fixed-point solution of a soliton are able to drive the system away from the soliton solution exponentially or whether the soliton solution is stable in the sense that small perturbations are exponentially damped.

Solitons in nonlinear optical media are extremely attractive candidates for applications in waveguiding, including the potential to create waveguide couplers or switches for optical telecommunication applications. Owing to the ease of soliton generation in the nonlinear photorefractive crystals that we consider in our book, all these effects are now experimentally accessible. Thereby, we can exploit the advantage of optics that the control parameters are easily adjustable. The possibility of guiding light by light is not a vision any longer – the generation of highly parallel, light-induced waveguide couplers and dividers that we present in this book shows the capability of nonlinear optical materials to guide light just by the action of light itself.

Besides their challenging fundamental aspects, studies of nonlinear dynamics in optics are undoubtedly of enormous practical and technological relevance. Devices based on nonlinear effects are key elements of modern technology. The enormous communication needs of the modern information society require complex optical transmission schemes. A larger energy flux is always associated with an increasing importance of nonlinearities, which are capable of exhibiting pattern formation effects under certain circumstances. The ongoing research on pattern formation effects is motivated by the information capacity of extended two-dimensional laser beams. Despite the interest in information aspects – the information capacity of a nonlinear system is determined by the number of different solutions of the system – systematic investigations of means to actively steer the pattern formation process are still rare. Our aim is to actively select a pattern of a specific geometry; doing so allows the operator to choose from the whole wealth of different spatial patterns. The pattern formation process can be controlled and steered in order to favor desired patterns – a basic condition for realizing information-processing systems by the use of pattern formation processes.

In the following chapters, various different scenarios are investigated that allow one to realize the mechanisms described here in a rather general way. Spatial solitons form because of a balance of nonlinear coupling and diffractive effects in the material. We increase the degree of complexity in successive stages; we go from one-dimensional to two-dimensional solitons and describe the dynamics of soliton formation. We then move on to the interaction of two solitons, which shows the fascinating possible applications of soliton fusion and energy exchange. In doing so we investigate the waveguiding properties of scalar solitons and the generation of soliton arrays. A higher degree of complexity is provided by multicomponent or vector solitons, which serve as a basis for multimodal soliton-based waveguide structures.

Spontaneous patterns are formed because of an instability of certain wave numbers, leading to a modulation instability that is driven by the beam-coupling features of the interacting beams in the nonlinear medium. In the description in this book, we start with a basic feedback system to learn about the principles of pattern formation in nonlinear optics. With a knowledge of the rich variety of steady-state and dynamic solutions, we then focus on the active manipulation and selection of a specific pattern from the manifold of different solutions, first by choosing a suitable system parameter in order to choose from different solutions in a bistable situation, and then by introducing a spatial control technique that works in the Fourier domain.

Finally, multiple feedback allows one to realize patterns by the construction of borders in oscillatory systems, leading to significant effects of mode pattern dynamics and competition. Methods for realizing spatio-temporal pattern control can be applied most effectively, in these configurations.

All these different systems have in common the fact that they are realized by use of a nonlinearity that has features of the standard Kerr nonlinearity but is saturable – the *photorefractive nonlinearity*. Therefore, some basic theoretical predictions based on an intensity-dependent refractive-index modulation can be verified for a certain range of experimental parameters. The photorefractive crystals used in our experiments throughout this book provide the key to the various nonlinear effects obtained with low cw powers, in the milliwatt or even microwatt range. As we shall show, the photorefractive nonlinearity offers a number of advantages and complies with our requirements for a robust and versatile system to explore in detail the formation of spatial solitons and of extended spatio-temporal patterns – two of the most challenging topics in the field of *Transverse Nonlinear Optics*.

Summary

The interplay between diffusion or diffraction processes and a nonlinearity leads, under certain circumstances, to astonishing effects in the transverse plane of an extended physical system that is driven far from equilibrium. This nonlinear coupling leads to a fascinating wealth of different spatial patterns, such as localized states, boundary domains, symmetric patterns and spirals, to name only a few examples of the possible results of the interaction of a nonlinearity with a spatial coupling process. The generic mechanism is due to the growth of pertubations, which result in the formation of novel wave numbers, which in turn form solitary, oscillating or stationary symmetric patterns. When this mechanism is combined with effects of boundaries, defects and inhomogeneities, many different patterns appear, which in the case of a photorefractive nonlinearity, can be studied at very low laser powers in a temporal region that is easy to access with standard cameras. For the above reasons, one of the most attractive challenges in the field of nonlinear optics is the use and exploitation of transverse effects for applications in photonics and information processing.

References

1. M.C. Cross and P.C. Hohenberg, *Pattern formation outside of equilibrium*, Rev. Mod. Phys. **65**, 851–1112 (1993).
2. H. Haken. *Synergetics. An Introduction. Nonequilibrium Phase Transitions & Self-Organization in Physics, Chemistry and Biology.* Springer, Berlin, Heidelberg, 1978.
3. H. Haken. *Advanced Synergetics.* Springer, Berlin, 1983.
4. H. Haken and A. Wunderlin. *Die Selbststrukturierung der Materie.* Vieweg, Wiesbaden, 1991.

2 Light Propagation in Nonlinear Optical Media

The propagation of a light beam through a nonlinear medium is the simplest scenario that one could imagine for light–matter interaction, but it is accompanied by a series of dramatic and fascinating changes in the spatio-temporal structure of the beam. Among these are light-induced scattering which may result in asymmetric beam fanning; the formation of spatial distributions of the electromagnetic fields that results in back reflection; phase-conjugation; and the possibility to prevent the beam from spreading and to enable self-trapping of the beam profile. Therefore, the spatial evolution of a single light beam is a central topic of nonlinear optical dynamics. A theoretical interpretation of this complex phenomenon and its experimental verification are of key importance for gaining insight into the nonlinear properties of a particular medium, on the one hand, and for contributing to our understanding of more general questions about the complex spatio-temporal behavior of nonlinear systems on the other hand. This is the reason why we have chosen this simple propagation of light through a nonlinear medium as the starting point for our investigations.

The topic is not a completely new one. Initial investigations on the self-trapping of light beams in nonlinear media were already performed in the early 1960s [1, 2]. The most important task, since the beginning of this research, was to find the conditions for the stability and nonlinear evolution of solitary-wave solutions of the nonlinear propagation equations associated with the propagation of the beam in the nonlinear material. This is one of the most crucial parts of the problem of self-trapping of optical beams and promises applications in waveguiding, information processing and, most simply, propagation without spreading losses.

In general, these effects may occur in all materials that display a nonlinear response that may act in such a way that diffraction effects due to the propagation of the beam can be compensated exactly. The balance between beam broadening on the one hand and self-focusing on the other hand gives rise to the formation of a solitary wave. In this case the envelope of the light wave does not change its profile during propagation. This wave is often called a *soliton* for short.

Solitons are waves that do not spread or disperse like all familiar waves, but retain their size and shape indefinitely – they are dynamically and structurally stable objects. The wave can be regarded as a quantity of energy that is permanently confined to a defined region of space. It can be set in motion, but it cannot dissipate by spatial or temporal spreading. In general, when two such waves collide, each comes away from the encounter with its identity intact. If a wave meets an "antiwave", both can be annihilated. A behavior of this kind is extraordinary, even strange in the context of wave physics, but it is familiar in another context. Given a description of an object with these properties, we would call it a particle.

2.1 Self-Focusing in Nonlinear Kerr Media – Spatial Soliton Generation

An initial theoretical investigation of the phenomena connected with self-trapping of an optical beam was performed in 1962 by Askaryan [1]. Only two years later, following the first experimental observations by Hercher [3], who observed filamentation and interpreted it initially as a damaging effect, Chiao et al [2] showed theoretically that the governing equation of self-focusing is nothing but the nonlinear wave propagation equation with a cubic potential, and that an optical beam propagating in such a one-dimensional nonlinear Kerr medium can indeed trap itself. Chiao et al considered the diffraction of a circular optical beam of uniform intensity propagating along the direction z in a material for which the index of refraction n may be expressed in terms of the field strength E as

$$n = n_0 + \Delta n(|\boldsymbol{E}|) = n_0 + \sum_i n_i |\boldsymbol{E}|^i \, . \tag{2.1}$$

In the case of a cubic ($\chi^{(3)}$), or Kerr, nonlinearity, this leads to

$$n = n_0 + \Delta n = n_0 + n_2 |\boldsymbol{E}|^2 \, , \tag{2.2}$$

where n_0 is the linear and n_2 is the nonlinear contribution to the refractive index. Δn represents the optically induced nonlinear variation of the index due to the intensity $I = |\boldsymbol{E}|^2$ of the electromagnetic wave.

In principle, a single beam can be represented by a superposition of several plane waves (spatial frequencies), with wave vectors $\boldsymbol{k}$ of equal magnitude but slightly different directions with respect to the optical axis. Since each plane-wave component can be characterized by a different projection of its wave vector $\boldsymbol{k}$ on the optical axis, each component propagates at a different phase velocity with respect to that axis. In this way, the component that propagates "on" the optical axis propagates faster than a component i that propagates at some angle Θ_i, whose propagation constant is proportional to $\cos(\Theta_i)$. The narrowest width of the spatial beam is obtained at a particular plane in space at which all components are in phase. However, as the beam propagates a distance z away from that plane, each plane-wave component acquires a different phase, equal to $2\pi n_0 z \cos(\Theta_i)/\lambda$, where λ is the wavelength in vacuum. This causes the spatial-frequency components to differ in phase and the beam broadens, i.e. diffracts. In general, the narrower the initial beam, the broader is its plane-wave spectrum (spatial spectrum) and the faster it diffracts with propagation along the z axis. Therefore, a beam of diameter d experiences an expansion due to diffraction with an angular divergence of $\Theta \approx 1.22\lambda/(n_0 d)$. The only method to eliminate the spatial spreading by means of linear optics is provided by waveguiding techniques. In a waveguide, the propagation behavior of a beam in a high-index medium is modified by total internal reflection from boundaries with media of lower refractive index, and under conditions of constructive interference between the reflections, the beam becomes trapped between these boundaries and forms a *guided mode.*

In nonlinear optics, the effect of light-induced lensing or self-focusing eliminates the spatial spreading of a beam and allows a nondiffracting optical wave to propagate. Intuitively, this can be explained when two distinct requirements are fulfilled. First of all, the optical beam has to modify the refractive index in such a way that it generates an effective positive lens, i.e. the refractive index in the center of the beam becomes larger than at the margins of the beam. The medium then resembles a graded-index waveguide. The second condition requires that the initial beam is also a guided mode of this self-induced waveguide. If these two conditions are fulfilled, the propagation of the beam becomes stationary, which means it propagates with a single propagation constant. All its plane-wave constituents now propagate with the same phase velocity, and as a result the beam becomes *self-trapped.* Its spatial divergence is eliminated, keeping it at a very narrow diameter, which can be as small as ten vacuum wavelengths. In other words, if the term $n_2|\boldsymbol{E}|^2$ in (2.2) produces a dielectric constant such that the nonlinear refractive-index change in the vicinity of the beam results in an internal-reflection angle exceeding Θ, then spreading by diffraction will no longer occur. For $\Theta \ll 1$, this requires that the total beam power P equals

$$P = \frac{\pi d^2}{4}\frac{n_0 E^2 c}{8\pi} \geq (1.22\lambda)^2 \frac{c}{64 n_2}. \tag{2.3}$$

This simple approximation indicates that a beam above a certain critical power may self-trap for any arbitrary beam diameter. The power level itself decreases with with λ^2. For conventional dielectric materials, the constant n_2 gives a critical self-focusing power P_{crit} in the range of one or two orders of magnitude of 10^6 W for visible light – a power level commonly obtained in laser beams.

In the case of a Kerr nonlinearity, the nonlinearity is proportional to the power of the laser beam ($\Delta n \propto |\boldsymbol{E}|^2$). In the case of a beam that is extended in one transverse dimension and propagates in one direction, a so-called (1+1)-dimensional beam, exact (1+1)-dimensional self-trapped or solitary waves were found, soon after the discovery of the self-trapping effect, in the early

1970s by Zakharov and Shabat [4]. The nonlinear coefficient n_2 may be associated with high-frequency Kerr effects involving molecular orientation or with nonlinearities due to the electronic polarizability of the nonlinear material, for example. Self-trapping can also occur in the case of a light pulse propagating in an optical fiber. Due to its enormous potential in applications and to the fact that this effect can be described analytically, nonlinear pulse compression in ultrafast optics and all-optical switching became the major field of theoretical and experimental investigations of solitons (see e.g. [5, 6, 7, 8]).

Moreover, these early investigations of pulse propagation also gave useful insight into the physics of beam propagation in a nonlinear medium. If the time variable t is exchanged with the spatial transverse coordinate x, the governing equation changes to describe the propagation of a spatially extended beam instead of a temporal pulse. To distinguish between these two types of solitons, self-trapped optical beams are referred to as *spatial solitons*, whereas nondispersive optical pulses in fibers are called *temporal solitons*. Remarkably the theoretical model of a Kerr-nonlinearity is universal for both temporal and spatial solitons in real physical systems.

In recent years, optical spatial solitons have become an exemplary model for the exploration of the deep and rich phenomena of solitons, which are found in almost all nonlinear systems in nature. This is due to the ease of formation and observation of solitons in complex dynamic situations and interaction scenarios. In two transverse dimensions, which is the most important case for spatial solitons, *photorefractive spatial solitons* are the most attractive candidates for exploring soliton physics. They allow solitary beam formation at low laser intensities and easy tuning of the nonlinearity by externally applied electric fields. Moreover observation of the formation and dynamics of solitons is simple and easy due to the reduced speed of the photorefractive refractive-index modulation.

An example of self-focusing of a beam in a photorefractive medium is shown in Fig. 2.1. The image at the top (Fig. 2.1**a**) shows a beam propagating from the left to the right, diffracting naturally, when the nonlinearity is not operating. In contrast, Fig. 2.1**b** shows the same beam when it is self-guided because of the nonlinearity of the material. In the case of this photorefractive nonlinearity, its strength can be tuned by changing an externally applied electric field, thus allowing one to realize and compare linear diffraction and nonlinear self-focusing.

In this book, we shall concentrate on this distinct type of spatial soliton, which is, more precisely, a representative of the class of *scalar bright solitons*. Several other types of solitons exist that are mathematically equivalent to these bright solitons. Their direct counterpart are *dark solitons*, which can form in a defocusing medium. In this case a dark notch propagates in a nondiffracting way on a bright background. In addition *vector*, or *multicomponent solitons*, consisting of several optical fields, have attracted appreciable

Fig. 2.1. Top-view photograph of a 10 μm wide beam propagating in a 5 mm long photorefractive crystal, diffracting naturally when the nonlinearity is not operating (**a**), and being self-guided when the nonlinearity is applied (**b**) (from [9] printed by kind permission of Mordechai Segev)

attention recently. They will be the main subject of Chap. 7, in the case of bright solitons. Solitons that form owing to a supplementary balance of gain or loss, known as *dissipative* and *cavity solitons*, and also as *auto-solitons*, are described in Sect. 2.4.2. Details of several other types of spatial solitons can be found in [10, 11], for example.

2.2 The Nonlinear Schrödinger Equation

In general, the properties of spatial optical solitons in one transverse dimension can be described by a single equation for the propagation of a beam in a nonlinear medium. The electric field of the optical beam

$$|\boldsymbol{E_{opt}}| = \frac{1}{2} A(x,z) \exp\left[i(kz - \omega t)\right] + \text{c.c.} \tag{2.4}$$

obeys the wave equation, where $k = \omega n/c$ is the wave number, ω is the frequency of the optical field and $A(x,z)$ represents the complex slowly varying amplitude of the electric field spreading in the coordinate x transverse to the propagation direction z; which is governed by the standard equation

$$\left(\mathrm{i}\frac{\partial}{\partial z} + \frac{1}{2k}\frac{\partial^2}{\partial x^2} + \frac{kn_2}{n_0} f(|A|^2) \right) A = 0 \; . \tag{2.5}$$

The function $f(|A|^2)$ describes the intensity-dependent refractive-index change. Its most convenient and simplest form is given by

$$f(|A|^2) = |A|^2 \; , \tag{2.6}$$

which characterizes a Kerr nonlinearity.

If we compare (2.5) with the well-known Schrödinger-equation

$$\left(\mathrm{i}\frac{\partial}{\partial t}+\frac{\partial^2}{\partial x^2}+V\right)\Phi=0\;, \tag{2.7}$$

and replace V by $|A|^2$ and t by z, an astounding similarity of the two equations can easily be recognized. The role of the potential V in the Schrödinger equation is represented by $|A|^2$, which in turn is proportional to the nonlinear change of the refractive index. For this reason, (2.5) is called the *nonlinear Schrödinger equation*. Since in the Schrödinger equation V expresses the potential which traps the quasiparticle represented by the wave function Φ, $|A|^2$ represents a trapping potential for a light beam which is proportional to the refractive-index change Δn given in (2.2). Therefore, if n_2 in (2.2) is positive, the potential, which is proportional to $|A|^2$ has the effect of trapping the wave energy, which otherwise tends to spread owing to diffraction. This is exactly the self-trapping process described above.

Stable self-trapping can only be achieved if the beam broadening due to diffraction is exactly counterbalanced by the nonlinear self-trapping effect. It is well known in linear optics that the diffraction process itself is inversely proportional to the initial width of a light beam. A beam which is very small in diameter spreads much faster during propagation compared to a beam with a larger spatial extent. Therefore the initial beam size has to be carefully chosen for a given nonlinearity in order to obtain a stationary self-focused beam profile.

2.2.1 Symmetry Considerations

Mathematically, the nonlinear Schrödinger equation (2.5) is often represented in normalized coordinates as

$$\mathrm{i}\frac{\partial}{\partial z'}A'+\frac{1}{2}\frac{\partial^2}{\partial x'^2}A'+|A'|^2A'=0\;, \tag{2.8}$$

which can easily be derived from (2.5) by introducing the coordinates $z' = z/(kw_0^2)$, $x' = x/w_0$ and $A' = \sqrt{kn_2/n_0}A$, where w_0 is a defined spatial width. If we choose w_0 to be the initial beam waist, then $kw_0^2 = L_\mathrm{d}$ can be considered as the diffraction length, which is a familiar expression in the framework of linear Gaussian optics. L_d defines the propagation distance at which the peak intensity of a Gaussian beam has dropped to half of its initial value. Equation (2.8) is integrable and it has soliton solutions of the form

$$A' = \mathrm{sech}(x')\,\mathrm{e}^{\mathrm{i}z'/2}\;. \tag{2.9}$$

The number of solutions of (2.8) is infinite. Moreover, from knowledge of a single solution given by (2.9), more complicated solutions can be derived by some simple symmetry considerations.

Scaling Transformation

By adding a simple factor that changes the scaling of our soliton solution, we can generate a complete family of solutions. Suppose $U(x, z)$ is a solution to (2.8); we can then obtain a one-parameter family of solitons by the simple transformation

$$U'(x, z|a) = aU(xa, za^2) \ , \tag{2.10}$$

by introducing a real parameter a. This *scaling transformation* can be applied to any arbitrary solution of the nonlinear Schrödinger equation. It expands a single solution into a one-parameter family of soliton solutions, which can be written explicitly as

$$U'(x, z|a) = a \operatorname{sech}(ax) \, \mathrm{e}^{\mathrm{i}(a^2 z)/2} \ . \tag{2.11}$$

The parameter a now defines the amplitude of the soliton, its transverse width and its period in z.

Galilean Transformation

The solution space can be expanded further to even more complicated solutions by applying the *Galilean Transformation*

$$U'(x, z|b) = U(x - bz, z) \exp(\mathrm{i}bx - \mathrm{i}b^2 z/2) \ , \tag{2.12}$$

where the real parameter b can be interpreted as a transverse velocity of the soliton. By combining (2.12) and (2.11), we end up with a two-parameter family of soliton solutions

$$U(x, z|a, b) = a \operatorname{sech}\left[(a(x - bz))\right] \exp\left[(\mathrm{i}bx + \mathrm{i}(a^2 - b^2)z/2\right] \ . \tag{2.13}$$

The two parameters a and b are mutually independent and can be chosen arbitrarily. They expand a single soliton solution into a two-dimensional parameter space of solutions. Furthermore, a whole variety of other solutions can be generated by applying the *inverse scattering technique* described below. This technique not only is a very powerful tool for solving the nonlinear Schrödinger equation but also provides a superposition principle that can be used to derive a manifold of complicated soliton solutions.

2.2.2 The Inverse Scattering Method

The general procedure for obtaining the solution to the nonlinear Schrödinger equation (2.8) is based on the property, that any completely integrable nonlinear equation can be associated with the superposition of a pair of linear differential equations containing a free parameter. The integrable equation is therefore equivalent to the requirement that these two linear equations are compatible for all allowed values of the free parameter.

The approach used to derive these two equations is based on the idea that a complex soliton solution can be constructed out of simpler solutions. Therefore, the original problem is transformed into a function space with a relatively simple temporal dependence. The data for a later time are calculated in this space. A subsequent inverse transformation then gives the required solution. This procedure is a typical one in solving differential equations and is well known in the analysis of optical systems for the linear propagation of beams. In this case, the Fourier transform is used to apply this technique of solving the propagation equation. It does nothing more than represent the starting distribution of the equation as a superposition of harmonic waves. Here, the differential equation is transformed into an algebraic equation, with Fourier components representing the spectrum of the starting distribution. The most important feature of such a procedure is that every step of the transformation is linear, even though, in the case of the inverse scattering transform that we want to discuss here, the starting equation may be strongly nonlinear.

In the case of a nonlinear wave problem, we can consider the equation as a problem of scattering on a potential. The potential is completely described by the starting distribution. Scattering on the potential will create stationary or bound states and a continuous spectrum. This sounds familiar to the case of the soliton solution problem: a wave may evolve into a soliton solution and a continuous, decaying wave.

Now let us consider the soliton equation again. We want to derive two linear equations from it that are connected to the original equation by a defined integration condition. The structure of the equations is defined by the spectrum of the so-called *inverse scattering technique* as described above. The procedure is shown in a simplified schematic in Fig. 2.2 in comparison with the well-known procedure of the Fourier transform. A consequence of this approach is that any integrable equation can be solved (i.e. integrated) exactly, as an initial-value problem.

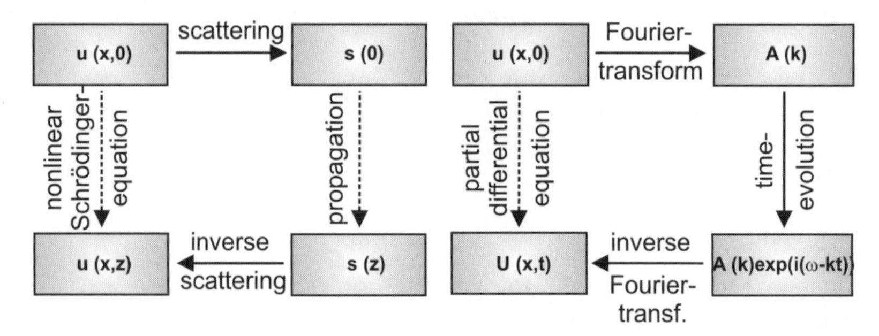

Fig. 2.2. Principle procedure of the inverse scattering method (*left*) in comparison with the Fourier transform algorithm (*right*)

In the following discussion we use the normalized nonlinear Schrödinger equation (2.8) and introduce κ as a tuning parameter for the nonlinearity; $\kappa > 0$ indicates the self-focusing regime, whereas $\kappa < 0$ represents the self-defocusing regime. The equation that we consider is

$$\mathrm{i}\frac{\partial}{\partial z}U + \frac{\partial^2}{\partial x^2}U + \kappa|U|^2U = 0 . \tag{2.14}$$

The inverse scattering method is applicable to equations of the type

$$\frac{\partial U}{\partial z} = S(V) , \tag{2.15}$$

where S, generally speaking, is a nonlinear differential operator in z, which can be represented in the form

$$\frac{\partial L}{\partial z} = \mathrm{i}[L, A] . \tag{2.16}$$

Here, L and A are linear differential operators containing the function $U(x, z)$ in the form of a coefficient. The commutator $[L, A] = LA - AL$ is nothing more than the integrability condition that connects the linear and nonlinear equations associated with a partial differential equation. Therefore, the step from nonlinearity to linearity is to associate a set of linear problems with the nonlinear differential equation, provided that the condition (2.16) is satisfied. The pair of linear differential equations is then found, and the spectrum of the operator L does not depend on z. The asymptotic characteristics of its eigenfunctions can be calculated for any value of z. Finding an eigenvalue equation which has such a property, for example that the eigenfunction S becomes independent of z when the potential satisfies certain nonlinear evolution equations, and the reconstruction of the function $U(x, z)$, is a common feature of the process of finding soliton solutions for nonlinear propagation equations.

The inverse scattering method was developed by Gardner and coworkers [12], who first applied this technique to the well-known Korteweg–de Vries equation describing nonlinear waves in weakly dispersive media. Gardner et al. revealed the fundamental role played by particular solutions of the Korteweg–de Vries equation. These soliton solutions are directly connected with the discrete spectrum of the operator: it was established that the asymptotic state for $z \to \pm\infty$ for many initial conditions is a finite set of solitons.

Lax [13] showed, soon after this discovery, that this method is applicable to all equations which can be represented in the form (2.16), in the case when the eigenfunctions of the operator can be reconstructed from its scattering matrix. Therefore, the solution has the same characteristics for most integrable equations. Smooth initial data in the x direction that vanishes rapidly for $x \to \pm\infty$ evolves as $z \to \infty$ into a finite number of N solitons and an oscillating wave train. The solitons typically travel at N different speeds, while the wave train decays in amplitude and spreads out in space. The oscillatory wave train is qualitatively similar to the solution of the linearized problem obtained by simply omitting the nonlinear terms.

Solitons have no counterpart in the linearized problem. Because, more than a century before solitons were discovered, Liouville proved that certain finite-dimensional Hamiltonian systems could be integrated completely if there exists a canonical transformation from the original coordinates to what are called *action-angle variables*, we may assert that equations that admit solitons are Hamiltonian, and that the inverse scattering transform is a canonical transformation of the action-angle variables of the problem. Therefore, in a strict mathematical sense, equations that admit solitons are nontrivial examples of infinite-dimensional Hamiltonian systems that are completely integrable in the sense of Liouville.

Zakharov and Shabat [4, 14] applied the inverse scattering method in 1970 to the nonlinear Schrödinger equation (2.14). They discovered that the eigenvalue of the Dirac-type eigenvalue equation

$$L\Phi = \lambda\Phi \;, \tag{2.17}$$

where

$$\Phi = \begin{pmatrix} \Phi_1 \\ \Phi_2 \end{pmatrix} \tag{2.18}$$

and

$$L = \mathrm{i}\begin{bmatrix} 1+\beta & 0 \\ 0 & 1-\beta \end{bmatrix}\frac{\partial}{\partial x} + \begin{bmatrix} 0 & U^* \\ U & 0 \end{bmatrix} \tag{2.19}$$

becomes invariant in z if U evolves in accordance with the nonlinear Schrödinger equation. Here, $\beta^2 = 1 - 2/\kappa = \text{const.}$ represents a propagation constant, and the spatial evolution along the propagation direction of the eigenfunction Φ is given by

$$\mathrm{i}\frac{\partial\Phi}{\partial z} = \mathcal{A}\Phi \;, \tag{2.20}$$

where

$$\mathcal{A} = -\beta\begin{bmatrix} 1 & 0 \\ 0 & 1 \end{bmatrix}\frac{\partial^2}{\partial x^2} + \begin{bmatrix} \frac{|U|^2}{1+\beta} & \mathrm{i}\frac{\partial U^*}{\partial x} \\ -\mathrm{i}\frac{\partial U}{\partial x} & \frac{-|U|^2}{1-\beta} \end{bmatrix} . \tag{2.21}$$

Without loss of generality, it can be assumed that $\kappa > 2$ and $\beta^2 > 0$.

For the nonlinear Schrödinger equation, the appropriate structure of the eigenvalues of (2.20) can be written as [5]

$$\mathrm{i}\frac{\partial\Phi_1}{\partial x} + U_0(x)\Phi_2 = \zeta\Phi_1 \;, \tag{2.22}$$

$$-\mathrm{i}\frac{\partial\Phi_2}{\partial x} - U_0^*(x)\Phi_1 = \zeta\Phi_2 \;. \tag{2.23}$$

If we write the eigenvalue in the form

$$\zeta = b + \mathrm{i}a \,, \tag{2.24}$$

the particular solution of (2.14) derived by the inverse scattering method as described above is given by [5]

$$U(x,z) = \sqrt{2\kappa}\, b \exp\left[-4\mathrm{i}(b^2 - a^2)z - 2\mathrm{i}bx + \mathrm{i}z_0\right] \times \mathrm{sech}[2a(x - x_0) + 8abz] \,. \tag{2.25}$$

We shall refer to this solution as a *soliton*, as well. Generally speaking, a soliton solution to the nonlinear Schrödinger equation is characterized by four constants. Two arbitrary constants, a and b, determine its amplitude and direction, respectively. These two constants are independent of each other, allowing one to form solitons of different amplitudes and directions. In addition, x_0 is the starting coordinate and z_0 a phase constant. The amplitude and angular direction of the soliton are characterized by the imaginary and real parts of the eigenvalue. The solution in (2.25) is the simplest representative of an extensive family of exact solutions of (2.14), which can be expressed in explicit form. In the self-focusing problem, a soliton has the meaning of a homogeneous waveguide channel inclined at an angle $\Theta = \arctan(1/4b)$ to the z axis (or to the direction of the k-vector of the initial wave). More details of the inverse scattering method can be found in [14, 15].

2.2.3 Stability Analysis

The stability properties of a soliton are of huge importance for their physical feasibility. One can easily derive these properties by using a stability criterion for soliton solutions of the general nonlinear Schrödinger equation (2.5). This criterion was first derived by Kolokolov and coworkers [16, 17] (for a general review see [18]). The important information about the stability of a solution can be extracted from the dependence of the beam power P on the nonlinear shift of the constant of integration, which in our case is the propagation constant β.

A necessary and sufficient criterion for the stability of soliton solutions of (2.14) with an arbitrary nonlinearity $f(|u|^2)$ is given by

$$\frac{\partial \mathcal{E}}{\partial \beta}(\beta) = \frac{\partial}{\partial \beta}\left(\int_V |\mathcal{E}(x,z')|^2 \,\mathrm{d}x\right) > 0 \,, \tag{2.26}$$

where $\mathcal{E} = \int |U(x)|^2 \mathrm{d}x$ is the soliton energy in normalized units. Therefore, the function $\mathcal{E}$ needs to be a monotonically increasing function of β.

The validity of this so-called *Vakhitov–Kolokolov criterion* is based on the specific properties of the eigenvalue problem that appears after linearizing the nonlinear Schrödinger equation near the solitary wave solution. The condition (2.26) is associated with the existence of only one negative eigenvalue of that problem. If this condition is fulfilled in the whole parameter space, the soliton solution is stable against small fluctuations. If the condition is not fulfilled, the stability criterion may not be formulated directly in terms of the beam power P, but becomes more complicated and depends on the particular structure and type of nonlinearity.

A linear stability analysis does not typically allow one to predict the subsequent evolution of unstable solitons. Therefore, the analysis of the stability or instability can be extended with more complex dynamic studies of the nonlinear wave equation, giving qualitative results about the possible deviation of the initial beam shape from the soliton solution. Recently, the *nonlinear theory of soliton instabilities* has been suggested in order to obtain quantitative results about the evolution of unstable solitons [19]. This theory is based on the multiscale asymptotic technique and allows one to describe soliton evolution near the stability threshold, i.e. the marginal stability curve. According to this approach, unstable bright solitons display three general scenarios for their evolution: they diffract, collapse or switch to a stable state with long-lived oscillations of their amplitude. These *oscillating solitons* can exist owing to the presence of soliton internal modes, which is one of the major properties that separate integrable from nonintegrable models.

In (2.25), the integrated intensity of the radiation in the soliton is given by

$$I = \int |U|^2 \, \mathrm{d}x = 4a \; . \tag{2.27}$$

This means that a soliton can exist for any amplitude and direction.

2.2.4 Conservation Laws

The nonlinear Schrödinger equation is integrable owing to the fact that it has an infinite number of conserved quantities. The three lowest-order integrals are defined as

$$Q = \int_{-\infty}^{+\infty} |U(x,z)|^2 \mathrm{d}x \; , \tag{2.28}$$

$$M = \int_{-\infty}^{+\infty} \left(|U^*| \frac{\partial U}{\partial x} - U \frac{\partial U^*}{\partial x} \right) \mathrm{d}x \; , \tag{2.29}$$

$$H = \int_{-\infty}^{+\infty} \left(\left| \frac{\partial U}{\partial x} \right|^2 - |U|^4 \right) \mathrm{d}x \; , \tag{2.30}$$

and are called as energy (Q), momentum (M) and Hamiltonian (H). All of these quantities remain constant as the soliton propagates. Even in for nonintegrable and hence nonconservative systems, these lowest-order integrals can be calculated. They are a very effective tool for investigating the formation of solitary waves and their stability.

One of the most important consequences of the integrability of the nonlinear Schrödinger equation and these conservation laws is that not only do the solitons conserve power and velocity upon collision, but they also retain their shape and their absolute number. Generally speaking, they display particle-like properties, since all collision processes display fully elastic features.

2.3 Self-Focusing in Non-Kerr Media – Hamiltonian Systems

The idealized situation of a Kerr-like nonlinear medium does not include a wealth of phenomena that appear during self-trapping of optical beams in real nonlinear optical media, owning to two main factors: higher-order nonlinearities and higher degrees of freedom of the nonlinear medium. In many cases, higher intensities give rise to a field-induced change in the refractive index that is influenced by higher-order nonlinearities. Often, the nonlinearity becomes a cubic–quintic one such as,

$$\Delta n(|A|^2) = n_2|A|^2 + n_4|A|^4 \, . \tag{2.31}$$

As an example, if the medium nonlinearity is cubic, wave-coupling processes may give rise to resonant four-wave-mixing effects, which in turn result in a strong coupling of the envelope of the fundamental field to a secondary field. In this case, the simple nonlinear Schrödinger equation is no longer valid. Instead, a coupling between the fundamental and higher-order modes may support multicomponent solitary waves which differ drastically from the conventional solitons of the scalar nonlinear Schrödinger equation. As long as the higher-order nonlinearity is stronger than cubic, e.g. a power-law focusing, the nonlinear Schrödinger equation has localized solutions which blow-up, so that a singularity appears at finite z. This blow-up or *collapse* means that the model of the nonlinear Schrödinger equation fails as an envelope equation, since it breaks the scales on which it was derived in the framework of the multiscale asymptotic technique. For spatial solitons, this condition means that if the transverse dimension $D = 2$, then the cubic nonlinearity $|A|^2 A$ is already sufficient to induce collapse. In the case of $D = 1$, quintic or even higher-order nonlinearities are necessary to induce a collapse. Blow-up also indicates nonparaxiality in the self-focusing of the beam.

However, catastrophic beam self-focusing can be prevented and corrected for by any kind of competing nonlinearity, for example when competition between self-focusing ($n_2 > 0$) at smaller intensities and self-defocusing ($n_4 < 0$) at larger intensities exists. In a more general case, models with competing nonlinearities can be described by a power-law dependence on the beam intensity

$$\Delta n_{\rm nl}(I) = n_{2p}(|A|^2)^p + n_{4p}(|A|^2)^{2p} \, , \tag{2.32}$$

where p is a positive constant andi, usually, $n_{2p}n_{4p} < 0$.

As one of the typical consequences of these competing higher-order terms, the optically induced refractive index change may become saturated at higher field strengths. From an experimental point of view such an effect is quite natural, because for any real material there is an upper limit to the refractive index change that can be induced optically. The field strength at which saturation occurs depends on the particular process that causes the nonlinear index change. In all cases, a physical limit is imposed onto the case of soliton propagation, and we have to consider these effects in order to investigate deviations from solitary wave behavior. Effects that lead to saturation include third-order dispersion, nonlinear dispersion, self-steering effects and the delayed response of the nonlinearity of a fiber in the case of temporal solitons. In the case of spatial solitons, a saturable nonlinearity may, for example, be provided by a photorefractive nonlinear response of the material, which will be subject of Chap. 3.

Introducing saturation into a Kerr-like system implies that the equation of propagation (2.5) is no longer integrable, which brings about some dramatic consequences for its solitary solutions. In general, no analytical solutions can be found, and the number of conserved quantities is no longer infinite. Although there exist solitary solutions that might be written in an analytic form, their generic behavior differs remarkably from that of the solutions of an integrable system. First of all, the stability of the soliton solutions differs drastically from that of solitons in a pure Kerr system. Collisions between two stable solitons can be characterized as inelastic, since the particle-like waves undergo a shape transformation and eventually exchange energy. The solitons do not maintain their profile and shape during a collision. Further, the number of solitons is not necessarily conserved in a collision process, which gives rise to soliton fission and fusion effects.

An additional and very important aspect is the transfer of energy into radiation. Starting from arbitrary initial conditions, a stable soliton might evolve in a self-focusing saturable nonlinear medium while part of the energy is being transformed into radiation. The total energy in the system is still conserved, but is no longer concentrated entirely within the soliton. This radiation is typical of solitons in saturable media and is especially important in collision processes.

From a general point of view, all physical systems that support the generation of any kind of solitary structure can be classified into three main categories: integrable, Hamiltonian and dissipative systems [21]. Integrable systems have analytic solutions, with an infinite number of conserved quantities and therefore represent a subclass of Hamiltonian systems, as depicted in Fig. 2.3. The latter lack integrability and possess only a limited number of conserved quantities. Nevertheless, solitary solutions exist and comprise at least a one-parameter family of solutions. Hamiltonian systems themselves, finally can be regarded as a subclass of dissipative systems, which include effects of gain and loss. Solitary solutions of dissipative systems differ drastically from those of Hamiltonian or even integrable system. Their main features are illustrated in Sect. 2.4.

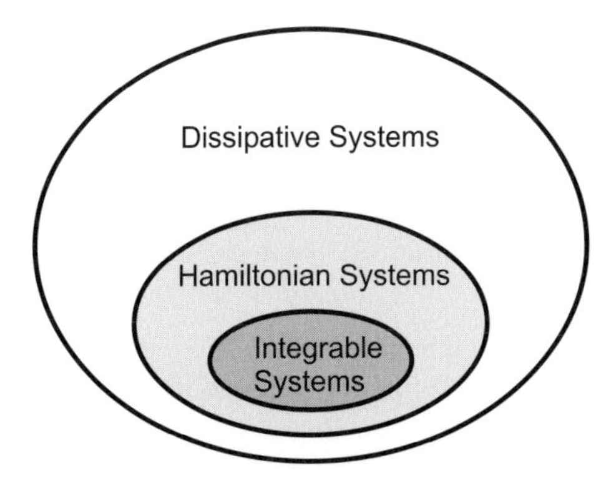

Fig. 2.3. Schematic classification of soliton-supporting systems. Integrable systems can be interpreted as a subclass of Hamiltonian systems, which in turn represent a subclass of dissipative systems (after [20], printed by kind permission of Nail Akhmediev)

Many of the above-mentioned aspects contradict the strict mathematical definition of a soliton as a very robust, confined entity of energy in time or space which displays elastic collision characteristics. Nevertheless, the term "soliton" has been widely used for several kinds of solitary waves that result from self-trapping. Since we deal with self-focused light structures in saturable photorefractive nonlinear media in Chap. 4, we also make use of the term "soliton" in this broader sense. However, the reader should keep in mind the essential difference between solitons in integrable and nonintegrable systems.

Even though a saturable medium is not integrable, it remains conservative in many respects. A distinct number of conserved quantities such as the energy Q, momentum M and Hamiltonian H (see (2.28–2.30)) can still be defined and exploited for the purpose of finding soliton solutions and their subsequent characterization. Despite the fact that the system is not integrable, its solitary solutions display similar features to solitons in integrable systems. In particular, when the deviations from an exact Kerr nonlinearity are considerably small, many properties of an integrable Kerr system can still be observed. Among the lowest-order conserved quantities, the Hamiltonian H plays a crucial role in the investigation of integrable and nonlinear wave systems in general [22], because the Hamiltonian formalism is extremely useful for the stability analysis of solitary solutions in nonintegrable systems. Hamiltonian systems comprise a one-parameter family of soliton solutions. From a two-dimensional representation in the form of a Hamiltonian versus energy $(H-Q)$ diagram [21], stationary solutions can easily be characterized.

The definition of the energy Q is given by (2.28), whereas the Hamiltonian for an arbitrary $f(|A|^2)$ in (2.5) is defined by

$$H = \int_{-\infty}^{+\infty} \left(\left| \frac{\partial u}{\partial z} \right|^2 - |f(|A|^2)|^2 \right) \mathrm{d}z \, . \tag{2.33}$$

From a knowledge of the Hamiltonian, the behavior of the stationary solutions of the system can be defined to a large extent. Stationary soliton solutions can generally be found by using the ansatz

$$A(x, z) = a(x)\mathrm{e}^{\mathrm{i}\beta z} \, , \tag{2.34}$$

where β again represents a propagation constant. The amplitude $a(x)$ is independent of the propagation coordinate z, and $\mathrm{e}^{\mathrm{i}\beta z}$ is a phase term which is independent of the transverse coordinate x. By use of to the symmetry considerations derived in Sect. 2.2.1, a one-parameter family of solutions

$$A(x, z, q) = a(x, q)\mathrm{e}^{\mathrm{i}\beta z} \, , \tag{2.35}$$

where q is a real soliton parameter, can be generated. The equation for stationary solutions can then be expressed in a variational formulation as

$$\delta(H - qQ) = 0 \, . \tag{2.36}$$

This formalism does not only derive stationary solutions. Its greatest benefit is the characterization of stationary solutions, which can be determined to be stable or unstable. For any fixed Q, a soliton solution is stable if the corresponding H has a local minimum, with q being a Lagrange multiplier [21].

Another source of phenomena beyond Kerr-like solitons is associated with the surroundings of the beam, which may include more dimensions of freedom than the soliton needs for stable propagation, giving rise to a broader class of perturbations and therefore instabilities of the solitary wave solution.

In the following, we shall discuss these aspects and the changes they introduce into the formation of spatial solitary beams compared with the exact soliton solution obtained in the case of a pure Kerr nonlinearity.

2.3.1 One-Dimensional Self-Focusing in Saturable Media

The influence of saturation in the self-focusing problem was examined analytically quite early on by use of the paraxial ray approximation [23, 24], the moment theory [25] and the variational approach [26]. All these approaches are based on the constant-shape approximation and predict an oscillating beam radius that produces periodic foci of the beam with propagation. Numerical simulations [27, 28] confirmed some of these approximate results. On the other hand, these simulations revealed that in general the constant-shape approximation can be violated, and under certain conditions the propagating beam can develop into a series of concentric ring structures. Later, numerical simulations [29, 30] of self-focusing in a saturable nonlinear medium predicted azimuthal-symmetry breaking and beam breakup under certain conditions.

Here, we introduce spatial-soliton formation for saturable nonlinearities with respect to its potential to realize higher-dimensional spatial solitons that cannot be stable in the case of a pure Kerr nonlinearity. This model is extremely useful for describing soliton formation processes in a series of experimental systems, such as off-resonant interaction of an optical beam in an atomic vapor system, or self-trapping due to saturation of the AC Stark shift of two-photon states. In addition, we shall focus in Chap. 4 on a photorefractive system which also displays a nonlinearity of saturable type.

In order to take into account the saturation behavior, the nonlinear Schrödinger equation (2.5) can be modified in a simple way by setting

$$f(|A|^2) = \frac{|A|^2}{1 + |A|^2/I_{\mathrm{sat}}} , \tag{2.37}$$

where I_{sat} is the saturation intensity. By making use of the same scaling formalism as in (2.8), we derive the normalized propagation equation for a saturable nonlinearity

$$\mathrm{i}\frac{\partial}{\partial z'}A' + \frac{1}{2}\frac{\partial^2}{\partial x'^2}A' + \frac{|A'|^2 A}{1 + \sigma|A'|^2} = 0 . \tag{2.38}$$

Here, $\sigma = n_0/(kn_2 I_{\mathrm{sat}})$ is the only parameter in (2.38) and is referred to as the saturation parameter. It contains the saturation intensity on the one hand and the strength of the nonlinearity, given by n_2, on the other hand. When the field intensity, given by $|A'|^2$, is much smaller than the saturation intensity, i.e. $|A'|^2 \ll I_{\mathrm{sat}}$, the system operates in a Kerr regime and the last term of (2.38) becomes proportional to $|A'|^2$. In the other limit, in which $|A'|^2 \gg I_{\mathrm{sat}}$, the nonlinear term in (2.38) approaches zero and the impact of the nonlinear effect decreases. In this way, the nonlinearity displays an upper threshold, a feature which becomes very important when it comes to self-focusing in both transverse dimensions (see Sect. 2.3.3).

2.3.2 One-Dimensional Spatial Solitons

In close analogy to the case of Kerr-type solitons, a single known solution $U(x, z, \sigma)$ of (2.38) can be extended to a one-parameter family of solutions by applying the scaling transformation (2.11):

$$U'(x, z, \sigma) = aU(ax,\ a^2 z, \sigma/a^2) , \tag{2.39}$$

where a can be any real parameter.

In the following, we seek a solution of (2.38) that describes a localized beam propagating undisturbed through a saturable self-focusing medium. Therefore, we make use of the general ansatz according to (2.34) and follow the approach of Gatz and Herrmann [31].

The complex amplitude can be expressed in terms of the real variables ϱ and φ by

$$A'(x',z') = \sqrt{\varrho(x',z')}\,\exp\left[\mathrm{i}\varphi(x',z')\right] . \tag{2.40}$$

Substituting this expression into (2.38) separates it into two linear equations [31]:

$$\frac{\partial \varrho}{\partial z'} + \frac{\partial}{\partial x'}\left(\varrho \frac{\partial \varphi}{\partial x'}\right) = 0 , \tag{2.41}$$

$$\frac{1}{4\varrho}\frac{\partial^2 \varrho}{\partial x'^2} - \frac{1}{8\varrho}\left(\frac{\partial \varrho}{\partial x'}\right)^2 + \frac{\varrho}{1+\sigma\varrho} = \frac{\partial \varphi}{\partial z'} + \frac{1}{2}\left(\frac{\partial \varphi}{\partial x'}\right)^2 . \tag{2.42}$$

The bright fundamental soliton solution of this system of partial differential equations is a localized solution of ϱ that is stationary in z' and has a phase φ that does not depend on x'. In other words, we require $\partial\varrho/\partial z' = 0$. From (2.41), we obtain

$$\varrho(x')\frac{\partial \varphi}{\partial x'}(x',z') = C(z') , \tag{2.43}$$

where $C(z')$ is an arbitrary function resulting from the integration. In analogy to the case of the nonlinear Schrödinger equation for a Kerr nonlinearity, one can show that the condition for a localized solution that is stationary in z' can only be satisfied when $C = 0$. Consequently, we obtain

$$\varphi = \beta z' + \varphi_0 , \tag{2.44}$$

where β is a constant of integration. By substituting this expression into (2.42), we obtain the following by integration with respect to ϱ [31]:

$$\left(\frac{\partial \varrho}{\partial x'}\right)^2 = 8\left(\beta - \frac{1}{\sigma}\right)\varrho^2 + \frac{8}{\sigma^2}\varrho \ln\left(1+\sigma\varrho\right) . \tag{2.45}$$

Since we are interested in a localized solution with a maximum $\varrho(x' = 0) = \varrho_0$, we can find from (2.45) with this initial condition that

$$\beta = \frac{1}{\sigma} - \frac{\ln\left(1+\sigma\varrho_0\right)}{\sigma^2\varrho_0} . \tag{2.46}$$

The differential equation (2.45) can then be transformed into an integral equation for ϱ [31]:

$$\varrho(x') = \varrho_0 \exp\left(\int_0^{x'} 2\sqrt{2}\left\{\frac{\ln\left(1+\sigma\varrho(\xi)\right)}{\sigma\varrho(\xi)} - \frac{\ln\left(1+\sigma\varrho_0\right)}{\sigma\varrho_0}\right\}^{1/2} \mathrm{d}\xi\right) . \tag{2.47}$$

The solution of (2.47) can no longer be found analytically, which is the major difference from the pure Kerr-type nonlinear Schrödinger equation (2.8). The procedure for the numerical integration can be found in [31] and

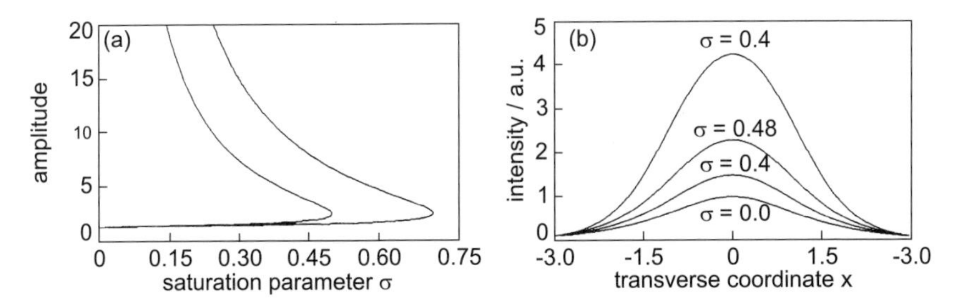

Fig. 2.4. Dependence of the soliton amplitude $\sqrt{\varrho_0}$ (**a**) and of the soliton shape (**b**) on the saturation parameter σ (after [31], printed by kind permission of Joachim Herrmann)

will not be repeated here. We describe only the most interesting features of the solutions.

For various different saturation parameters σ, the shapes of soliton solutions (or soliton amplitudes) can be found. For small σ ($\sigma < 0.2$), the soliton amplitude remains small and stationary and thus is comparable to the solution of the nonlinear Schrödinger equation for a Kerr nonlinearity (2.8). For larger values of σ, the soliton amplitude increases slightly, up to a limit $\sigma_{\max}$. For $\sigma > \sigma_{\max}$, no soliton solution exists. The most striking phenomenon is that the soliton amplitude becomes a two-valued function of the parameter σ depending on the soliton beam width. This situation is depicted in Fig. 2.4a. For illustration, the shapes of two solitons corresponding to the upper and lower solution branches for $\sigma = 0.4$ are depicted in Fig. 2.4b. The two limiting cases given by $\sigma = 0$ (no saturation) and $\sigma = \sigma_{\max} = 0.48$ are also shown, where only one soliton solution exists.

Therefore, from the two-valued dependence, we find that two soliton states with different peak intensities or different powers exist for the same saturation paramater σ. This feature will be discussed extensively in Sect. 2.3.3 for the (2+1)-dimensional geometry which is closer to the real experimental system. Such a property is similar to a property of optical bistable systems: there exist two possible (output) soliton states for the same value of an input parameter w. Therefore, we denote the soliton solution of (2.38) as *bistable solitons.*

2.3.3 Two-Dimensional Self-Focusing Effects in Saturable Media

We now extend the problem of solitary-beam generation to propagation in the full three-dimensional geometry, in which the beam propagates along the z coordinate and diffracts in the transverse x–y plane. This configuration is commonly referred to as (2+1)-dimensional to distinguish between the transverse and the longitudinal directions.

Catastrophic Collapse in Kerr Media

In contrast to the (1+1)-dimensional case in a Kerr system described in Sect. 2.2, soliton solutions do not exist at all in the (2+1)-dimensional geometry, and the propagation of a circular beam is generally unstable. Instead of forming a circular soliton, the beam undergoes catastrophic self-focusing and breaks up [32]. Indeed, many early experiments in nonlinear optics demonstrated this catastrophic self-focusing, which frequently led to damage [3].

The reason for this is that the solution now belongs to a subclass of all possible solutions and has a lower dimensionality than the basic set of equations in the (2+1)-dimensional geometry. It contains a free, but "hidden", coordinate, and is at the same time independent of this coordinate. Therefore, it is now subject to a broader class of perturbations. Perturbations that affect the hidden coordinate are of particular importance. The corresponding *symmetry-breaking instability*, which is due to the growth of the perturbations along the hidden, initially homogeneous coordinate, is called a *transverse modulation instability*. It characterizes a breakup of the continuous wave (or plane wave) and results in the filamentation of a one-dimensional stripe or two-dimensional beam during propagation, which prevents stable solitary-wave propagation. Often, this instability is connected with the generation of trains of localized beams. The study of this instability goes back to papers by Bespalov and Talanov [33] as well as Benjamin and Feir [34, 35], who discussed its manifestations for a homogeneous or plane-wave ground state. Subsequently, several analyses of the stability of two-dimensional self-bound solutions in a Kerr medium [36, 37], which is described by the cubic Schrödinger equation, predicted a catastrophic unlimited self-focusing, which leads to a random filamentation of an optical beam if the initial power is larger than the critical power for self-focusing $P_{\text{crit}} = c\lambda^2/(\pi n_0 n_2)$ given by (2.3). For a power smaller than this critical power, the beam diffracts.

One important feature of this instability is that P_{crit} is independent of the beam size, which means that variations in the peak power result in instability and not in self-trapping. The main result of these investigations is that Kerr-type solitons are stable only in (1+1) dimensions – they can be observed in single-mode waveguides, but not in a bulk material. These instability properties are universal and are exhibited by all Kerr solitons in nature.

Influence of Saturation

For other nonlinearities, the behavior may be different. In order to examine the stability of two-dimensional solitons in saturable media, we refer once again to the investigations described in Sect. 2.3.1 for the case of one-dimensional solitons in saturable media. Saturation allows one to form solitons that differ from those created in Kerr-type media. They have two solution branches that are sensitive to perturbations. If the initial beam differs drastically from the solitary wave solution, instability occurs, and filamentation processes can be observed. This suggests ways to prevent or at least weaken the catastrophic collapse and generate (2+1)-dimensional self-guided waves or even spatial solitons in a bulk medium.

Although the feasibility of a scheme for self-trapping in a three-dimensional medium using resonant enhancement in an off-resonant atomic vapor was demonstrated by Bjorkholm and Ashkin [38], no attempts to apply these effects to the formation of two-dimensional spatial solitons were made. The main problem here is the strong absorption in an off-resonant interaction, which counteracts the formation of a soliton. This discouraging experimental situation finally led to an almost complete stop of investigations of how to use saturation to realize higher-dimensional optical spatial solitons in the 1970s. The concept of self-trapping of beams was completely lost from view for almost twenty years.

It was only recently that the problem of (2+1)-dimensional spatial solitons was revisited beginning with investigations of beam filamentation. As a result, various approaches for the formation of spatial solitons in bulk media have been studied theoretically and demonstrated experimentally.

In the case of extremely high-power laser pulses of the order of 10 TW, relativistic self-channeling in a dense plasma has been predicted [39, 40] and demonstrated [41, 42]; here, nearly all neutral molecules of the medium are ionized, and saturable self-focusing develops because of relativistic effects, owing to the increase of electron inertia under the influence of an intensive light wave. For a power in the nonrelativistic region of the order of 20 GW, self-channeling of a small-scale filament formed from a femtosecond laser pulse has recently been observed [43]. In this region of laser power, the self-induced photoionization leads to a counteracting self-defocusing against the Kerr nonlinearity of the neutral molecules and therefore induces a saturable effect. The formation of spatial solitons by this mechanism was recently described in [44]. In addition, the formation of (2+1)-dimensional solitons in quadratic nonlinear crystals has been proposed [45, 46], and recently such an effect was demonstrated in a KTP crystal with optical beams with a power of the order of 100 W [47].

The most promising new type of spatial solitary waves are spatial solitons obtained by use of photorefractive nonlinearity in an electro-optic crystal. Their main features and the principle of their generation will be subject of Chap. 4. In order to understand their general behavior, and to understand how to exploit them for investigations of the basic nature of (2+1)-dimensional spatial solitons, we shall describe in the following section the prediction of (2+1)-dimensional soliton solutions for saturable media in general.

2.3.4 Two-Dimensional Spatial Solitons

In 1997, Gatz and Herrmann [48] studied theoretically the propagation of beams and the properties of spatial solitons in two transverse dimensions in a bulk medium for the general case of a local saturable nonlinearity. Although a detailed investigation of the photorefractive effect shows that this type of spatial soliton is not exactly equivalent to that present in photorefractive materials, it gives a general insight into the properties of solitons in two transverse dimensions.

In the case of two transverse dimensions x and y, the normalized nonlinear Schrödinger equation (2.38) changes to

$$\mathrm{i}\frac{\partial}{\partial z'}A + \frac{1}{2}\nabla_{\perp'}A + \frac{|A|^2}{1+\sigma|A|^2} = 0\,, \tag{2.48}$$

where $\nabla_{\perp'}$ represents the transverse Laplace operator $\partial^2/\partial x'^2 + \partial^2/\partial y'^2$. Corresponding to the (1+1)-dimensional case, a new solution of (2.48) can be found from a known solution $U(x,y,z)$ by use of the symmetry relation

$$U'(x,y,z,\sigma) = aU(ax, ay, za^2, \sigma/a^2)\,. \tag{2.49}$$

Two-Dimensional Soliton Solution

Considering a radially symmetric problem, Gatz and Herrmann derived a solution for a localized beam propagating undisturbed through a nonlinear medium by expressing the complex amplitude A in terms of the real variables ϱ and φ and introducing the radial coordinate $s = \sqrt{x^2+y^2}$, such that

$$A(s,z) = \sqrt{\varrho(s,z)}\,\exp\left[\mathrm{i}\varphi(s,z)\right]\,. \tag{2.50}$$

By substitution into (2.48), they obtained [48]

$$\frac{\partial\varphi}{\partial z'} + \frac{1}{8}\left[-\frac{1}{\varrho^2}\left(\frac{\partial\varrho}{\partial s}\right)^2 + \frac{2}{\varrho}\frac{\partial^2\varrho}{\partial s^2} - 4\left(\frac{\partial\varphi}{\partial s}\right)^2 + \frac{2}{\varrho s}\frac{\partial\varrho}{\partial s}\right] + f(\varrho) = 0\,, \tag{2.51}$$

$$-\frac{\partial\varrho}{\partial z} + \frac{\partial}{\partial s}\left(\varrho\frac{\partial\varphi}{\partial s}\right) + \frac{\varrho}{s}\frac{\partial\varphi}{\partial s} = 0\,, \tag{2.52}$$

where $f(\varrho) = \varrho^2/(1+\sigma\varrho^2)$.

Equations (2.51) and (2.52) have, in principle, the same structure as (2.41) and (2.42). Consequently, they can be solved in a similar way to the equations describing the one-dimensional case. For fundamental bright soliton solutions of (2.51) and (2.52), we require that ϱ is stationary in z, with a phase φ that does not depend on s. The further steps to affiliate a single integral equation are described in [48]. The resulting equation can be solved with the help of a numerical relaxation procedure. As in the one-dimensional case, the

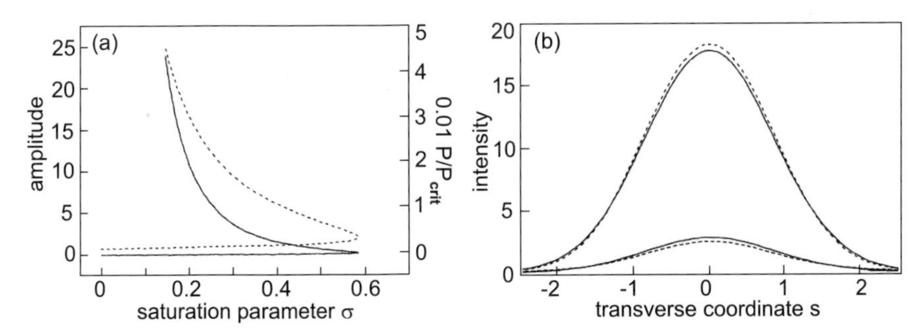

Fig. 2.5. Dependence of the amplitude (*d*ashed curve) and the normalized soliton power (*s*olid curve) on the normalized saturation parameter σ (**a**) and the corresponding shapes of the solitons (*d*ashed curves) for $\sigma = 0.5$ in comparison with the corresponding Gaussian beams with the same power (*s*olid curves) (**b**), (after [48], printed by kind permission of Joachim Herrmann)

solutions are again a two-valued function of the saturation parameter σ. It is remarkable, that for $\sigma > \sigma_{\text{crit}} = 0.59$, no soliton solution exists.

From the two-valued curve shown in Fig. 2.5a it is evident that for the a given parameter σ, two soliton states with different peak intensities exist. In Fig. 2.5b, examples of the shapes of these two solitons for the upper and lower solution branches are shown. The figure shows clearly that the shapes of the solitons differ from those of a Gaussian beam with equal power, illustrated by the solid curves in Fig. 2.5b; this is an important feature of (2+1)-dimensional solitons in saturable media. This fact is an inherent problem in experimental soliton investigations, since the initial beam profile typically features a Gaussian shape. Therefore, in experiments it is very unlikely that the initial light intensity distribution matches an exact soliton profile.

The solid curve in Fig. 2.5a refers to the right-hand scale and shows the normalized power P of the soliton in units of the critical power P_{crit} for self-focusing in a classical Kerr medium without saturation, given by

$$P = \int_0^\infty \varrho(s,\sigma)\, 2\pi s \, \mathrm{d}s \,. \tag{2.53}$$

This curve shows a similar two-valued behavior to the amplitude, and it approaches unity when σ approaches zero. It represents a relation between the power P of the (2+1)-dimensional soliton, the beam radius and the nonlinear saturation coefficient σ.

Stability of Two-Dimensional Solitons

To analyze the stability of the solutions obtained, we make use of the power P as a conserved quantity and apply the Vakhitov–Kolokolov criterion given by (2.26). The soliton power can be expressed in terms of the propagation constant β [48] and it can be shown that the required criterion $\partial P/\partial\beta > 0$ is valid over the whole parameter range, since P is a monotonically increasing function of β. Therefore both solution branches depicted in Fig. 2.5 are stable against small fluctuations.

Conservation Laws for Two-Dimensional Solitons

Analogously to the one-dimensional Kerr case (2.28), one can define some conserved quantities for the equations of the two-dimensional system. By multiplying (2.48) by A^* and integrating over the transverse plane, we obtain the first invariant, the power

$$P = 2\pi \int_{-\infty}^{+\infty} \varrho(x,y)\,\mathrm{d}x\,\mathrm{d}y\ , \tag{2.54}$$

where we have used the representation $A = \sqrt{\varrho}\exp(\mathrm{i}\varphi)$.

Two further invariants are given by

$$\Theta_x = \int_{-\infty}^{+\infty} \varrho\frac{\partial\varphi}{\partial x}\,\mathrm{d}x\,\mathrm{d}y\ , \tag{2.55}$$

$$\Theta_y = \int_{-\infty}^{+\infty} \varrho\frac{\partial\varphi}{\partial y}\,\mathrm{d}x\,\mathrm{d}y\ . \tag{2.56}$$

Multiplying by $\partial\varphi^*/\partial z$, we obtain the Hamiltonian [48]

$$H = \int_{-\infty}^{+\infty} \left(\left|\frac{\partial\varphi}{\partial x}\right|^2 + \left|\frac{\partial\varphi}{\partial y}\right|^2 - 2F(\varrho) \right) \mathrm{d}x\,\mathrm{d}y\ , \tag{2.57}$$

where $F(\varrho)$ is given by

$$F(\varrho) = \int_0^{\varrho} f(I)\,\mathrm{d}I = \varrho/\sigma - [\ln(1+\sigma\varrho)]/\sigma^2\ . \tag{2.58}$$

The Hamiltonian can, finally, be represented in the form $H = H_{\mathrm{I}} + H_{\mathrm{kin}}$, where

$$H_{\mathrm{I}} = \int_{-\infty}^{+\infty} \left\{ \frac{1}{4\varrho}\left[\left(\frac{\partial\varrho}{\partial x}\right)^2 + \left(\frac{\partial\varrho}{\partial y}\right)^2 \right] - 2F(\varrho) \right\} \mathrm{d}x\,\mathrm{d}y\ , \tag{2.59}$$

$$H_{\mathrm{kin}} = \int_{-\infty}^{+\infty} \varrho \left[\left(\frac{\partial\varphi}{\partial x}\right)^2 + \left(\frac{\partial\varphi}{\partial y}\right)^2 \right] \mathrm{d}x\,\mathrm{d}y\ . \tag{2.60}$$

These invariants are essential for an understanding of the properties of solitons. The invariant P can be considered as a mechanical analogue of a mass, Θ_x and Θ_y are the components of the momentum, and H is the energy, which is the sum of the kinetic energy H_{kin} and the negative internal energy H_{I} which describes an analogue of a binding force. In particular, for solitons, (2.57) is given by $H = H_{\mathrm{I}} + (\Theta_x^2 + \Theta_y^2)/4P$. H_{I} is a decreasing function of P/P_{crit}, which can be interpreted as an analogue of the mass defect in nuclear fusion. A single fused soliton possesses a smaller internal energy H_{I} than that of two separated solitons. The dramatic change in the interaction of solitons in saturable media that is due to this effect is the most important and exciting topic in Chap. 5 and leads to soliton fusion and repulsion, which is not known in classical soliton interaction.

From this perspective, it is clear that there is no longer a simple mapping between temporal and spatial optical solitons. Spatial solitons can exist in (2+1)-dimensional media and are therefore a much richer and more complicated phenomenon. Often, stationary beam propagation in planar waveguides has been considered somewhat similar to pulse propagation in fibers. This is a direct manifestation of the *spatio-temporal analogy* in wave propagation: the propagation coordinate z is treated as the evolution variable and the spatial beam profile along the transverse direction, in the case of a waveguide, is similar to the temporal pulse profile, in the case of a fiber. This analogy has already been cited above for the case of (1+1)-dimensional solitons, and is based on the simple notion that both beam evolution and pulse propagation can be described by the cubic nonlinear Schrödinger equation.

However, there is a crucial difference between these two phenomena that manifests itself strongly as soon as we go to higher dimensions. In the case of nonstationary pulse propagation in a fiber, the operating wavelength is usually selected to be near the zero of the group velocity dispersion. Therefore, the absolute value of the fiber dispersion is small enough to be compensated by a weak nonlinearity such as that produced by the (very weak) Kerr effect in an optical fiber which leads to a relatively small nonlinearity-induced change in the refractive index. Consequently, the nonlinearity in such a system is always weak, and it should be modeled well by the same form of the cubic nonlinear Schrödinger equation that is known to be integrable by means of the inverse scattering technique. However, for very short pulses, the cubic nonlinear Schrödinger equation describing the long-distance propagation of the pulse needs to be corrected to include some additional, though small, effects, such as higher-order dispersion and Raman scattering. All these corrections can be taken into account by a perturbation theory. Thus, in fibers, nonlinear effects are weak and become important only when dispersion is small (which means near the zero-dispersion point), in which case they affect pulse propagation over large distances of the order of hundreds of meters or kilometers. In contrast to pulse propagation in optical fibers, the physics underlying stationary beam propagation in planar waveguides and bulk media is completely different. In this case the nonlinear change in the refractive index must compensate for the beam spreading caused by diffraction, which is not a small effect. This is why much larger nonlinearities are usually required in order to observe spatial solitons, and very often such nonlinearities are not of the Kerr type; for example, they saturate at higher intensities. This leads to models of generalized nonlinearities where the properties of solitary waves are different from those described by the integrable cubic nonlinear Schrödinger equation (2.48).

Dynamical Behavior of Two-Dimensional Spatial Solitons in Saturable Media

One of the consequences of the different conservation laws of two-dimensional solitons is the different beam propagation behavior that can arise for more general initial conditions in the case of a saturable medium. Gatz and Herrmann investigated this question in detail [48], by numerically solving the nonlinear wave equation (2.48) with the help of a beam propagation method [49], for different saturation parameters σ, different initial powers and different initial beam shapes.

For a rather large saturation parameter ($\sigma = 0.5$), the field profile $|A(0,z)|^2$ obtained from an initial Gaussian beam shape with a power in the lower solution branch of the soliton solution has an unexpected evolution compared with the one-dimensional case. The beam does not converge to a perfect soliton shape during propagation, but shows small oscillations around the soliton parameters, as illustrated in Fig. 2.6. This remarkable oscillation is due to the fact that the initial Gaussian profile does not match the exact soliton shape. To provide a deeper insight into this behavior, the evolution of the on-axis intensity $\varrho(z) = |A(0,z)|^2$ for different powers is shown in Fig. 2.7. For initial soliton shapes in the lower and upper solution branches, the beam parameters show an exact steady-state behavior, whereas Gaussian input beams with the same power display small oscillations. In the intermediate regions between the two-valued soliton branches, solitons cannot exist, but Gaussian beams show a similar oscillating behavior; sometimes the oscillations are even smaller than in the case of a beam in the soliton existence region. When the saturation parameter σ is decreased, the amplitude of the oscillations or recurrent motions become much larger. An impressive example is shown in the right column of Fig. 2.7a. An initially Gaussian beam with a power in the lower solution branch now undergoes a dominant initial reshaping process with a remarkable change of beam parameters and profile, which is indicated by the solid curve, whereas a beam with a soliton field profile of the same power remains unchanged, as illustrated by the dashed line. For larger powers in the upper solution branch, the Gaussian input beam shows an even more complicated recurrent behavior, in which the initial beam parameters return very close to their initial values quasi-periodically with propagation.

Note that the evolution of the beam in the case of solitons in the upper and lower solution branches differs considerably from the properties of beam propagation in the lower (1+1)-dimensional case, described by the nonlinear

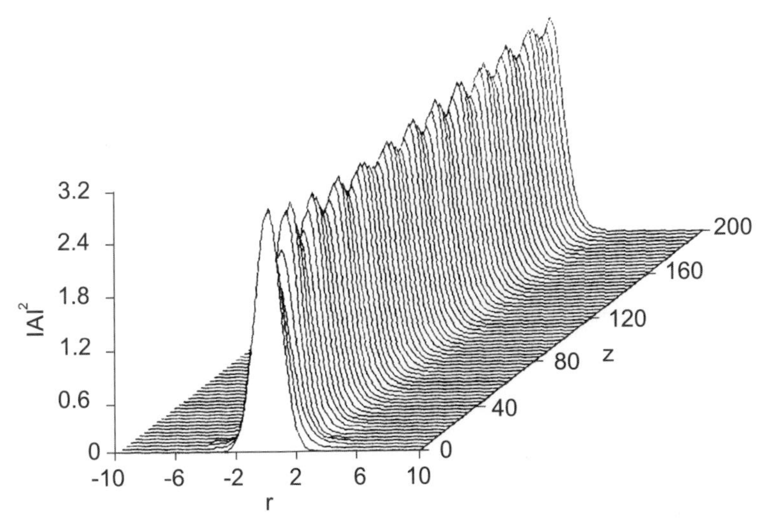

Fig. 2.6. Evolution of a Gaussian input beam with a soliton power in the lower solution branch for $\sigma = 0.5$ (after [48], printed by kind permission of Joachim Herrmann)

Schrödinger equation for a Kerr nonlinearity. In that case, the solutions converge to the soliton solution for an arbitrary initial shape, if the pulse power is equal to or up to 50% larger than the soliton power. When the power is located in the intermediate region or above the upper soliton branch, the strong deviation of the power from the soliton condition does not lead to a larger amplitude of the periodic variation of the beam parameters compared with the case of the upper or lower solution branch.

When the saturation parameter approaches zero ($\sigma = 0.01$, see Fig. 2.8), the behavior shows some qualitative differences compared with the larger-saturation cases discussed above. In the case of an initially Gaussian beam with a power corresponding to the lower solution branch, the initial reshaping process does not evolve to a recurrent motion, but instead diffraction prevails with propagation and leads to a broadening and decreasing beam profile. The reason for this somewhat unexpected behavior is connected with the deviation of the soliton shape from the Gaussian shape, analogous to the deviations visible in Fig. 2.5b. The deviation of the shape leads to diffraction of a small part of the power, ΔP. Consequently, the evolution of the beam does not lead to a new equilibrium state because the reduced power is below the soliton existence curve depicted in Fig. 2.5a. In the case of a Gaussian input beam with an initial power located in the intermediate region between the two solution branches, the amplitude of the oscillations of $\varrho(z)$ becomes very large.

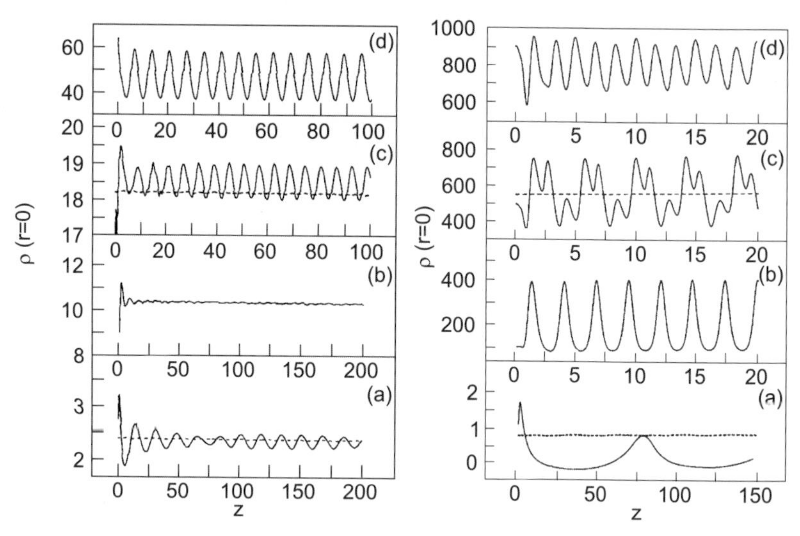

Fig. 2.7. Evolution of the on-axis intensity $\varrho(z)$ of Gaussian input beams (*s*olid curves) or soliton input beams (*d*ashed curves) for different powers Q and different values of σ. *Left*: $\sigma = 0.5$. (**a**) $P = 12.4$ (lower solution branch), (**b**) $P = 40.9$, (**c**) $P = 79.7$ (upper solution branch), (**d**) $P = 291$. *Right*: $\sigma = 0.16$. (**a**) $P = 6.8$ (lower solution branch), (**b**) $P = 454$, (**c**) $P = 2.24 \times 10^3$ (upper solution branch), (**d**) $P = 4.08 \times 10^3$ (after [48], printed by kind permission of Joachim Herrmann)

When the intensity is further enhanced, a strong initial beam narrowing due to the low saturation parameter occurs, leading to an intensity that grows to a factor of 10^3 times its initial value after a propagation distance of $z = 0.25$. For an even larger propagation length, the intensity of the central focal spot decreases, but a spatial ring is formed on the outside. Upon further propagation, the azimuthal symmetry of this ring is broken, and the modulation grows with propagation until it leads to a complete filamentation of the beam. Thus, we obtain again a case of filamentation of a self-focused beam, as already well known for a Kerr nonlinearity, where no stable solitary solutions can exist in the (2+1)-dimensional geometry.

Some of the results found by Gatz and Herrmann have also been reported for other physical systems with a saturable nonlinearity, which underlines the universality of this behavior. A recurrent motion with multiple quasi-periodic foci and the formation of rings were also found for saturable nonlinear media by Marburger and Dawes [27]. However, the critical dependence of the evolution of the beam on its initial shape, as well as on the saturation parameter, and especially on the type of dynamics of slightly perturbed initial soliton beams is rather unexpected, because the soliton beam itself was found to be stable in the framework of the standard approach of linear stability theory (see Sect. 2.2.3).

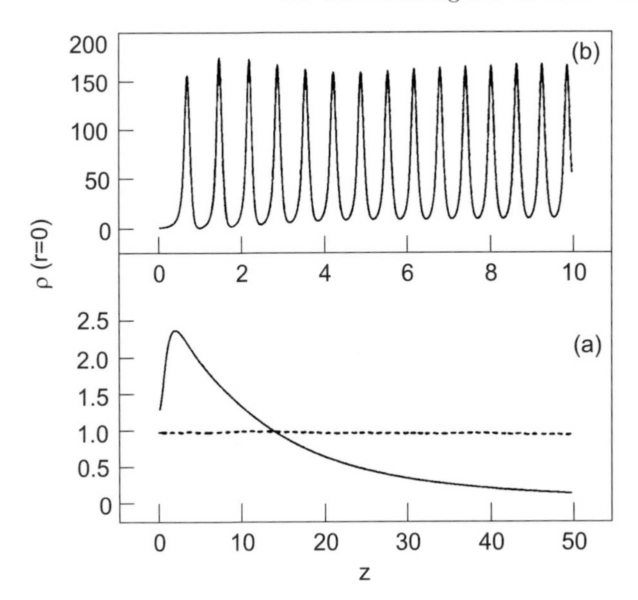

Fig. 2.8. Evolution of the on-axis intensity $\rho(z')$ of Gaussian input beams (*solid* curves) and a soliton input beam (*dashed* curve) for different powers P, for $\sigma = 0.01$. (**a**) $P = 5.9$ (lower solution branch), (**b**) $P = 18.1$ (after [48], printed by kind permission of Joachim Herrmann)

To summarize these results, we can say that on the one hand, catastrophic self-focusing leads to a collapse of the beam for a vanishing saturation parameter σ. On the other hand, an arbitrarily small parameter σ larger than zero cannot stabilize this collapse; as could be concluded from the linear stability analysis alone. Therefore, the general conclusion that can be drawn from the above dynamic study is that spatial solitons in a three-dimensional medium with a local saturable refractive index change can be observed only in a very restricted range of the parameter σ, which imposes certain limitations on the experimental conditions.

We shall show in Chap. 4 that most of the relations that have been described in this chapter also apply for the most interesting and realistic experimental example of (2+1)-dimensional spatial solitons – the generation of photorefractive spatial screening solitons. In a photorefractive medium with an external bias electric field oriented along the x axis and an optical beam propagating along the z axis, polarized along the x axis, the change of the refractive index is caused by the electro-optic effect. The generation of free charge carriers by the optical beam screens the space charge field inside the beam in the illuminated region, and the refractive index becomes a nonlinear function of the beam intensity. There are some pecularities of the photorefractive nonlinearity that make it an attractive candidate not only for experimental verification of soliton formation and interaction effects in saturable media, but also for theoretical investigations of soliton behavior in these materials.

2.4 Self-Focusing in Dissipative Systems

In the preceding sections, we focused on the influence of saturation in a nonlinear system, which provides bright soliton solutions. The most striking characteristics that arise from saturation are probably that the system is no longer integrable and therefore that no analytic solutions can exist. Nevertheless, the system remains Hamiltonian, and there are still some conserved quantities that can be exploited for the characterization of solitary beams. In contrast to this situation, dissipative systems are non-Hamiltonian systems. They generally feature a flux of energy instead of energy conservation. These systems are subject to permanent gain and loss processes and are therefore not in an equilibrium state. However, solitary structures exist in a manifold of physical systems, including systems with convection instabilities, binary fluid convection [50] and even phase transitions [51]. In optics, systems with gain and loss can be realized in experiments that include any kind of cavity or feedback. Among those systems, we want to mention the optical parametric oscillator, and lasers that show spatial soliton effects.

Generally speaking, solitary solutions in dissipative systems differ qualitatively from those in Hamiltonian systems. In Hamiltonian systems, soliton solutions comprise a one or even two-parameter family of solutions, and the soliton exists because of a balance between linear diffraction and nonlinear self-focusing. This scenario is roughly depicted in Fig. 2.9a. For any given value of the diffraction, the nonlinearity can be adjusted in order to generate a solitary structure. In dissipative systems, an additional balance between

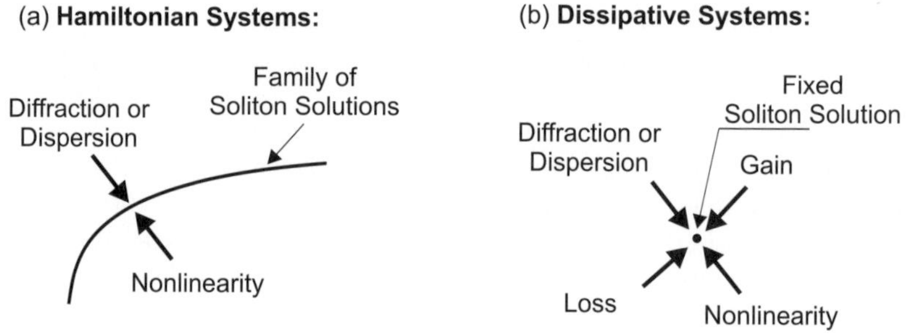

Fig. 2.9. Principal differences between a Hamiltonian system (**a**) and a dissipative system (**b**). In Hamiltonian systems, solitons exist because of a balance between diffraction and nonlinearity, resulting in a whole family of solutions. Dissipative systems support only isolated solutions, owing to the additional balance of loss and gain (after [20], printed by kind permission of Nail Akhmediev)

gain and loss is required to generate a solitary solution. In this case, there exist only isolated solutions, and a one-parameter family of solutions cannot be found. The detuning of one parameter cannot be counterbalanced by simply adjusting one of the other parameters. Instead, the single solitary solution has to be considered as an attractor, as illustrated in Fig. 2.9b. Dissipative systems can be described by a generalized form of the nonlinear Schrödinger equation (2.5) with additional terms that represent the effects of gain and loss. The *complex cubic–quintic Ginzburg–Landau* equation can be used to describe almost every kind of dissipative system in general; this equation can be written as

$$\mathrm{i}\frac{\partial}{\partial t}A + \frac{1}{2}\nabla_\perp^2 A + |A|^2 A = \mathrm{i}\delta A + \mathrm{i}\varepsilon |A|^2 A + \mathrm{i}\xi \nabla_\perp^2 A + \mathrm{i}\mu |A|^4 A - \nu |A|^4 A. \tag{2.61}$$

For the specific case of an externally driven optical cavity, t is the time and $\nabla_\perp$ represents the diffraction in the transverse plane. Depending on their sign δ and ε are linear and nonlinear loss or gain terms, respectively. ξ represents a diffraction term and μ is a higher-order loss/gain term. Finally, ν is a parabolic correction term of the nonlinearity.

If the real constants δ, ε, β, μ equal zero, (2.61) becomes Hamiltonian, and in the case when $\nu = 0$ also, the system represents the two-dimensional nonlinear Schrödinger equation.

2.4.1 Influence of Losses

Equation (2.61) is nonintegrable and only a few particular solutions can be obtained. Before we consider a complete dissipative system, we shall first describe the impact of losses on soliton propagation in a material with a saturable nonlinearity. We apply (2.61) to a one-dimensional spatial system and set $\varepsilon = \xi = \mu = \nu = 0$. In this case, only linear losses are taken into account. Equation (2.61) then reduces to

$$\mathrm{i}\frac{\partial}{\partial z}A + \frac{1}{2}\frac{\partial^2}{\partial x}A + \frac{|A|^2}{1+\sigma|A|^2}A = \mathrm{i}\delta A\,. \tag{2.62}$$

If the linear losses are small, a perturbative approach can be applied starting from a numerical solution of the appropriate Hamiltonian system given by (2.38) [31]. For small losses ($\delta \approx 0.05$), a quasi-soliton behavior can still be observed, but the beam profile becomes slightly broader during propagation as the peak power decreases. However, because the relation between the width of the beam and its peak power is preserved, this structure can still be considered as a solitary wave. In contrast, the beam profile does not maintain such a quasi-soliton property for larger losses ($\delta > 0.07$). In this case, the beam slowly broadens until the beam profile is damped completely down to zero.

The standard soliton perturbation theory discussed above is usually applied to analyze soliton dynamics under the action of a small external perturbation, ensuring a behavior near the soliton existence curve. However, if an unstable soliton undergoes profile oscillations, it may evolve under the action of its own, self-induced perturbation. This is a qualitatively different physical problem that can be investigated in a similar way using a variation of the Vakhitov–Kolokov criterion (see Sect. 2.2.3). Kivshar and coworkers have investigated the criterion for this kind of perturbation in analogy to the stability analysis of the case of a one-dimensional Kerr nonlinearity.

Here, we summarize the main features of this investigation. Further details can be found in [19]. The most remarkable result of this asymptotic analysis is in the dynamics of solitons near the instability threshold. These dynamics can be described by an equation which is equivalent to the equation of motion of an effective inertial, conservative particle moving under the action of a potential force which is proportional to the difference between the beam amplitude and the soliton amplitude. This gives rise to instability dynamics, which may lead either to convergence to a soliton solution or to complete decay.

In the case of a general two-power-law nonlinearity, $f(|A|^2) = -|A|^p + \sigma|A|^{2p}$, which possesses an explicit soliton solution for any $p > 0$, three different regimes of the soliton dynamics exist [52]:

- If the amplitude of the unstable perturbed soliton is smaller than the amplitude of the (unstable) stationary solution, the instability leads to a decrease of the soliton amplitude. This results in a spreading of the soliton into small-amplitude waves, which finally decay into linear waves owing to diffraction.
- If the initially perturbed unstable soliton has an amplitude slightly larger than that of the stationary soliton, the exponentially growing instability pushes the soliton into the stability region where a stable stationary soliton solution exists. However, owing to the inertial nature of the soliton evolution, the transition from an unstable to a stable state is accompanied by long-lived periodic oscillations of the soliton amplitude. These oscillations can be explained by the existence of a nontrivial internal mode of a bright soliton in the generalized nonlinear-Schrödinger-equation model [52].
- For larger deviations from the stable equilibrium state, the periodic oscillations of the soliton amplitude disappear again, because as soon as the soliton enters the unstable region, it spreads out owing to diffraction.

2.4.2 Dissipative Solitons

While (2.62) represents a perturbed Hamiltonian system with soliton-like solutions, dissipative solitons require gain and loss as described by (2.61). Since a dissipative system does not conserve energy, the complex Ginzburg–Landau equation (2.61) does not have any conserved quantities. Instead one can define two rate equations for the energy Q and momentum M, which are given here for a one-dimensional system:

$$\frac{\partial}{\partial t}Q = F(A) \,, \tag{2.63}$$

$$\frac{\partial}{\partial t}M = J(A) \,, \tag{2.64}$$

where M and Q are defined by (2.28), and the functions $F(A)$ and $J(A)$ are given by

$$F(A) = 2\mathrm{Im}\int_{-\infty}^{\infty}\left[\delta|A|^2 + \varepsilon|A|^4 + \mu|A|^6 - \xi\left|\frac{\partial}{\partial x}A\right|^2\right]\mathrm{d}x \,, \tag{2.65}$$

$$J(A) = 2\int_{-\infty}^{\infty}\left[\delta + \varepsilon|A|^2 + \mu|A|^4 + \xi\left|\frac{\partial^2}{\partial x^2}A\right|^2\right]\frac{\partial}{\partial x}A^*\,\mathrm{d}x \,. \tag{2.66}$$

Equations (2.63) and (2.64) are generally a quite effective tool for solving various problems related to the Ginzburg–Landau equation.

Assuming that the coefficients on the right-hand side of (2.61) are small, one can again apply a perturbative approach, starting from an analytical two-parameter solution of the nonlinear Schrödinger equation (2.13). By inserting the analytical solutions of the (1+1)-dimensional Schrödinger equation into (2.61), one can derive two differential equations from (2.13) that include the soliton parameters a and b:

$$\begin{aligned} \frac{\partial a}{\partial t} &= 2a\left[\delta - \xi b^2 + \frac{1}{3}(2\varepsilon - \xi)a^2 + \frac{8}{15}\mu a^4\right] \,, \\ \frac{\partial b}{\partial t} &= -\frac{4}{3}\xi b a^2 \,. \end{aligned} \tag{2.67}$$

Now, for some specific values of the coefficients δ, ξ, ε, μ and ν, the parameters of (2.67) can be expressed in a two-dimensional plot. Depending on the set of coefficients, (2.67) displays one or even two singular points on the semiaxis $b = 0$, $a > 0$, which may represent a stable or unstable solitary solution. By finding the roots of the expression in the square brackets in (2.67), one can make statements about the stability. When a is imaginary, and hence the roots of a^2 are negative, there are no singular points, and no soliton solutions exist at all. In contrast, when a is real, there exist two fixed points and two corresponding soliton solutions. For a more detailed analysis and details of the coefficients and parameters, we refer to [20, 21, 53].

2.4.3 Cavity Solitons

An optical cavity or resonator generally provides a dissipative system in the sense that there exists a permanent flow of energy. Such a system never reaches a state of equilibrium; instead, we have to consider a balance between losses and external driving or intracavity gain effects. Therefore, a cavity soliton needs two kinds of balance, namely between spatial spreading (diffraction) and nonlinearity, and between dissipation and external pumping. This is the most obvious difference to solitons in Hamiltonian systems that propagate in a bulk material. A cavity soliton does not propagate; rather, it is trapped between the two mirrors of the cavity and may move only in the transverse plane. Another striking feature is that there is no unique effect but a whole variety of different effects that support cavity soliton generation. It is not really surprising that a nonlinear medium allows the generation of solitary structures when placed inside a cavity. But even a linear medium inside a cavity – e.g. a saturable absorber that does not provide a light-intensity-dependent refractive index – can lead to the generation of a robust solitary structure. Among the various materials that are suitable for the generation of cavity solitons, there are Kerr-like media, two-level systems (saturable absorbers), quadratic media and semiconductors.

The photorefractive nonlinearity is one of the most suitable effects for the study of the formation of cavity solitons, because it allows easy observation of their formation and stabilization owing to its slow dynamics. Moreover, photorefractive two-beam coupling allows to realize cavities with gain, which represent active nonlinear resonators.

Cavity solitons can be excited by localized pulses of light and can in principle be used to store information [54]. Once excited, a solitary structure does not disappear as long as the driving field is present. It can actually only be erased by a deliberate action, for example with an appropriate phase-reversed pulse [55]. This feature, in combination with the property that cavity solitons never exist for a complete range of their control parameters but only for distinct values of them, they are easily distinguishable and may serve as all-optical bits of information. Additionally, dissipative solitons are attractors and can evolve from arbitrary conditions.

A cavity that supports the generation of solitary structures can be implemented with a laser resonator containing a nonlinear element. In that case a manifold of different soliton types such as vortices, bright solitons, phase solitons, and pairs of bright and dark solitons can in principle be demonstrated [56]. The first experimental observation of a solitary structure in an optical resonator made use of a nematic liquid-crystal cell as a nonlinear element [57]. Remarkably, dissipative solitons can be generated even when one of the cavity mirrors is removed. In that case, the cavity transforms into a single feedback system which allows solitary structures. Such a single feedback system has been successfully applied by making use of a liquid-crystal light valve [58] or an Na-vapor cell [59] as the essential nonlinear element.

Combining such a single feedback system and a photorefractive crystal gives rise to even more complicated and fascinating phenomena. If the system is operating above a certain power threshold, spontaneous transverse pattern formation may arise from modulation instabilities. The second part of this book is dedicated to this topic, the main aspects of which are extensively described in Chaps. 8–10.

Summary

An optical beam traversing a nonlinear medium may experience self-trapping under certain conditions. When a balance between the spreading of the beam due to propagation and the effect of focusing due to the nonlinearity is reached, a self-stabilizing beams that can be described in the framework of soliton theory is obtained. In this chapter, we introduced a mathematical description of a soliton on the basis of the nonlinear Schrödinger equation for a Kerr nonlinear medium, a medium with an intensity-dependent nonlinear refractive index. We demonstrated the unique features of these particle-like waves in a one-dimensional transverse geometry and presented the most typical analytic family of solutions, obtained by applying symmetry considerations. Subsequently, the possibility of forming two-dimensional spatial solitons was addressed. A beam that is extended in both transverse dimensions with respect to its propagation direction will experience catastrophic self-focusing in a Kerr medium, thereby being destroyed into filaments. By introducing a saturable nonlinear response, the catastrophic collapse can be prevented, allowing again a balance between nonlinearity and diffraction. However, the propagation equation describing such a system is no longer integrable, and hence no analytic solution can be formulated. Because, for an isotropic nonlinear response, the description remains Hamiltonian, the Hamiltonian formalism can be used to find solitary solutions and test their stability. Owing to the nonintegrability, the soliton characteristics change remarkably. Radiation losses, oscillations of the beam profile and inelastic collision processes take place. Finally, if we take the effects of gain and losses into account, the system becomes dissipative, allowing a new mechanism of stabilization in soliton formation that involves balance of nonlinearity, diffraction, gain and losses. In nonlinear optical dissipative systems, solitary structures can be generated in a variety of configurations, including configurations with feedback.

References

1. G.A. Askaryan, *Effect of gradient of high-power electromagnetic field on electrons and atoms*, Zh. Eksp. Teor. Fiz. **42**, 1567 (1962). (Sov. Phys. JETP **15**, 1088 (1962)).
2. R.Y. Chiao, E. Garmire and C.H. Townes, *Self-trapping of optical beams*, Phys. Rev. Lett. **13**, 479 (1964). see also Phys. Rev. Lett. **14**, 1056 (1965).
3. M. Hercher, *Laser-induced damage in transparent media*, J. Opt. Soc. Am. **54**, 563 (1964).

4. V.E. Zakharov and A.B. Shabat, *Exact theory of two-dimensional self-focusing and one-dimensional self-modulation of waves in nonlinear media*, Zh. Eksp. Teor. Fiz. **61**, 118 (1971). (Sov. Phys. JETP **34**, 62 (1972)).
5. A. Hasegawa. *Optical Solitons in Fibers*, volume 116 of *Tracts in Modern Physics*. Springer, Berlin, Heidelberg, 1989.
6. A. Hasegawa and Y. Kodama. *Solitons in Optical Communications*. Oxford University Press, Oxford, 1995.
7. A. Hasegawa and F.D. Tappert, *Transmission of stationary nonlinear optical pulses in dispersive dielectric fibers. I. Anomalous dispersion*, Appl. Phys. Lett. **23**, 142 (1973).
8. L.F. Mollenauer, R.H. Stolen and J.P. Gordon, *Experimental observation of picosecond pulse narrowing and solitons in optical fibers*, Phys. Rev. Lett. **45**, 1095 (1980).
9. *http : //physics.technion.ac.il/ ~ msegev/sel_pub.html.*
10. M. Segev, *Optical spatial solitons*, Opt. Quantum Electron **30**, 503 (1998).
11. S. Trillo and W. Torruellas. *Spatial Solitons*, volume 82 of *Springer Series in Optical Sciences*. Springer, Berlin, Heidelberg, 2001.
12. C.S. Gardner, J.M. Greene, M.D. Kruskal and R.M. Miura, *Method of solving the Korteweg–deVries equation*, Phys. Rev. Lett. **19**, 1095 (1967).
13. P.D. Lax, *Integrals of nonlinear equations of evolution and solitary waves*, Commun. Pure Appl. Math. **21**, 467 (1968).
14. V.E. Zakharov, S.V. Manakov, S.P. Novikov and L.P. Pitaevsky. *Theory of Solitons*. Consultants Bureau, New York, 1984.
15. M.J. Ablowitz and H. Segur. *Solitons and the Inverse Scattering Method*. SIAM, Philadelphia, 1981.
16. N.G. Vakhitov and A.A. Kolokolov, *Stationary solutions of the wave equation in a medium with nonlinearity saturation*, Radiophys. Quantum Electron. **16**, 783 (1973).
17. A.A. Kolokolov, *Stability of stationary solutions of nonlinear wave equations*, Izv. Vyssh. Uchebn. Zaved. Radiofiz. **17**, 132 (1974). (Radiophys. Quantum Electron. **17**, 1016 (1976)).
18. J.J. Rasmussen and K. Rypdal, *Blow up in nonlinear Schrödinger equations: a general review*, Phys. Scr. **33**, 481 (1986).
19. D.E. Pelinovsky, V.V. Afanasjev and Yu.S. Kivshar, *Nonlinear theory of oscillating, decaying, and collapsing solitons in the generalized nonlinear Schrödinger equation*, Phys. Rev. E **53**, 1940 (1996).
20. N. Akhmediev. *Soliton Driven Photonics*, 371. Kluwer, Amsterdam, 2001.
21. N.N. Akhmediev and A. Ankiewicz. *Solitons – Nonlinear Pulses and Beams*. Chapman & Hall, London, 1997.
22. L.D. Fadeev and L.A. Takhtadjan. *Hamiltonian Methods in the Theory of Solitons*. Springer, Berlin, Heidelberg, 1987.
23. J.H. Marburger, *Self-focusing: theory*, Prog. Quant. Electron. **4**, 35 (1975).
24. W.G. Wagner, H.H. Haus and J.H. Marburger, *Large-scale self-trapping of optical beams in the paraxial ray approximation*, Phys. Rev. **175**, 256 (1968).
25. S.N. Vlasov and E.V. Sheinina, *On the theory of interaction of counterpropagating waves in a nonlinear cubic medium*, Radiophys. Quantum Electron. **27**, 15 (1983).
26. M. Karlson, *Optical beams in saturable self-focusing media*, Phys. Rev. A **46**, 2726 (1992).

27. J.H. Marburger and E.L. Dawes, *Dynamic formation of a small-scale filament*, Phys. Rev. Lett. **21**, 556 (1968).
28. E.L. Dawes and J.H. Marburger, *Computer studies of self-focusing*, Phys. Rev. **179**, 862 (1969).
29. V. Konno and H. Suzuki, *Self-focusing of laser beam in nonlinear media*, Phys. Scr. **20**, 382 (1979).
30. J.M. Soto-Crespo, E.M. Wright and N.N. Akhmediev, *Recurrence and azimuthal-symmetry breaking of a cylindrical gaussian beam in a saturable self-focusing medium*, Phys. Rev. A **45**, 3168 (1992).
31. S. Gatz and J. Herrmann, *Soliton propagation in materials with saturable nonlinearity*, J. Opt. Soc. Am. B **8**, 2296 (1991).
32. P.L. Kelley, *Self-focusing of laser beams*, Phys. Rev. Lett. **15**, 1005 (1965).
33. V.I. Bespalov and V.I. Talanov, *Filamentary structure of light beams in nonlinear liquids*, Pis'ma Zh. Eksp. Teor. Fiz. **58**, 903 (1970). (Sov. Phys. JETP **31**, 486 (1970)).
34. T.B. Benjamin and J.E. Feir, *The disintegration of wave trains on deep water*, J. Fluid Mech. **27**, 417 (1967).
35. T.B. Benjamin and J.E. Feir, *The disintegration of wave trains on deep water. Part I. Theory*, J. Math. Phys. **46**, 133 (1967).
36. N.N. Akhmediev, V.I. Korneev and R.F. Nabiev, *Modulation instability of the ground state of the nonlinear wave equation: optical machine gun*, Opt. Lett. **17**, 393 (1992).
37. G.S. McDonald, K.S. Syed and W.J. Firth, *Dark spatial soliton break-up in the transverse plane*, Opt. Commun. **95**, 281 (1993).
38. J.E. Bjorkholm and A. Ashkin, *Cw self-modulation and self-focusing of light in sodium vapour*, Phys. Rev. Lett. **32**, 129 (1974).
39. C.E. Max, J. Arona and A.B. Langdon, *Self-modulation and self-focusing of electromagnetic waves in plasmas*, Phys. Rev. Lett. **33**, 209 (1974).
40. A.B. Borisov, A.V. Borovskij, O.B. Shiryaev, V.V. Korobkin, A.M. Prokhorov, I.C. Solem, T.S. Luk, K. Boyer and C.K. Rhodes, *Relativistic charge-displacement self-channeling of intense ultrashort laser pulses in plasmas*, Phys. Rev. A **45**, 2309 (1992).
41. A.B. Borisov, A.V. Borovskij, V.V. Korobkin, A.M. Prokhorov, O.B. Shiryaev, X.M. Shi, T.S. Luk, A. McPherson, I.C. Solem, K. Boyer and C.K. Rhodes, *Observation of relativistic and charge-displacement self-channeling of intense subpicosecond uv-radiation in plasmas*, Phys. Rev. Lett. **68**, 2309 (1992).
42. P. Monot, T. Auguste, P. Gibbon, F. Jokober, G. Mainfray, A. Duliev, M. Louis-Jacquet, G. Malka and I.L. Miguel, *Experimental demonstration of relativistic self-channeling of a multiterawatt laser pulse in an underdense plasma*, Phys. Rev. Lett. **74**, 2953 (1975).
43. A. Braun, G. Korn, X. Liu, D. Du, J. Squier and G. Mourou, *Self-channeling of high-peak-power femtosecond laser pulses in air*, Opt. Lett. **20**, 73 (1973).
44. S. Henz and J. Hermann, *Two-dimensional spatial optical solitons in bulk Kerr-media stabilized by self-induced photoionization*, Phys. Rev. E **53**, 4092 (1996).
45. K.N. Karamzin and A.P. Sukhorukov, Zh. Eksp. Teor. Fiz. **20**, 774 (1974). (JETP Lett. **20**, 339 (1974)).
46. K. Hayata and M. Koshiba, *Multidimensional solitons in a quadratic nonlinear medium*, Phys. Rev. Lett. **71**, 3275 (1993).

47. W. Torruellas, Z. Wang, D.J. Hagan, E.W. van Stryland, G.I. Stegeman, L. Torner and C.R. Menyuk, *Observation of two-dimensional spatial solitary waves in a quadratic medium*, Phys. Rev. Lett. **74**, 5036 (1995).
48. S. Gatz and J. Herrmann, *Propagation of optical beams and the properties of two-dimensional spatial solitons in media with a local saturable nonlinear refractive index*, J. Opt. Soc. Am. B **14**, 1795 (1997).
49. T.R. Taha and M.J. Ablowitz, *Analytical and numerical aspects of certain nonlinear evolution equations. II. Numerical, nonlinear Schrödinger equation*, J. Comput. Phys. **55**, 203 (1984).
50. P. Kolodner, *Collisions between pulses of traveling-wave convection*, Phys. Rev. A **44**, 6466 (1991).
51. R. Graham. *Fluctuations, Instabilities and Phase Transitions.* Springer, Berlin, Heidelberg, 1975.
52. Yu.S. Kivshar, *Bright and dark spatial solitons in non-Kerr media*, Opt. Quantum Electron. **30**, 571 (1998).
53. N. Akhmediev and A. Ankiewicz. *Solitons of the complex Ginzburg–Landau equation. In: Spatial Solitons*, 311. Springer, Berlin, Heidelberg, 2001.
54. W.J. Firth and A.J. Scroggie, *Optical bullet holes: robust controllable localized states of a nonlinear cavity*, Phys. Rev. Lett **76**, 1623 (1996).
55. D.V. Skryabin, *Energy of the soliton internal modes and broken symmetries in nonlinear optics*, J. Opt. Soc. Am. B **19**, 529 (2002).
56. C.O. Weiss, G. Slekis, V.B. Taranenko, K. Staliunas and R. Kuszelewicz. *Spatial Solitons in Resonators. In: Spatial Solitons*, 395. Springer, Berlin, Heidelberg, 2001.
57. M. Kreuzer, H. Gottschling and T. Tschudi, *Structure formation and self-organization phenomena in optical bistable elements*, Mol. Cryst. Liq. Cryst. **207**, 219 (1991).
58. A. Schreiber, *Experimental investigation of solitary structures in a nonlinear optical feedback system*, Opt. Commun. **136**, 415 (1997).
59. B. Schäpers, M. Feldmann, T. Ackemann and W. Lange, *Interaction of localized structures in an optical pattern-forming system*, Phys. Rev. Lett. **85**, 748 (2000).

3 The Photorefractive Nonlinearity

The photorefractive effect – representing a photoinduced change of the index of refraction via the electro-optic effect – has become the nonlinear optical mechanism of choice for optical image processing. This is primarily because propagation of a light beam in a photorefractive medium is accompanied by a series of dramatic and fascinating changes in its spatial structure.

The photorefractive effect can be regarded as a by-product of the intensive early research on nonlinear optical effects – it was discovered in 1966 when it ruined the phase-matching condition in a second-harmonic generation experiment [1]. The $LiNbO_3$ crystal used in that experiment was reported to be "optically damaged" after irradiation by a laser beam. This turned out to be the result of a basic phenomenon, the *photorefractive effect*, which caused a distortion of the wavefront through a spatial index variation. The photorefractive effect results in a *local optically induced change in the refractive index of a nonlinear electro-optic material.* One of the main features of photorefractive materials is the large effective nonlinearity even at low laser powers together with inherent optical real-time information-processing properties using all three dimensions in space. Consequently, the potential for optical storage applications [2] boosted the intensive search for photorefractive materials, effects and devices.

A wide range of electro-optic materials were found to show photorefractive properties, among them ferroelectric crystals such as lithium niobate ($LiNbO_3$), strontium barium niobate (SBN), barium titanate ($BaTiO_3$), and potassium niobate ($KNbO_3$), paraelectric crystals such as bismuth silicon oxide (BSO) and bismuth germanium oxide (BGO), and well-known semiconductors such as gallium arsenide (GaAs). Considerable attention has been paid recently to the investigation of photorefractive polymers as low-cost materials for possible applications.

All these materials have a large nonlinear response at low intensities in the milliwatt range, an intensity that can be provided by standard continuous-wave laser sources. The response time of the materials varies from microseconds for semiconductor materials to seconds for some electro-optic crystals. Photorefractive materials are, typically, sensitive over a broad range of wavelengths. Doping with metal ions during the growth process may influence the speed of the grating formation process and increase the photorefractive sensitivity in a certain spectral region.

Photorefractive crystals are nowadays used for a number of applications. Apart from volume holographic data storage, which was suggested early and has been improved in recent years using a number of multiplexing and encoding techniques [3], photorefractive materials can be used for the correlation and convolution of optical images [4]. Closely related to correlators are associative memories based on photorefractive crystals, which are able to recover a complete stored image from partial input information [5], showing the ability of a photorefractive nonlinear material to perform parallel information processing. In addition to the holographic response, an important property of a number of photorefractive crystals (among them $BaTiO_3$ and $KNbO_3$) is the beam-coupling effect – a transfer of energy between two coherently intersecting beams inside the crystal. This unique property of nonreciprocal energy transfer has led to a number of applications in the general context of image processing, namely self-pumped phase conjugators [6], photorefractive novelty filters [7], image amplification [8] and optical neural networks [3, 9]. In particular, the sluggish temporal response of several photorefractive materials and their large nonlinearity even at low laser powers render them ideal subjects for the topic on which we focus throughout this book – the investigation and characterization of transverse spatial light phenomena such as spatial solitons and pattern formation. The topic of photorefractive materials and their applications is covered by a number of review articles and books [10, 11, 12, 13, 14, 15]. An excellent overview of the historical developments is provided in [10].

3.1 Principle of the Photorefractive Effect

The principle of the photorefractive mechanism, which will be discussed later on in detail, can be summarized as follows. The intensity profile of a beam incident on a photorefractive medium creates an energy gradient. In the regions of high light intensity, charge carriers from impurity atoms are liberated from the valence into the conduction band of the crystal, where they become mobile. As a consequence, a migration process starts, which originates from three different charge carrier transport processes. First, migration may result from diffusion, in which the charges move towards a place of lower charge concentration owing to their thermal mobility. Second, it may occur due to the action of an external field, which causes drift. The third possibilty is migration due to the photovoltaic effect, a drift effect caused by an internal electric field which has, in turn been created by the light pattern itself. The most distinctive feature of the photovoltaic effect is that photoelectrons which are excited into the charge transfer band migrate in a preferential direction, which is determined by the direction of the polar axis of the crystal. Finally, the electrons (or holes) can recombine with free sites in the lattice of the crystal, i.e empty donors (or acceptors), thus creating a new distribution of the charge. Owing to the movement of the charge carriers, a spatial difference in the excitation rate of the ionized donors and in the trapping rate of the electrons emerges, resulting in a spatially modulated electric charge density. Consequently, an electric field builds up – the *space charge field* – which may be shifted with respect to the intensity distribution that generated it, which leads to a nonlocal effect. This field in turn modulates the refractive index via the linear electro-optic effect. Depending on the dominant transport mechanism, an overall phase shift between the intensity modulation and the refractive index arises, which is generally time- and space-dependent. To provide insight into the relevant details, we explain qualitatively the separate influences of drift and diffusion effects on the generation of the space charge field and hence the refractive index modulation. Figure 3.1 demonstrates the effect of a drift-dominated charge carrier transport process on the refractive

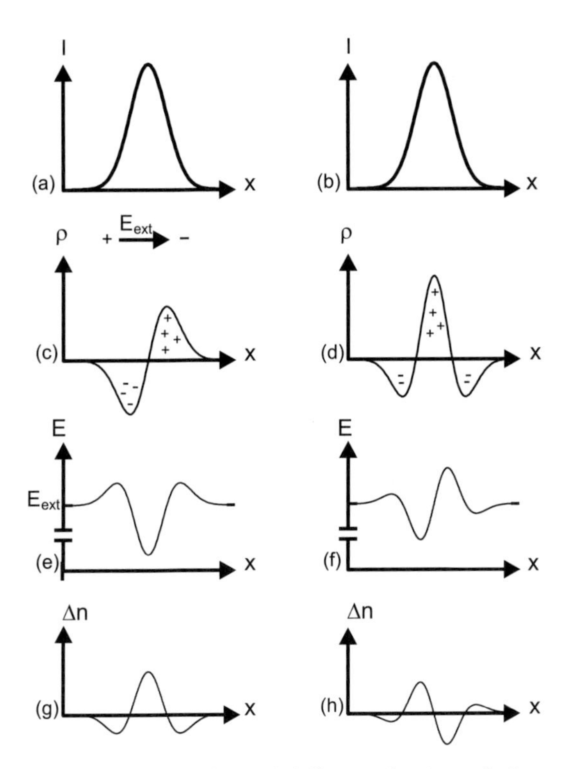

Fig. 3.1. Simplified sketch of drift- and diffusion-dominated charge carrier transport processes and their impact on the induced refractive index modulation. Light intensity pattern (**a**), (**b**), charge density (**c**), (**d**), space-charge field (**e**), (**f**), and refractive index modulation (**g**), (**h**) for a drift (*left column*) and for a diffusion process (*right column*)

index modulation in the left column (Fig. 3.1a, c ,e, g) and the effect of a diffusion-dominated process in the right column (Fig. 3.1b, d, f, h). Starting from a Gaussian intensity distribution as shown in Fig. 3.1a, electrons from the central area are excited into the conduction band. Owing to the presence of the external electric field $E_{\rm ext}$ as depicted in Fig. 3.1c, the electrons drift in a defined direction, and the resulting charge density ρ becomes asymmetric. By integrating $\rho(x)$ one can derive the optically induced space charge electric field $E_{\rm sc}$ (Fig. 3.1e), which transforms into a symmetric, elevated refractive index modulation $\Delta n_{\rm pr}$ (Fig. 3.1g) via the Pockels effect. In this case the light intensity pattern in Fig. 3.1a induces a positive refractive index modulation that resembles a focusing structure. When the external electric field in Fig. 3.1c points in the other direction, the opposite effect takes place, and the net refractive index change is negative, leading to a defocusing refractive index profile.

In the right column of Fig. 3.1, the impact of the induced refractive index change in the case of diffusion is illustrated. Starting from the same intensity distribution (Fig. 3.1b), the electrons diffuse symmetrically to areas of lower light intensity in the vicinity of the optical beam. The resulting space charge field (Fig. 3.1d) is asymmetric, and the net refractive index change shown in Fig. 3.1h represents more an asymmetric, deflecting structure than a symmetric, focusing one. From this simple illustration it is clear that the two effects, of diffusion and drift induce different types of refractive index changes, which can never be separated in a real physical system. The induced refractive index change acts as a bulk diffractive element for the beam itself, changing its propagation characteristics dramatically. In the case of a drift-dominated transport mechanism (see Figs. 3.1a, c, e, g), the refractive index modulation mimics the intensity distribution, thus creating an effective focusing lens, which may lead to self-induced waveguiding of the incident beam. In the case of a diffusion-dominated charge transport mechanism, the refractive index modulation results in an asymmetric lensing effect, which leads to fanning of a beam in the direction of the symmetry axis of the material (see Fig. 3.1h). This fanning effect causes a strong bending of the trajectory of an optical beam in a photorefractive crystal [6, 16, 17, 18, 19]. This is obvious from studies of beam propagation in a purely diffusion-dominated material; a theoretical simulation [17] is illustrated in Fig. 3.2a, and an experimental observation in Fig. 3.2b.

This short survey shows that photorefractive media have a type of nonlinearity that differs in many important ways from the conventional Kerr-type nonlinearity. First, Fig. 3.1 indicates that it is intrinsically nonlocal in nature. Second, many features of nonlinear beam mixing in these media are almost independent of the intensities of the interacting beams, which determine only the characteristic relaxation timescales. This is an essential feature of nonlinear saturable mechanisms. In addition, the photorefractive response can also

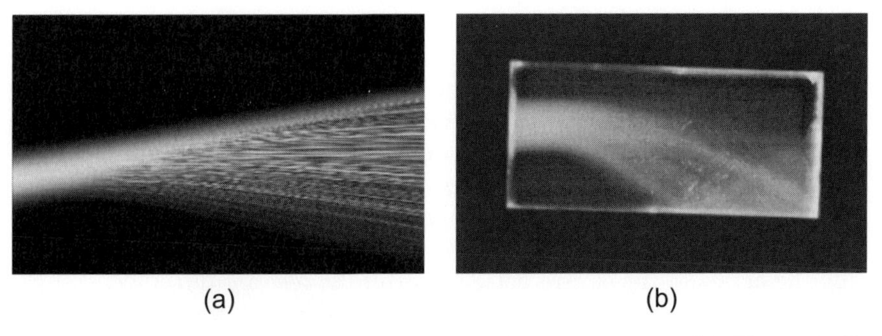

Fig. 3.2. Fanning of a beam inside a photorefractive medium. (**a**) Theoretically calculated distribution of the light intensity (after [17], printed by kind permission of Dana Z. Anderson), (**b**) experimental observation in $BaTiO_3$

be controlled by an externally applied AC or DC electric field, thus enabling a direct control of the nonlinearity.

In the following paragraphs, the essential equations necessary for the understanding of the photorefractive effect are given. All nonlinear processes derived from the photorefractive effect can be traced back to these fundamental equations. They explicitly take into account the anisotropic nature of the medium, as well as the presence of externally applied control fields.

The structure of the equations is essentially that of a nonlinear electrostatic problem. Their solution involves calculation of a long-range electrostatic potential that is created by the space charge field in the material, with asymptotic boundary conditions. The spatial variation of the electrostatic potential, in turn, determines the refractive index modulation that is formed in the bulk of the material. This process is solely due to the linear electro-optic effect, which is described in the following section for the case of a general electrostatic space charge field. The derivation of the space charge field itself for various beam configurations will be the subject of the subsequent sections.

3.2 Refractive Index Modulation in Electro-Optic Crystals

The variation of the refractive index results from the action of the linear electro-optic effect, which is due to a low-frequency optically induced space charge electric field. The linear electro-optic effect (also referred to as the Pockels effect) is described by the change of the impermeability tensor [11]

$$\delta\left(\frac{1}{n^2}\right)_{ij} = \sum_k r_{ijk} E_{\mathrm{sc},k} , \tag{3.1}$$

and represents a second-order nonlinear effect. Here, r_{ijk} is a third-rank tensor with $3^3 = 27$ elements, called *electro-optic coefficients.* $E_{sc,k}$ represents the kth component of the space charge field ($k = x, y, z$). The strength of the electro-optic effect depends on the orientation of the space charge field E_{sc} within the crystal and the polarization direction of the light wave. By taking into account the optical properties and symmetries of the photorefractive crystal, the electro-optic coefficients can be reduced to an effective electro-optic coefficient r_{eff}. By using this contracted notation we take account of, only the action of the electro-optic tensor in the relevant direction.

The large majority of photorefractive media are anisotropic electro-optic crystals. In these media, the dielectric tensor ε represents a tensor of rank 2 and can be written in its principal coordinates as

$$\varepsilon = \begin{pmatrix} \varepsilon_x & 0 & 0 \\ 0 & \varepsilon_y & 0 \\ 0 & 0 & \varepsilon_z \end{pmatrix} = \varepsilon_0 \begin{pmatrix} {n_x}^2 & 0 & 0 \\ 0 & {n_y}^2 & 0 \\ 0 & 0 & {n_z}^2 \end{pmatrix} , \tag{3.2}$$

where ε_x, ε_y, ε_z are the principal dielectric constants, and n_x, n_y and n_z represent the principal indices of refraction. In the most general case, the principal indices of refraction are all different; this involves two optical axes. $KNbO_3$, the material of choice for the pattern formation experiments described in this book, belongs to this group of biaxial crystals and possesses three different indices of refraction. However, many crystals, such as SBN and $BaTiO_3$, are uniaxial because they have two equal principal indices of refraction as a result of their crystal symmetry. Symmetry considerations reduce the number of independent coefficients of the impermeability tensor to 18 owing to a permutation invariance, $i = j$, of the first two indices. The electro-optic coefficients are often written in terms of a 6 times 3 matrix, using a contracted notation [11]. In fact, as a result of the crystal point group symmetries, the number of independent coefficients decreases further because many coefficients become zero. As a general rule, an increase in the crystal symmetry leads to a decrease in the number of independent electro-optic coefficients. It is important to note that all r_{ijk} are zero for crystals possessing inversion symmetry, i.e. the linear electro-optic coefficients vanish in centrosymmetric crystals.

Let us focus on the symmetries of the photorefractive materials SBN, $BaTiO_3$ and $KNbO_3$, as examples. These crystals belong to the point groups $4mm$ ($BaTiO_3$ and SBN) and $2mm$ ($KNbO_3$); their electro-optic tensors in contracted notation therefore reduce to [11]

$$r_{SBN} = \begin{pmatrix} 0 & 0 & r_{13} \\ 0 & 0 & r_{13} \\ 0 & 0 & r_{33} \\ 0 & r_{42} & 0 \\ r_{42} & 0 & 0 \\ 0 & 0 & 0 \end{pmatrix} , \quad r_{KNbO_3} = \begin{pmatrix} 0 & 0 & r_{13} \\ 0 & 0 & r_{23} \\ 0 & 0 & r_{33} \\ 0 & r_{42} & 0 \\ r_{51} & 0 & 0 \\ 0 & 0 & 0 \end{pmatrix} . \tag{3.3}$$

Depending on the crystal chosen, in the case of beam-coupling experiments, the largest electro-optic coefficient can be selected by choosing the polarization and direction of the incident light appropriately. For all crystals used in the work, the extraordinary polarization gives the largest effective electro-optic coeffient, and was consequently used for the realization of transverse pattern formation and soliton generation. Since SBN, $BaTiO_3$ and $KNbO_3$ – the predominant crystals for our purposes – have different point geometries, we have derived an ideal configuration for each crystal, as described below. For all three crystals, we assume a space charge field E_{sc} along z, which is also the crystallographic c axis, so that

$$\boldsymbol{E}_{sc} = \begin{pmatrix} 0 \\ 0 \\ 1 \end{pmatrix} |\boldsymbol{E}_{sc}| . \tag{3.4}$$

3.2.1 Effective Electro-Optic Coefficient of $KNbO_3$

For $KNbO_3$ inserting (3.4) and (3.3) into (3.1) reduces the change of the impermeability tensor to

$$\delta\left(\frac{1}{n^2}\right) = \begin{pmatrix} r_{13} & 0 & 0 \\ 0 & r_{23} & 0 \\ & 0 & r_{33} \end{pmatrix} |\boldsymbol{E}_{sc}| . \tag{3.5}$$

Assuming further an extraordinary polarization vector

$$\boldsymbol{p} = \begin{pmatrix} \cos\vartheta \\ \sin\vartheta \\ 0 \end{pmatrix} , \tag{3.6}$$

where ϑ is the angle with respect to the crystallographic a axis, the refractive index distribution can be written as

$$\Delta n_{pr} = -\frac{1}{2} n_0^3 \langle p | \delta\left(\frac{1}{n^2}\right) | p \rangle = -\frac{1}{2} n_0^3 r_{eff} |\boldsymbol{E}_{sc}| , \tag{3.7}$$

which implies an effective electro-optic coefficient for $KNbO_3$ equal to

$$r_{eff} = r_{13}\cos^2\vartheta + r_{23}\sin^2\vartheta . \tag{3.8}$$

Since $r_{13} \gg r_{23}$ for $KNbO_3$ [11], the largest refractive index change Δn can be achieved in this configuration by choosing a polarization parallel to the crystallographic a axis, i.e. $\vartheta = 0$; the setup is therefore usually adjusted to achieve this in experiments.

3.2.2 Effective Electro-Optic Coefficient of SBN

An expression for the electro-optic coefficient of SBN can be derived in a similar way to that for $KNbO_3$. Since we are most interested in exploiting the relatively large electro-optic coefficient r_{33} of SBN, the configuration described above for $KNbO_3$ is not directly applicable for the present purpose. Instead we choose the polarization to be in the x–z plane, i.e.

$$\boldsymbol{p} = \begin{pmatrix} \cos\vartheta \\ 0 \\ -\sin\vartheta \end{pmatrix} . \tag{3.9}$$

In this case, the refractive index change can be written as

$$\Delta n_{\mathrm{pr}} = -\frac{1}{2} n_0^3 r_{\mathrm{eff}} |\boldsymbol{E}_{\mathrm{sc}}| , \tag{3.10}$$

where

$$r_{\mathrm{eff}} = r_{13}\cos^2\vartheta + r_{33}\sin^2\vartheta . \tag{3.11}$$

Since $r_{33} \gg r_{13}$ in SBN [20], we choose $\vartheta = \pi$ in order to achieve a maximum refractive index change. Consequently, the light wave then has to be linearly polarized parallel to the crystallographic c axis. In the notation used throughout this chapter, the z axis denotes the direction of the crystallographic c axis. The reader should not be confused when we refer to light propagation along z in SBN crystals in Chaps. 4–7. In this case, the crystal is rotated by 90° and the c axis points in the horizontal x direction.

3.2.3 Effective Electro-Optic Coefficient of $BaTiO_3$

Owing to the similarity of the crystal structures of SBN and $BaTiO_3$, the electro-optic coefficient of $BaTiO_3$ can be analyzed in almost the same way as in the preceding section. Again, we we want to exploit the largest electro-optic coefficient, which is r_{42} in the case of $BaTiO_3$, requiring an extraordinary beam polarization in the direction of the c axis of the material. Using (3.9) and (3.10), the effective electro-optic coefficient can again be formulated. Because we refer to this material mainly in Chap. 11, where we describe the case of two interacting beams, we shall consider here the effective electro-optic coefficient for the interaction of two beams – a signal beam and a pump beam – with the incidence angles α_1 and α_2, measured relative to the plane perpendicular to the crystallographic c axis. This coefficient is given by

$$\begin{aligned} r_{\mathrm{eff}} &= \frac{1}{n_0 n_3^3} \cos\left(\frac{\alpha_1+\alpha_2}{2}\right) \\ &\times \left[n_0^4 r_{13} \sin\alpha_1 \sin\alpha_2 + 2 n_0^2 n_{\mathrm{e}}^2 r_{42} \sin^2\left(\frac{\alpha_1+\alpha_2}{2}\right) + n_{\mathrm{e}}^4 r_{33} \cos\alpha_1 \cos\alpha_2 \right] . \end{aligned} \tag{3.12}$$

For all crystals discussed here, i.e. SBN, $BaTiO_3$ and $KNbO_3$, the effective electro-optic coefficients are in the range of 10 to 1700 pm/V, resulting in a space charge field of $|\boldsymbol{E}_{\mathrm{sc}}| \approx 2-6$ kV/cm. With an undisturbed refractive index for the extraordinary polarization of $n_0 \approx 2.3$, the resulting refractive index change is in the range of $\Delta n_{\mathrm{pr}} \approx 10^{-5} - 10^{-4}$.

3.3 Creation of the Space Charge Field – the Band Transport Model

The physical description of the charge transport mechanism relies basically on the band transport model, which was first developed by Kukhtarev and coworkers [21] and is based on a three-step process. In the first step, electrons or holes are liberated from donors or traps via photoionization. These charges may be provided by lattice defects or dislocations or – in the case of nominally undoped crystals – by traces of impurities. In most of the ferroelectrics that have a pronounced photorefractive effect, iron impurities are the most important donor and trapping centers. After that liberation process, charge migration due to diffusion, drift or the photovoltaic effect (photoinduced stimulation) takes place, as the second step, in the valence or conduction band, causing the charges to move to areas of lower energy and charge density. At these places, they finally recombine with empty donors or traps and form the space charge field that is essential to the process. In many photorefractive materials, it turns out that simultaneous electron and hole conductivity is responsible for the creation of the space charge field.

However, in some cases with a short time constant for the buildup of the grating or a large energy band gap between the valence and the conduction band, the *charge-hopping model* proposed by Hall and coworkers [22, 23, 24] is more appropriate. This model substitutes the three-step process of Kukhtarevs model by a one-step mechanism of thermal or photoinduced tunneling of excited states. Both models aim to explain the basis of the mechanism that leads to the creation of the refractive index grating. They naturally simplify the real situation of light–particle interaction in photorefractive materials. Both models describe special cases of simple trap distributions, whereas in reality the distribution of traps is much more complex because several different dopants or impurity ions interact. Moreover, the existence of traps near the band edge of the conduction band (*shallow traps*) or of the valence band (*deep traps*) suggests that processes of higher order, with two or more transitions between traps, may take place [25, 26, 27, 28]. In order to explain the basic features of the photorefractive effect, we refer in this chapter to the simple band transport model of Kukhtarev and Vinelski [21].

We start from the rate equations for electrons and holes – which are commonly known in semiconductor physics – which arise because an effective current density j acts on the crystal when it is illuminated with a light intensity pattern. That current density is composed of three constituents arising from well-known effects – drift in an electric field, diffusion and the bulk photovoltaic effect. If we assume a photorefractive material of the electron-conducting type, the current density of the holes can be neglected. For a material of the hole-conducting type, similar equations can be derived straightforwardly, in this case neglecting the current density of the electrons. For an electron-conducting material, the resulting current density of the electrons is given by

$$\boldsymbol{j} = \boldsymbol{j}_{\text{drift}} + \boldsymbol{j}_{\text{diff}} + \boldsymbol{j}_{\text{pv}} \tag{3.13}$$

$$= e\mu n_{\text{e}}\boldsymbol{E}_{\text{sc}} + eD\boldsymbol{\nabla} n_{\text{e}} + \beta_{\text{pv}}(N_{\text{D}} - N_{\text{D}}^{+})\boldsymbol{c}I \ , \tag{3.14}$$

where $D = \mu k_{\text{B}}T/e$ is the diffusion constant, e and μ are the charge and the mobility, of the electrons, respectively; N_{D}, N_{D}^{+}, and n_{e} are the densities of donors, ionized donors and free electrons, respectively; k_{B} is the Boltzmann constant; T is the absolute temperature; I is the intensity of the electromagnetic radiation; and $\boldsymbol{E}_{\text{sc}}$ is the amplitude of the static space charge electric field. $\boldsymbol{c}$ represents the dependence of the bulk photovoltaic effect on its direction, assuming that the largest component of the photogalvanic tensor is β_{pv}, which generates a current along the direction of the axis of spontaneous polarization of the medium (the c axis), whereas other components are neglected.

With the assumption that all acceptors are ionized, the space charge density depends only on the number densities of the acceptors N_{A}, the ionized donors N_{D}^{+}, and of the free electrons n_{e}, and we obtain

$$\rho = e(N_{\text{D}}^{+} - N_{\text{A}} - n_{\text{e}}) \ . \tag{3.15}$$

The change of the concentration of the electrons with time depends on the two competing effects of charge carrier generation G on the one hand and of recombination R on the other hand. The generation of electrons is proportional to the density of nonionized donors $(N_{\text{D}} - N_{\text{D}}^{+})$ and to the probability of thermal excitation and photoexcitation. In a similar way, the recombination rate R increases with the electron density n_{e} and the number of recombination centers N_{D}^{+}:

$$G = (sI + \beta_{\text{th}})(N_{\text{D}} - N_{\text{D}}^{+}) = \beta_{\text{th}}(I/I_{\text{sat}} + 1)(N_{\text{D}} - N_{\text{D}}^{+}) \ , \tag{3.16}$$

$$R = \zeta N_{\text{D}}^{+} n_{\text{e}} \ . \tag{3.17}$$

In these equations, s is the constant for photoinduced excitation, β_{th} is the constant for thermal excitation and ζ is the recombination coefficient for linear recombination. It is convenient to introduce the saturation intensity $I_{\text{sat}} = \beta/s$. This important parameter determines the ratio of the thermal-excitation coefficient to the photoionization coefficient which is expressed by a characteristic intensity. This formulation is useful because not only does the optical beam intensity $I = |A|^2$ control the nonlinear interaction, but also a quantity equivalent to an intensity arises from the thermal excitation. Moreover, an artificially created contribution to the thermal excitation due to a supplementary, incoherent background illumination is often used in experimental situations, because it allows one to control the nonlinear behavior of the photorefractive material. In the following we refer to the normalized intensity $\hat{I} = I/I_{\mathrm{sat}}$ and omit the $\hat{}$ - symbol. In some publications [29] the intensity is normalized by a characteristic intensity I_0 (e.g. in the center of the beam) and can then be written as

$$I = (I + \beta_{\mathrm{th}}/s)/I_0 \equiv \tilde{I} + I_{\mathrm{d}} \,. \tag{3.18}$$

Note that in this notation the so-called dark intensity I_{d} is proportional to I_{sat}. In the following, we shall use mainly the notation based on I_{sat}, keeping in mind the analogy with the formulation based on I_{d}.

It is important to note that only the electrons are mobile charge carriers here whereas the ionized donor or acceptor atoms are stationary within the crystal lattice. Therefore, we arrive at the following continuity equation for the change of the concentration of the electrons:

$$\frac{\partial \rho}{\partial t} = e\left(\frac{\partial N_{\mathrm{D}}^{+}}{\partial t} - \frac{\partial n_{\mathrm{e}}}{\partial t}\right) = \boldsymbol{\nabla} \boldsymbol{j} \,. \tag{3.19}$$

The change of the density of the ionized donors depends, for those immobile charge carriers only on the difference between the generation and recombination rates:

$$\frac{\partial N_{\mathrm{D}}^{+}}{\partial t} = G - R = \beta_{\mathrm{th}}(I+1)(N_{\mathrm{D}} - N_{\mathrm{D}}^{+}) - \zeta N_{\mathrm{D}}^{+} n_{\mathrm{e}} \,. \tag{3.20}$$

Since the densities of electrons and ionized donors vary at the same rate when the current is independent of spatial position, the equation for the average electron density is given by

$$\frac{\partial n_{\mathrm{e}}}{\partial t} = \beta_{\mathrm{th}}(I+1)(N_{\mathrm{D}} - N_{\mathrm{A}}) - \zeta N_{\mathrm{A}} n_{\mathrm{e}} \,. \tag{3.21}$$

The equations (3.14), (3.19), and (3.20) or (3.21), combined with Gauss's law (Poisson's equation)

$$\varepsilon\varepsilon_0 \boldsymbol{\nabla} \boldsymbol{E}_{\mathrm{sc}} = \rho = e(N_{\mathrm{D}}^{+} - N_{\mathrm{A}} - n_{\mathrm{e}}) \,, \tag{3.22}$$

are referred to as the fundamental equations of the *band transport model*, after Kukhtarev et al. [21]. Here ε_0 is the permittivity of a vacuum and ε is the dielectric constant of the material along the c axis. The equations describe the propagation of light waves in the medium and the properties of the photorefractive effect for that particular material completely. These equations are the basis of all our further derivations of the various material–light interaction processes in photorefractive nonlinear materials.

3.4 Space Charge Field for a Single Beam

Transforming the equations of the band transport model into a single equation for the amplitude of the static electric space charge field in a photorefractive crystal is a very involved task that is beyond the scope of our description. However, using several assumptions and approximations that are realistic for typical experimental situations, it is possible to extract some simplified expressions for the space charge field. Here, we concentrate on two different approaches by Christodoulides and Carvalho [30] and by Zozulya and Anderson [29], which are referred to here as isotropic and anisotropic, respectively.

3.4.1 Isotropic Approximation to the Space Charge Field

Following Christodoulides and Carvalho [30], we assume a (1+1)-dimensional geometry with a quasi-homogeneous illumination of the crystal, which is equivalent to the assumption that the width of the light beam exceeds by far the characteristic spatial scale of the electric field, given by the Debye wave number $k_\mathrm{D} = (e^2 N_\mathrm{A}/k_\mathrm{B}T\varepsilon_0\varepsilon)^{1/2}$. Further, we assume that the system is in a steady state ($\partial/\partial t = 0$) and that the photovoltaic term of (3.14) can be neglected. Therefore the current density consists only of the two constituents associated with drift and diffusion. We suppose that the optical wave propagates along the z direction and that an external electric field $\boldsymbol{E}_\mathrm{ext}$ is applied in the transverse x direction (the direction of the optical c axis). Supposing that the electron density n_e is negligible compared with the density of impurity ions, i.e. $N_\mathrm{D}^+, N_\mathrm{A} \gg n_\mathrm{e}$, we can derive the following expression for the density of ionized donor impurities from (3.22):

$$N_\mathrm{D}^+ = N_\mathrm{A}(1 + \frac{\varepsilon_0\varepsilon}{eN_A}\boldsymbol{\nabla}\boldsymbol{E}_\mathrm{sc}) \, . \tag{3.23}$$

With this expression in hand, we can also derive an equation for the density of free electrons from (3.20):

$$n_\mathrm{e} = \frac{\beta_\mathrm{th}(N_\mathrm{D} - N_\mathrm{A})}{\gamma N_\mathrm{A}}(1+I)(1 + \frac{\varepsilon_0\varepsilon}{eN_\mathrm{A}}\boldsymbol{\nabla}\boldsymbol{E}_\mathrm{sc})^{-1}. \tag{3.24}$$

If we assume that the light intensity I asymptotically approaches a constant value in the border region of the crystal, i.e. $I(x, y \to \pm\infty, z) = I_\infty$, the space charge field is independent of the transverse coordinates in this region and equals the externally applied field $\boldsymbol{E}_\mathrm{ext}$. Therefore, the space charge field $|\boldsymbol{E}_\mathrm{sc}(x \to \pm\infty)|$ can be expressed in terms of a constant voltage V and the crystal width L, resulting in

$$|\boldsymbol{E}_\mathrm{sc}(x \to \pm\infty)| \equiv |\boldsymbol{E}_\mathrm{ext}| = V/L \, . \tag{3.25}$$

Consequently, the free-electron density in this border region, where $n_0 = n_\mathrm{e}(x, y \to \infty)$, is given by

$$n_0 = \frac{\beta_{\text{th}}(1+I_\infty)(N_{\text{D}}-N_{\text{A}})}{\gamma N_{\text{A}}} \,. \tag{3.26}$$

Since the system is in a steady state, the current density $\boldsymbol{j}$ has to be constant everywhere, according to (3.19) and therefore $n_0 \boldsymbol{E}_{\text{ext}} = n_{\text{e}} \boldsymbol{E}_{\text{sc}} + (k_{\text{B}}T/e)\boldsymbol{\nabla} n_{\text{e}}$ or

$$\boldsymbol{E}_{\text{sc}} = \frac{n_0 \boldsymbol{E}_{\text{ext}}}{n_{\text{e}}} - \frac{k_{\text{B}}T}{e n_{\text{e}}} \boldsymbol{\nabla} n_{\text{e}} \,. \tag{3.27}$$

Substitution of (3.24) and (3.26) in (3.27) yields the following relation:

$$\begin{aligned} \boldsymbol{E}_{\text{sc}} &= \boldsymbol{E}_{\text{ext}} \frac{I_\infty + 1}{I+1}\left(1 + \frac{\epsilon\epsilon_0}{eN_{\text{A}}}\boldsymbol{\nabla}\boldsymbol{E}_{\text{sc}}\right) - \frac{k_B T}{e}\frac{\boldsymbol{\nabla} I}{I+1} \\ &+ \frac{k_B T}{e}\frac{\epsilon\epsilon_0}{eN_{\text{A}}}\left(1 + \frac{\epsilon\epsilon_0}{eN_{\text{A}}}\boldsymbol{\nabla}\boldsymbol{E}_{\text{sc}}\right)^{-1} \cdot \nabla^2 \boldsymbol{E}_{\text{sc}} \,. \end{aligned} \tag{3.28}$$

In the case of a drift-dominated charge carrier migration process, where the external field $|\boldsymbol{E}_{\text{ext}}|$ can reach an appreciable value, the diffusion ($k_{\text{B}}T/e$ terms) can be neglected in (3.28). Further, assuming that the beam intensity varies slowly with respect to the spatial coordinates, i.e. $\boldsymbol{\nabla}\boldsymbol{E}_{\text{sc}} \approx 0$, (3.28) can be simplified to get a simple expression for the space charge field,

$$\boldsymbol{E}_{\text{sc}} \approx \boldsymbol{E}_{\text{ext}} \frac{I_\infty + 1}{I+1} \,. \tag{3.29}$$

Note that (3.29) has an isotropic character since $\boldsymbol{E}_{\text{sc}}$ depends only on the intensity distribution and not on any spatial coordinates. If we also assume that $I_\infty = 0$, which is reasonable for studies of solitons in photorefractive crystals, the refractive index change resulting from (3.10), with $|\boldsymbol{E}_{\text{sc}}|$ given by (3.29), can be expressed as

$$\Delta n_{\text{pr}} = -\frac{1}{2} n_0^3 r_{\text{eff}} |\boldsymbol{E}_{\text{ext}}| \frac{1}{1+I} \,, \tag{3.30}$$

which reflects the saturable characteristics of the photorefractive nonlinearity. This model is quite appropriate for describing the formation and interaction of spatial solitons in generic saturable media, but it does not take into account the distinctly anisotropic characteristics of a biased photorefractive medium, which is the typical kind of medium utilized for spatial-soliton investigations.

3.4.2 Anisotropic Approximation to the Space Charge Field

To describe the space charge field of an optical beam in a biased photorefractive medium we follow the anisotropic approach developed by Zozulya and Anderson [29]. This approach is based on the simple assumption that the space charge field can be expressed by means of an electrostatic potential, i.e.

$$\boldsymbol{E}_{\mathrm{sc}} = -\boldsymbol{\nabla}\varphi \,. \tag{3.31}$$

As already explained for the isotropic case, we assume the system to be in a steady state and neglect the contribution of any possible photovoltaic term. Consequently, the current density (3.14) can be rewritten as

$$\boldsymbol{j} = -e\mu n_{\mathrm{e}}\,\boldsymbol{\nabla}\varphi + \mu k_{\mathrm{B}}T\boldsymbol{\nabla}n_e \,. \tag{3.32}$$

Since we are dealing with a single spatially extended beam, we assume that the intensity varies slowly with respect to the spatial coordinates, which yields the following from (3.24):

$$n_{\mathrm{e}} = \frac{\beta_{\mathrm{th}}(1+I)(N_{\mathrm{D}}-N_{\mathrm{A}})}{\gamma N_{\mathrm{A}}} \,. \tag{3.33}$$

Substitution of (3.33) into (3.32), assuming the steady-state condition $\boldsymbol{\nabla}\boldsymbol{j} = 0$ and using (3.19) then yields

$$\boldsymbol{\nabla}I\cdot\boldsymbol{\nabla}\varphi + (1+I)\boldsymbol{\nabla}\cdot(\boldsymbol{\nabla}\varphi) - \frac{k_{\mathrm{B}}T}{e}\boldsymbol{\nabla}^{\mathbf{2}}(1+I) = 0 \,. \tag{3.34}$$

If the external electric field is applied along the positive x direction, it is appropriate to assume that the electrostatic potential consists of a light-induced term φ_0 and an external bias term $-|\boldsymbol{E}_{\mathrm{ext}}|x$. Inserting $\varphi = \varphi_0 - |\boldsymbol{E}_{\mathrm{ext}}|x$ into (3.34) and dividing by $(1+I)$ leads to an expression for the potential equation for the electrostatic field,

$$\begin{aligned}\boldsymbol{\nabla}^{\mathbf{2}}\varphi_0 = &-\boldsymbol{\nabla}\ln(1+I)\boldsymbol{\nabla}\varphi_0 + |\boldsymbol{E}_{\mathrm{ext}}|\frac{\partial\ln(1+I)}{\partial x}\\ &-\frac{k_{\mathrm{B}}T}{e}\{\boldsymbol{\nabla}^{\mathbf{2}}\ln(1+I) + [\boldsymbol{\nabla}\ln(1+I)]^2\} \,.\end{aligned} \tag{3.35}$$

With this specific type of nonlinearity the space charge field $\boldsymbol{E}_{\mathrm{sc}} = -\nabla\varphi = \boldsymbol{E}_{\mathrm{ext}} - \nabla\varphi_0$ can be expressed in terms of the externally applied voltage V, where $V/L = |\boldsymbol{E}_{\mathrm{ext}}|$ and the light-induced electrostatic potential φ_0 can be calculated numerically from (3.35). The potential equation contains contributions to the space charge field from both drift and diffusion. The drift is represented by the first two terms on the right-hand side of (3.35), and the diffusion is described by the last term on the right-hand side, which is proportional to $k_{\mathrm{B}}T/e$. Both charge carrier migration effects have a distinct impact on the refractive index modulation. The drift of electrons (or holes) in an external electric field leads to the formation of an anisotropic, focusing refractive index lens, whereas diffusion results in a refractive index contribution that varies almost linearly across the beam. Both effects destroy the circular symmetry of the incident light beam. The latter leads to a transverse deflection of the beam towards the optical axis, which is referred to as beam bending [6], whereas drift induces a symmetric refractive index structure with respect to the horizontal and vertical transverse axes. Therefore, an incident beam with a circularly symmetric Gaussian profile will become elliptically shaped and laterally displaced as it propagates through the biased nonlinear crystal. The physics described by (3.35) can be exploited to form a *screening photorefractive spatial soliton*, which will be described in detail in Sect. 4.2.3.

3.5 Space Charge Field for Two Interfering Beams

A more complex situation arises in the presence of two beams intersecting in a photorefractive medium; this situation is commonly known as photorefractive two-wave mixing. Above a certain threshold for the nonlinear coupling strength, the energy and phase coupling between the two beams gives rise to complex nonlinear dynamics, including modulation instabilities and spatial pattern formation processes.

To obtain an explicit expression for the modulation of the refractive index, let us consider two monochromatic plane waves with wave vectors $\boldsymbol{k}_1$ and $\boldsymbol{k}_2$, incident on a photorefractive medium. The two prominent examples of two-wave-mixing geometries are depicted in Fig. 3.3: the codirectional geometry (Fig. 3.3a) and the contradirectional geometry (Fig. 3.3b). The two waves enter the crystal from the same side (Fig. 3.3a), or from opposite sides (Fig. 3.3b). If we ignore the vectorial nature of the electromagnetic fields and use a scalar representation, the interference of two plane waves

$$E_1(\boldsymbol{r},t) = A_1(\boldsymbol{r},t)\exp(\mathrm{i}\boldsymbol{k}_1\cdot\boldsymbol{r})\exp(-\mathrm{i}\omega t) + \mathrm{c.c.} \tag{3.36}$$

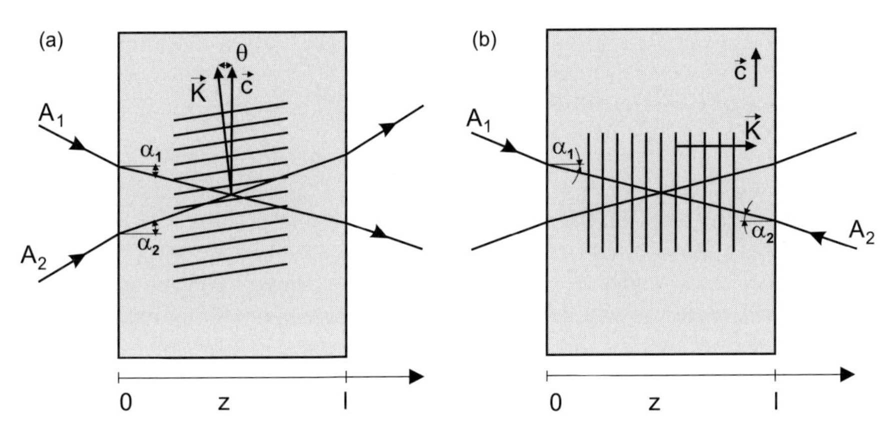

Fig. 3.3. Principle of two-wave mixing of two waves with amplitudes A_1 and A_2 and incidence angles α_1 and α_2. (**a**) Transmission grating configuration (codirectional geometry), (**b**) reflection grating configuration (contradirectional geometry). The grating direction is indicated by $\boldsymbol{K} = \boldsymbol{k}_2 - \boldsymbol{k}_1$, which represents the grating vector. $\boldsymbol{c}$ is the crystallographic c axis, and θ is the orientation of the grating vector relative to the crystallographic c axis

and

$$E_2(\boldsymbol{r},t) = A_2(\boldsymbol{r},t)\exp(\mathrm{i}\boldsymbol{k}_2 \cdot \boldsymbol{r})\exp(-\mathrm{i}\omega t) + \text{c.c.} \tag{3.37}$$

leads to an intensity modulation inside the medium given by

$$I = |E_1 + E_2|^2 = I_0\{1 + m\cos[(\boldsymbol{k}_2 - \boldsymbol{k}_1)\cdot \boldsymbol{r}]\}\,, \tag{3.38}$$

where

$$I_0 = |A_1|^2 + |A_2|^2\,. \tag{3.39}$$

The *modulation depth* is defined by

$$m = \frac{A_1^\star A_2 + A_1 A_2^\star}{I_0}\,, \tag{3.40}$$

which is nothing more than the "visibility" of interference fringes known from classical optics. In the case where both waves enter the crystal from the same side, a *transmission grating* is formed with a wave number

$$K_\mathrm{T} = \frac{4\pi n_0 \sin[(\alpha_1 - \alpha_2)/2]}{\lambda}\,, \tag{3.41}$$

where α_1 and α_2 denote the incidence angles of the two interfering beams. A *reflection grating* is formed if the beams enter the medium from opposite sides. This grating is described by a wave number

$$K_\mathrm{R} = \frac{4\pi n_0 \cos[(\alpha_1 - \alpha_2)/2]}{\lambda}\,. \tag{3.42}$$

An important special case with respect to the capability for exhibiting modulation instabilities is that of two exactly counterpropagating beams with wave vectors $\boldsymbol{k}_1 = -\boldsymbol{k}_2$. In this case, (3.38) reduces to

$$I(z) = I_0[1 + m\cos(2kz)] \tag{3.43}$$

and represents a scalar intensity grating inside the crystal with a grating wave number $2k$.

The incidence angles α_1 and α_2 of the beams then define the orientation of the wave vector of the grating relative to the c axis of the crystal (note the sign conventions for the two angles) as follows:

$$\theta = \frac{\alpha_1 + \alpha_2}{2}\,. \tag{3.44}$$

With this knowledge, we are now able to calculate the value of the space charge field from the charge density by using (3.22). The integral in (3.22) can be calculated if we assume that quadratic recombination can be neglected and that the modulation depth is small. This requirement corresponds to the case of small contributions from high diffraction orders in the decomposition of the diffraction pattern into a Fourier series or, in other words, to the assumption of a quasi-linear modulation. This linearization is justified because of the selectivity of the Bragg condition in volume materials, which gives the minimum angular distance between two beams that are required to write two different, separable gratings with the same pump wave. Consequently, the solution of Kukhtarev's differential equations can be written in the form

$$X(z,t) = X_0(t) + X_{\rm p}(z,t)\,, \tag{3.45}$$

where X_0 is independent of spatial position and $X_{\rm p}$ is the periodic perturbation. X represents any variable in Kukhtarev's equations, $N_{\rm D}^+$, n, j, or $E_{\rm sc}$. The corresponding nonlinear terms can be found in (3.14), or (3.20) and (3.22). By applying the linearization described above, we can rewrite the corresponding differential equations, separate the space-independent and space-dependent terms and derive from the rsulting equations the differential equation for the space charge field $E_{\rm sc}$ [10]:

$$\frac{\partial E_{\rm sc}}{\partial t_{\rm n}} + p = {\rm i}m\frac{E_{\rm D} + {\rm i}(E_0 + E_{\rm pv})}{1 + (E_{\rm D} + {\rm i}E_0)/E_{\rm M}}\,, \tag{3.46}$$

where $m = I_1/I_0$ is the modulation of the interference pattern, and the new variable $t_{\rm n}$ is given by $zt_{\rm n} = t/\tau_{\rm d}$, where $\tau_{\rm d} = \varepsilon/(e\mu n_0)$ is the dielectric relaxation time. The parameter p is defined by

$$p = \frac{1}{1 + [E_{\rm D} + {\rm i}(E_0 + E_{\rm pv})]/E_{\rm M}}\left(1 + \frac{E_{\rm D} + {\rm i}(E_0 + E_{\rm pv})}{E_{\rm q}}\right) \tag{3.47}$$

and will be used in the following derivation, following [10], in order to simplify the representation of the space charge field. In addition, we shall use a scalar notation in the following sections, thereby ignoring the vectorial nature of the space charge field, again for reasons of simplicity of the expressions.

The characteristic fields $E_{\rm D}$, $E_{\rm M}$, $E_{\rm q}$ and the photovoltaic field $E_{\rm pv}$ were introducted by Kukhtarev et al. [21] to describe the space charge field of two interfering beams and may be written in terms of the parameters defined earlier as

$$\begin{aligned} E_{\rm D} &= \frac{k_{\rm B}TK}{e}\,, \qquad E_{\rm M} = \frac{\zeta N_{\rm A}^-}{\mu K}\,, \qquad E_{\rm pv} = \chi\alpha I\sigma^{-1}\,, \\ E_{\rm q} &= E_{\rm q0}(1-a) = \frac{eN_{\rm A}^-}{\varepsilon K}\left(1 - \frac{N_{\rm A}^-}{N_{\rm D}}\right)\,, \end{aligned} \tag{3.48}$$

where χ is the anisotropic tensor for charge transport, α is the absorption coefficient, $\sigma = \sigma_{\rm d} + \alpha f(i)$ represents the conductivity and $\sigma_{\rm d}$ represents the dark conductivity.

The physical meaning of these fields is obvious. $E_{\rm D}$ is the *diffusion field*. It is inversely proportional to the grating spacing. The smaller the grating spacing, the larger is the gradient of the electron distribution and the stronger are the forces of diffusion. Under certain circumstances, the space charge field may be equal to the diffusion field. E_M can be recognized if we write it in the form

$$E_M = \frac{\Lambda}{2\pi\mu\tau_e} . \tag{3.49}$$

Thus, E_M is the electric field which will move an electron a distance $\Lambda/2\pi$ during its lifetime. E_{q0}, called the *saturation* or *maximum field*, can also be associated with a simple interpretation. It is the maximum space charge field that can exist in the material for a sinusoidal charge distribution in the case of $a \ll 1$.

The equation for the space charge field (3.46) is a second-order differential equation owing to its complex coefficients, and has solutions that may be combinations of exponential and oscillatory functions. With the initial conditions $E_{sc}(t = 0)$, the solution is [10]

$$E_{sc} = \mathrm{i}m \frac{E_D + \mathrm{i}E_0}{1 + [E_D + \mathrm{i}(E_0 + E_{pv})]/E_q} [1 - \exp(-p)] . \tag{3.50}$$

When no external electric field or drift field is operating ($E_0 + E_{pv} = 0$), all the coefficients are real and the function increases monotonically, but for $E_0 + E_{pv} \neq 0$, the imaginary part is different from zero and an oscillatory behavior occurs. The rise time is then given by the real part of the exponential term, i.e. that of p in (3.50).

A similar solution exists for the decline of the electric field. Assuming that it has a certain value $E_{sc}(0)$ at $t = 0$, and that the excitation is suddenly stopped (i.e. $m = 0$), the solution of (3.46) is

$$E_{sc} = E_{sc}(0) \exp(-p) . \tag{3.51}$$

The decay time is given again by the real part of p, and the imaginary part of p is responsible for the oscillations. However, it is important to note that the symmetry between the rising and decaying solutions exists only for the normalized temporal variable t_n. In practical cases it is far from obvious that the dielectric relaxation time is the same for both the rise and decay. This depends on the way the excitation comes to a stop.

In order to solve the equation for the space charge field, we shall concentrate on two aspects of the rising solution that are important for the subsequent investigations of pattern formation.

3.5.1 Short-Time Limit of the Space Charge Field

Solutions in the vicinity of $t = 0$ can be obtained from (3.50) by expanding the exponential term, or directly from the differential equation (3.46) by disregarding p. In either case, the space charge field obtained is

$$E_{sc} = \mathrm{i}m \frac{E_M[E_D + \mathrm{i}(E_0 + E_{pv})]}{E_M + E_D + \mathrm{i}(E_0 + E_{pv})} t_n . \tag{3.52}$$

The limiting value of $\partial E_{\mathrm{sc}}/\partial t_{\mathrm{n}}$ is $\mathrm{i}m[E_{\mathrm{D}}+\mathrm{i}(E_0+E_{\mathrm{pv}}]$ when E_{M} is large, and $\mathrm{i}mE_{\mathrm{M}}$ when E_{M} is small. Hence E_{M} represents the maximum rate of change of E_{sc}. This relation can also be explained in a more physical way. The space charge field is limited by the rise time needed to generate photocarriers. Because the space charge field depends on the ionized-donor density N_{D}^{+}, and because in order to ionize a donor at least one photon must be incident, the fastest rate at which N_{D}^{+} can increase is $(\alpha\eta_{\mathrm{q}}mI_0/\hbar\omega)t$, where η_{q} is the probability of photon ionization. Consequently, the fastest rate at which E_{sc} can increase is

$$E_{\mathrm{sc}}=\mathrm{i}m\frac{e}{K\varepsilon}\frac{\alpha\eta_{\mathrm{q}}I_0}{\hbar\omega}, \tag{3.53}$$

which can be reduced to

$$E_{\mathrm{sc}}=\mathrm{i}m\frac{E_{\mathrm{M}}}{\tau_{\mathrm{d}}}, \tag{3.54}$$

in agreement with the derivation above.

In our consideration of the application of these effects to pattern formation, the case of pure diffusion is most interesting. For $E_0+E_{\mathrm{pv}}=0$, we obtain

$$E_{\mathrm{sc}}=\mathrm{i}m\frac{E_{\mathrm{M}}E_{\mathrm{D}}}{E_{\mathrm{M}}+E_{\mathrm{D}}}t_{\mathrm{n}}\,. \tag{3.55}$$

In this case the space charge field is purely imaginary. Moreover, there is an optimum value of K, since E_{D} is proportional to K and E_{M} is inversely proportional to K. The optimum occurs when $E_{\mathrm{M}}=E_{\mathrm{D}}$, which occurs for $K^2=e/(\mu\tau_{\mathrm{e}}k_{\mathrm{B}}T)$. If we write $eD=\mu k_{\mathrm{B}}T$, and use the definition of the diffusion length $L_{\mathrm{D}}=\sqrt{D\tau_{\mathrm{e}}}$, we obtain $K_{\mathrm{opt}}=L_{\mathrm{D}}^{-1}$, which indicates that the optimum occurs when the grating spacing is 2π times the diffusion length. Because the diffusion lengths are significantly different in different photorefractive materials, this affects the optimum grating spacing strongly. Therefore, for crystals with a lower mobility–lifetime product, the optimum grating spacing is much larger than for those with a higher mobility–lifetime product.

If fields due to drift charge transport mechanisms are also involved in the creation of the space charge field, the situation is more complex. In this case the energy interaction between the two optical waves depends only on the imaginary part of the space charge field, whereas the real part gives rise to phase coupling. If the aim is to achieve a large modulus of the space charge field (i.e. to obtain high diffraction efficiency), the application of an electric field can help considerably.

3.5.2 Stationary Solution for the Space Charge Field

When the space charge field is allowed to rise to its saturation value, the following solution can be easily obtained from the differential equation (3.46):

$$E_{\mathrm{sc}}=\mathrm{i}m\frac{E_{\mathrm{q}}[E_{\mathrm{D}}+\mathrm{i}(E_0+E_{pv})]}{E_{\mathrm{D}}+E_{\mathrm{q}}+\mathrm{i}(E_0+E_{pv})}\,. \tag{3.56}$$

If E_{M} is replaced by E_{q}, the above expression becomes nearly identical in form to (3.52).

Owing to the common space charge field or refractive index modulation of the beams, they may exchange energy as well as phase. The amount of energy transfer between the two interacting beams depends critically on the phase shift between the interference grating and the refractive index grating. This phase shift, in turn, is determined by the ratio of the imaginary contribution to the space charge field (diffusion), which characterizes the maximum energy coupling, to the real contribution (drift), which is responsible for phase coupling; the phase shift is given by

$$\tan\phi = \frac{\mathrm{Im}(E_{\mathrm{sc}})}{\mathrm{Re}(E_{\mathrm{sc}})} = \frac{E_{\mathrm{d}}E_{\mathrm{q}} + E_{\mathrm{d}}^2 + (E_0 + E_{pv})^2}{E_{\mathrm{d}}(E_0 + E_{pv})} . \tag{3.57}$$

Thus, to achieve a stationary beam-coupling situation, one has to adjust the optimum phase shift to $\phi = 90°$, for example by using the crystal without any external or photovoltaic field or by adjusting those two fields in such a way that they compensate and satisfy the equation $E_0 + E_{\mathrm{pv}} = 0$. In that special case, the space charge field is given by

$$E_{\mathrm{sc}} = -im\frac{E_{\mathrm{d}}E_{\mathrm{q}}}{E_{\mathrm{d}} + E_{\mathrm{q}}} , \tag{3.58}$$

which again reaches its maximum value for $E_{\mathrm{d}} = E_{\mathrm{q}}$, or a pure diffusion-dominated refractive index modulation. Again, the optimum grating spacing corresponding to this case is given by

$$\Lambda_{\mathrm{opt}} = 2\pi\left(\frac{k_{\mathrm{B}}T\varepsilon}{e^2N_{\mathrm{A}}(1 - \frac{N_{\mathrm{A}}}{N_{\mathrm{D}}})}\right)^{1/2} = 2\pi L_{\mathrm{Deb}} , \tag{3.59}$$

where L_{Deb} is the Debye screening length, derived earlier for the case of a single beam–matter interaction, corresponding to an effective charge density of $N_{\mathrm{A}}(1 - N_{\mathrm{A}}/N_{\mathrm{D}})$. This indicates that the space charge field must be small when the grating spacing is much less than the Debye length. The limiting values are now E_{q} for a small grating spacing and E_{D} for a large grating spacing. In the opposite case of a sufficiently large drift component (E_0 and/or E_{pv} are large), the space charge field may be made equal to E_{q} for any grating spacing. The time it takes for the space charge field to rise to 63% of its final value – this is the usual definition of the rise time, which is often referred to as the photorefractive time constant τ – may be obtained from (3.46) in a normalized form relative to the dielectric relaxation time as [10]

$$\frac{\tau}{\tau_{\mathrm{d}}} = \frac{1}{\mathrm{Re}(p)} = \frac{E_{\mathrm{q}}}{E_{\mathrm{M}}} + \frac{(E_{\mathrm{M}} + E_{\mathrm{D}})^2 + (E_0 + E_{\mathrm{pv}})^2}{(E_{\mathrm{q}} + E_{\mathrm{D}})(E_{\mathrm{M}} + E_{\mathrm{D}}) + (E_0 + E_{\mathrm{pv}})^2} . \tag{3.60}$$

Equation (3.60) shows that all the material parameters may play a role in determining the value of the rise time: both E_{M} and E_{q} have to be taken into account. Moreover, the rise time is proportional to the dielectric time constant, resulting in "slow" materials as $BaTiO_3$ and "quicker" ones such as BSO. In the absence of an applied electric field and when $E_D \ll E_M$, E_q (i.e. for large Λ) Eq. (3.60) yields $\tau/\tau_d = 1$, but otherwise the rise time can vary over quite a wide range.

3.5.3 Strong Modulation of the Space Charge Field

One of the crucial approximations of the Kukhtarev model for deriving the space charge field that enables us to obtain analytic solutions, is that the modulation m of the interference pattern is assumed to be small. In practice, this condition is satisfied when the crystal is used as an amplifier (i.e. weak signal and strong pump). In contrast, when the aim is a high diffraction efficiency or if the two interacting beams are derived from the same beam (e.g. by reflection from a feedback mirror), the choice is undoubtedly to have incident beams of equal intensity, i.e. a modulation of or near unity. For a number of photorefractive applications, a high modulation is favorable, and a large diffraction efficiency of the photorefractive grating can be achieved by choosing incident beams of equal intensities. Consequently, this problem of measurement at large modulation depth has attracted considerable interest, and a number of articles have appeared on that subject [31, 32, 33, 34]. In particular, experiments have revealed that measurement of the gain coefficient at a large modulation depth reduces the apparent coupling constant [31, 33]. Such effects at large modulation depth, often referred to as "large-signal effects", are caused by higher harmonics of the space charge field [31, 32], which leads to a nonlinear response of the space charge field to the modulation depth m. As a consequence, (3.67) is no longer valid for large modulation depths. A common approach to overcome this problem is the introduction of an empirical correction function $f_i(m)$ to describe this nonlinear dependence [31]. Several different functional dependences $f_i(m)$ ve been used to account for the experimental results [10] (this is the motivation behind the subscript i). For small modulation $m \ll 1$, these functions provide a linear dependence on the modulation depth as assumed by Kukhtarev's, equations and yield sublinear behavior for larger m. In particular, Belić et al. obtained an analytic solution of the two-wave-mixing equations for an arbitrary correction function [34].

In the case of a large modulation depth, near unity, it is still true that the space charge field must be a periodic function of spatial position, and that the fundamental period is given by Λ, as calculated from the interbeam angle. Again, if the space charge field is periodic, it can be expanded into a Fourier series. The fundamental component will vary as $\cos(Kz)$, in general with some phase angle relative to the interference pattern. That is exactly what we have assumed so far about the spatially varying space charge field. However, how will the fundamental component change as $m \to 1$ and will higher harmonics be involved? Because most photorefractive materials are bulk materials, Bragg selectivity is always present and will suppress the contribution of higher harmonics.

Here, we consider only the case of a diffusion-dominated material under steady-state conditions, which will be the important case for our investigations of pattern formation. Assuming linear recombination but retaining the nonlinearity in the current equation, the electron density is given by

$$n(z) = sI_0(N_\mathrm{D} - N_\mathrm{A})\tau_\mathrm{e}[1 + m\cos(Kz)] . \tag{3.61}$$

In the absence of an applied electric field, the current density is zero, and therefore the conduction and diffusion currents must cancel each other in the steady state, yielding

$$e[1 + m\cos(Kz)]E_\mathrm{sc}(z) = k_\mathrm{B}TK\sin(Kz) . \tag{3.62}$$

The space charge field can be written as

$$E_\mathrm{sc}(z) = \frac{mE_\mathrm{D}\sin(Kz)}{1 + m\cos(Kz)} . \tag{3.63}$$

Here, the dependence on m is clearly nonlinear, but it still reduces to the previous result when m is small. The change in the space charge field for one period can be recognized in the deviation from the sinusoidal profile. The variation along z becomes much steeper and asymmetric compared with the previous profile. The magnitude of the fundamental component may be obtained in this case from a Fourier expansion [23], as

$$f_1(m) = \frac{2}{m}\left(1 - \sqrt{1 - m^2}\right) . \tag{3.64}$$

Thus $f(m)$ reduces to m in the limit of small m, and $f(m = 1) = 2$. The most important result is that large-modulation effects are favorable, leading to higher space charge fields. However, at the same time, large-modulation effects may also be harmful, especially in the case of a detuning of the two interfering waves. In this case, the maximum of the space charge field at the optimum detuning may be associated with a decrease in the maximum of the imaginary part of the space charge field to about 30% of its low-modulation value. This effect was found by Au and Solymar [32, 35] and was taken into account by Refregier et al. [31] simply replacing the modulation m by another function. Setting

$$f_2(m) = \frac{1}{a}[1 - \exp(-am)] , \tag{3.65}$$

where $a = 2.8$, gives a sublinear variation of m. For AC excitation, the reduction in gain due to large-modulation effects was first demonstrated by Stepanov and Sochova [36] and later calculated by Brost [37], who introduced a phenomenological function of the form

$$f_3(m) = \frac{1}{a}[1 - \exp(am)] \exp(m) . \quad (3.66)$$

From this equation, it can be concluded that under large-modulation conditions the space charge field cannot significantly exceed the applied field. The correction function $f_2(m)$ was used by Kwak et al. [33] to explain the results of two-wave-mixing gain experiments in a diffusion-dominated material, namely $BaTiO_3$. These authors found excellent agreement with their experiments for a fitting parameter $a = 3.0$, which yields $f(m = 1) = 1/3[1 - \exp(-3)] = 0.317$ for a modulation depth of unity.

The description of these higher-order terms is of importance when one is formulating the coupling between two waves that interact through the mediation of the space charge field. At large modulation depth, the resulting coupling strength has a value which is approximately one-third of the coupling strength for lower modulation depths. Especially in the case of geometries that are important in pattern formation experiments, this has to be kept in mind for measurements of the coupling strength in counterpropagating beam configurations, which imply measurements at large modulation depth.

3.6 Equations of Motion

3.6.1 Two-Beam Coupling and Amplification

One of the main results of the Kukhtarev model is the direct proportionality between the space charge field E_{sc} and the modulation depth m of the intensity grating [38] (see e.g. (3.50)),

$$E_{\mathrm{sc}} \propto m\mathrm{e}^{\mathrm{i}\phi} , \quad (3.67)$$

where the modulation depth is defined as in (3.40), and ϕ represents the relative phase shift between the interference grating, and the space charge field grating which depends on the relative strengths of the charge transport processes. A central condition for the derivation of the space charge field used to obtain (3.57) is the assumption of a small modulation depth $m \ll 1$. When we insert the expression (3.56) into the equation for the refractive index modulation (3.7), this leads to a refractive index profile given by [11]

$$\Delta n_{\mathrm{pr}} = \frac{n_{\mathrm{pr}}}{2} \left[m \exp(\mathrm{i}\phi) \exp(-\mathrm{i}\boldsymbol{K} \cdot \boldsymbol{r}) + \mathrm{c.c.} \right] , \quad (3.68)$$

where n_{pr} is a proportionality constant representing a measure of the strength of the photorefractive effect. In particular, it includes the complex space charge field amplitude (which depends on possible diffusion, drift and photovoltaic contributions) and the effective electro-optic coefficient, which depends on the geometry of incidence and the polarization state of the interacting waves. Thus, this constant represents (together with the phase $\mathrm{e}^{\mathrm{i}\phi}$)

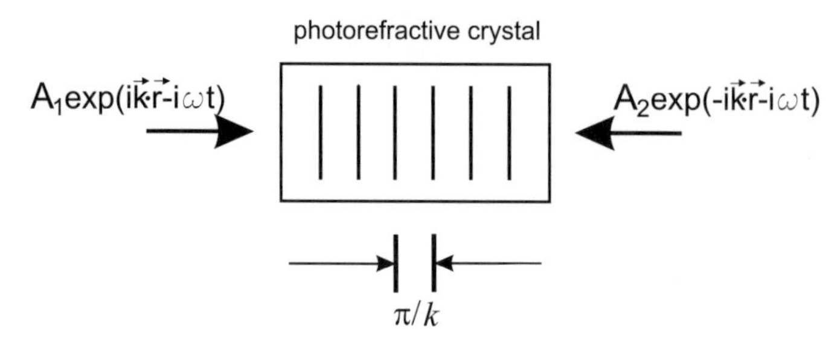

Fig. 3.4. Basic interaction geometry. Two plane waves with wave vectors $\boldsymbol{k}$ and $-\boldsymbol{k}$ counterpropagate in a photorefractive crystal and create a static interference pattern with a wavelength $\Lambda = \pi/|\boldsymbol{k}|$

the specific material properties. Equation (3.68) illustrates, in particular, the explicit dependence of n_{pr} on the modulation depth, which may vary across the crystal volume owing to beam-coupling effects. It is also apparent that the refractive index modulation due to the photorefractive effect does not depend on the total intensity, but on the *ratio* of the two incident intensities.

Now that we have found the explicit dependence of the refractive index modulation on the incident light, the next step is to find the impact of the material on the light. The question that arises is that of how this refractive index distribution modifies the motion and intensity of the light beam. A complete theory of coupled waves was first derived by Kogelnik [39], who suggested that, for the solution of the scalar wave equation, we may assume that the field varies only slowly with time, thus allowing us to neglect the second-order derivatives in the wave equation. This approximation is often known in nonlinear optics as the *slowly varying field envelope approximation* (SVEA), and can be applied to the case of photorefractive materials.

Our starting point is the wave equation in a medium with refractive index n, describing the propagation of an electromagnetic wave in a polarizable medium:

$$\nabla E + \frac{\omega^2}{c^2} n^2 E = 0 \,. \tag{3.69}$$

In order to refer to the configuration that is most important for pattern formation, we restrict ourselves in the following derivation to the case of two plane waves propagating in the positive and negative z directions as depicted in Fig. 3.4.

The total electromagnetic field is the sum of the two monochromatic waves:

$$E = E_1 + E_2 = A_1(x, y, z)e^{ikz} + A_2(x, y, z)e^{-ikz} \,. \tag{3.70}$$

Equation (3.70) is now inserted into the wave equation (3.69). Using the slowly varying envelope approximation

$$\left| \frac{\partial^2 A_j}{\partial z^2} \right| \ll \left| k \frac{\partial A_j}{\partial z} \right| , \tag{3.71}$$

the wave equation reads

$$\Delta_\perp A_1 e^{ikz} + \Delta_\perp A_2 e^{-ikz} + 2ik\frac{\partial A_1}{\partial z}e^{ikz} - 2ik\frac{\partial A_2}{\partial z}e^{-ikz}$$
$$-k^2 A_1 e^{ikz} - k^2 A_2 e^{-ikz} + \frac{\omega^2}{c^2}n^2(A_1 e^{ikz} + A_2 e^{-ikz}) = 0 \,. \tag{3.72}$$

Inserting the expression for the refractive index (3.68) into (3.72) and comparing the coefficients of e^{ikz} and e^{-ikz} yields

$$\frac{\partial A_1}{\partial z} - \frac{i}{2k_0 n_0}\Delta_\perp A_1 = i\frac{\pi n_{pr}}{\lambda}e^{-i\Phi}\frac{A_1 A_2^\star}{|A_1|^2 + |A_2|^2}A_2 \tag{3.73}$$

and

$$\frac{\partial A_2}{\partial z} + \frac{i}{2k_0 n_0}\Delta_\perp A_2 = -i\frac{\pi n_{pr}}{\lambda}e^{i\Phi}\frac{A_1^\star A_2}{|A_1|^2 + |A_2|^2}A_1 \,, \tag{3.74}$$

where $k = k_0 n_0$. The relative amplitude (modulation depth) of the refractive index grating is assumed to be small. Therefore, the quadratic term $(\Delta n)^2$ can be neglected. The nonlinearity is given by the right-hand side of these equations and motivates us to define the photorefractive coupling constant γ, by

$$\gamma = i\frac{\pi n_{pr}}{\lambda}e^{-i\phi} \,, \tag{3.75}$$

which is a measure of the strength of the photorefractive nonlinearity. This constant depends basically on the product $n_{pr}e^{-i\phi}$, into which all crystal properties are incorporated. The photorefractive coupling constant is by definition a complex quantity. For diffusion-dominated crystals such as $BaTiO_3$, γ becomes real, because the phase difference between the interference and the refractive index grating is $\phi = \pi/2$. With this definition of the photorefractive coupling constant, the two coupled wave equations for counterpropagating photorefractive two-wave mixing in the steady state given above can be written as

$$\frac{\partial A_1}{\partial z} - \frac{i}{2k_0 n_0}\Delta_\perp A_1 = \gamma\frac{A_1 A_2^\star}{|A_1|^2 + |A_2|^2}A_2 \tag{3.76}$$

and

$$\frac{\partial A_2}{\partial z} + \frac{i}{2k_0 n_0}\Delta_\perp A_2 = \gamma^\star\frac{A_1^\star A_2}{|A_1|^2 + |A_2|^2}A_1 \,. \tag{3.77}$$

These equations will be used as a starting point for our subsequent calculations. Equations (3.76) and (3.77) take into account the transverse shape of the incoming beams which is manifested by the appearance of the transverse Laplacian $\Delta_\perp$.

If we neglect the transverse variation of the beam profile, and take into account linear absorption in the material with an absorption coefficient α, the most general expression for the coupled wave equations for photorefractive two-wave mixing reads

$$\frac{\partial A_1}{\partial z} = \pm\gamma \frac{A_1 A_2^\star}{|A_1|^2 + |A_2|^2} A_2 - \alpha A_1 \ , \tag{3.78}$$

$$\frac{\partial A_2}{\partial z} = \gamma^\star \frac{\cos\alpha_1}{\cos\alpha_2} \frac{A_1^\star A_2}{|A_1|^2 + |A_2|^2} A_1 - \alpha A_2 \tag{3.79}$$

for the two waves A_1 and A_2. Here, the negative sign in (3.78) represents two copropagating waves, whereas the positive sign applies in the counter-propagating case. The general definition of the coupling constant used here incorporates an additional geometry factor as follows

$$\gamma = \mathrm{i}\frac{\pi n_{\mathrm{pr}}}{\lambda n_0 \cos\alpha_1} \exp(\mathrm{i}\phi) \ . \tag{3.80}$$

(3.78)tions and (3.79) represent a set of differential equations which describes completely the behavior of the beams propagating through the crystal. They describe the nonlinear coupling of the wave amplitudes. To calculate the intensity coupling, we factorize the complex amplitudes according to

$$A_1 = \sqrt{I_1} \exp(-\mathrm{i}\psi_1) \tag{3.81}$$

and

$$A_2 = \sqrt{I_2} \exp(-\mathrm{i}\psi_2) \ , \tag{3.82}$$

where ψ_1 and ψ_2 are the respective phases. Using this ansatz, we split (3.78) and (3.79) into two equations describing *energy coupling*,

$$\frac{\mathrm{d}}{\mathrm{d}z} I_1 = -2\gamma_{\mathrm{r}} \frac{I_1 I_2}{I_1 + I_2} - \alpha I_1 \tag{3.83}$$

$$\frac{\mathrm{d}}{\mathrm{d}z} I_2 = \pm 2\gamma_{\mathrm{r}} \frac{I_1 I_2}{I_1 + I_2} \mp \alpha I_2 \ , \tag{3.84}$$

and two equations describing *phase coupling* between the beams,

$$\frac{\mathrm{d}}{\mathrm{d}z} \psi_1 = \gamma_{\mathrm{i}} \frac{I_2}{I_1 + I_2} \ , \tag{3.85}$$

$$\frac{\mathrm{d}}{\mathrm{d}z} \psi_2 = \pm\gamma_{\mathrm{i}} \frac{I_1}{I_1 + I_2} \ . \tag{3.86}$$

Here, we have assumed a plane transverse amplitude profile ($\Delta_\perp A_1 = \Delta_\perp A_2 = 0$). The upper signs describe the transmission-grating (copropagating) case, whereas the lower signs describe the reflection-grating (counter-propagating) case. γ_{r} and γ_{i} denote the real and the imaginary part, respectively, of the photorefractive coupling constant:

$$\gamma = \gamma_{\mathrm{r}} + \mathrm{i}\gamma_{\mathrm{i}} \ . \tag{3.87}$$

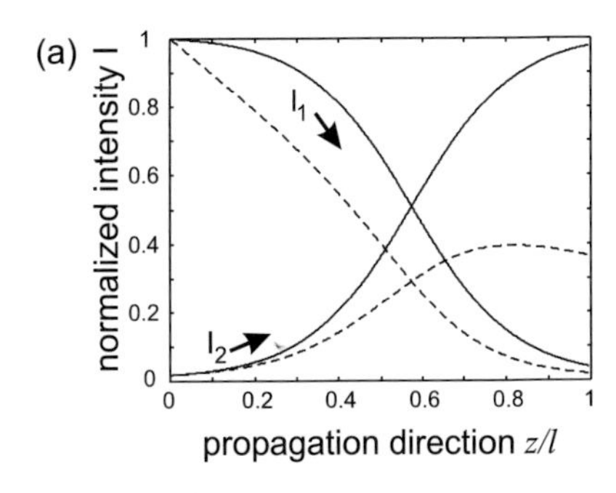

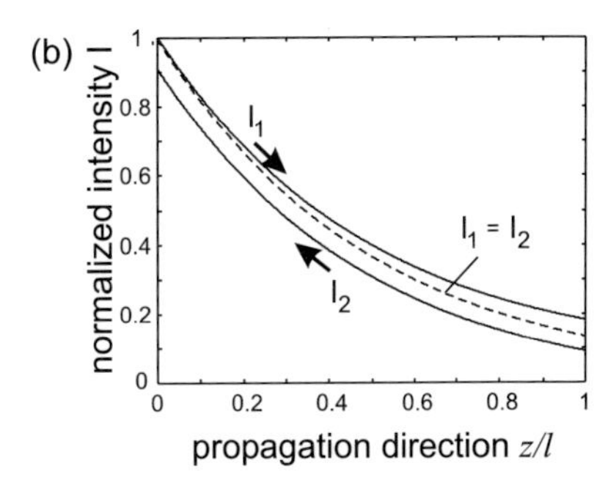

Fig. 3.5. Intensity variation across the crystal due to energy coupling between two beams I_1 and I_2. (**a**) Transmission (copropagating) geometry. *Solid curves*: input intensity ratio $I_1(0)/I_2(0) = 100$, coupling strength $\gamma l = 4$, $\alpha = 0$. *Dashed curves*: same parameters except $\alpha l = 0.5$. (**b**) Reflection (counterpropagating) geometry. *Solid curves*: mirror reflectivity $r = 0.5$ and $\gamma_{\mathrm{r}} l = 2$. *Dashed curves*: $r = 1$ and $\gamma_{\mathrm{r}} l = 2$

Apparently, the transfer of energy from one beam to another depends solely on the real part of the coupling constant, which is a maximum for a purely diffusion-dominated material. On the other hand, transfer of phase is related to the imaginary part, which emerges for drift and charge transport due to the photovoltaic effect. Transfer of phase also appears in the case of nondegenerate two-wave mixing, which is discussed later.

Equations (3.83) and (3.84) describe the z dependence of the intensities of the two co- or counterpropagating beams. We can directly conclude the following conservation laws:

$$I_1(z) + I_2(z) = \text{const.} \tag{3.88}$$

for the copropagating case, and

$$I_1(z) - I_2(z) = \text{const.} \tag{3.89}$$

for the counterpropagating case. Both are good enough implementations of the energy law in the cases under consideration.

Figure 3.5 shows the solutions of the differential equations (3.83) and (3.84). The figure shows typical curves for the amplification of one beam at the expense of the other for the two geometries under consideration. The counterpropagating case as depicted in Fig. 3.5b incorporates a feature which will be of considerable interest later on in the discussion of modulation instabilities: the second, counterpropagating beam is generated by reflection at a mirror at the end of the crystal.

3.6.2 Nondegenerate Wave Mixing

So far, we have considered only the case of degenerate wave mixing, where the incident waves have exactly the same optical frequency. A frequency detuning can be applied to one of the beams by an acousto-optical modulator (AOM) or by reflection at a moving mirror. Assuming that the mirror moves with a constant velocity v, the resulting frequency detuning Ω can be deduced by a direct application of the Doppler shift equations:

$$\Omega(v) = \frac{4\pi}{\lambda} v \cos\epsilon \,, \tag{3.90}$$

where ϵ is the angle of incidence relative to the mirror normal and v is the velocity of the mirror. Two frequency-shifted beams can be described by

$$E_1(r,t) = A_1(r,t)\exp(\mathrm{i}k_1 r)\,\exp(-\mathrm{i}\omega t) + \mathrm{c.c.} \tag{3.91}$$

and

$$E_2(r,t) = A_2(r,t)\exp(\mathrm{i}k_2 r)\,\exp(-\mathrm{i}\omega t)\exp(-\mathrm{i}\Omega t) + \mathrm{c.c.}\,, \tag{3.92}$$

where the second beam carries the frequency detuning Ω. One can directly conclude from Kukhtarev's equations [11, 21] that the influence of a frequency detuning is completely described by a modified coupling constant

$$\gamma = \frac{\gamma_0}{1 + \mathrm{i}\Omega\tau}\,, \tag{3.93}$$

where τ is the characteristic time constant of the photorefractive grating and γ_0 is the coupling constant in the degenerate case $\Omega = 0$. Equation (3.93) indicates that a purely real coupling acquires an imaginary contribution and vice versa. This has a strong influence on the two-beam coupling properties. The real and imaginary parts of the revised coupling constant now read

$$\gamma_\mathrm{r} = \frac{1}{1 + \Omega^2\tau^2}(\gamma_{\mathrm{r},0} + \gamma_{\mathrm{i},0}\Omega\tau)\,, \tag{3.94}$$

$$\gamma_\mathrm{i} = \frac{1}{1 + \Omega^2\tau^2}(\gamma_{\mathrm{i},0} - \gamma_{\mathrm{r},0}\Omega\tau)\,, \tag{3.95}$$

where $\gamma_{r,0}$ and $\gamma_{i,0}$ are the real and imaginary parts, respectively, of the photorefractive coupling constant in the absence of a frequency detuning. For a nonzero drift or photovoltaic contribution $\gamma_{\mathrm{i},0}$ to the photorefractive coupling constant, a frequency detuning can be used to enhance the energy transfer between two beams by amplification of the real part of the coupling constant. In particular, measurement of the two-wave-mixing gain as a function of a frequency detuning provides a means to measure any possible deviation from a $\phi = \pi/2$ phase shift between the interference and the refractive index grating [40].

Besides this frequency-detuning technique, a large number of methods for the enhancement of wave-mixing processes have been proposed, among them the use of tilted pump waves [41], and space charge field enhancement using an externally applied DC electric field, an applied AC field or combined techniques [10]. All of these methods are based on the manipulation of the real and imaginary parts of the complex coupling constant γ. In this sense, their modes of operation are similar, apart from the pronounced functional dependence of the coupling constant on the particular control parameter.

Now we have derived all material properties that are necessary to investigate in detail the propagation of one or several light beams in a photorefractive medium. This enables us to analyze in detail the different spatio-temporal effects that appear above a certain threshold in various different cases of interaction of ligth with the photorefractive saturable nonlinearity. In the case of formation of photorefractive spatial screening solitons, which will be presented in Chaps. 4 and 5, the light propagation is described by the generalized nonlinear Schrödinger equation. For the more complex situation of a single-feedback system, we shall use the two-beam coupling equations (3.76) and (3.76).

Summary

In this chapter we have provided a basic insight into the mechanism and the features of the photorefractive effect, focusing on those vital basic features, topics and equations, which are necessary as a starting point for a discussion of the topic of transverse pattern formation. The basic features of the formation of the refractive index modulation in a photorefractive material due to charge transport effects have been described using the formalism of Kukhtarev. These basic theoretical descriptions have subsequently been extended to the cases of single-beam propagation and two-beam interference in the material, which are the two essential cases for the investigation of soliton formation and of the interaction of counterpropagating beams.

References

1. A. Ashkin, G.D. Boyd, J.M. Dzeidzic, R.G. Smith, A.A. Ballman, J.J. Levinstein and K. Nassau, *Optically induced refractive index inhomogeneities in* $LiNbO_3$ *and* $LiTaO_3$, Appl. Phys. Lett. **9**, 72–74 (1966).
2. F.S. Chen, *Optically induced change of refractive indices in* $LiNbO_3$ *and* $LiTaO_3$, J. Appl. Phys. **40**, 3389–3396 (1969).
3. C. Denz. *Optical Neural Networks: an Introduction with Special Emphasize on Photorefractive Implementations.* Vieweg, Braunschweig,Wiesbaden, 1998.
4. H. Rajbenbach, S. Bann, P. Refregier, P. Joffre, J.P. Huignard, H.-S. Buchkremer, A.S. Jensen, E. Rasmussen, K.-H. Brenner and G. Lohmann, *Compact photorefractive correlator for robotic applications*, Appl. Opt. **31**, 5666–5674 (1992).
5. B.H. Soffer, G. Dunning, Y. Owechko and E. Marom, *Associative holographic memory with feedback using phase-conjugate mirrors*, Opt. Lett. **11**, 118–120 (1986).

6. J. Feinberg, *Self-pumped, continuous-wave phase conjugator using internal reflection*, Opt. Lett. **7**, 486–488 (1982).
7. D.Z. Anderson and J. Feinberg, *Optical novelty filters*, IEEE J. Quantum Electron. **25**, 635–647 (1989).
8. J.P. Huignard and A. Marrakchi, *Two-wave mixing and energy transfer in $Bi_{12}SiO_{20}$ crystal: application to image amplification and vibration analysis*, Opt. Lett. **6**, 622–624 (1981).
9. D. Psaltis, D. Brady and K. Wagner, *Adaptive optical networks using photorefractive crystals*, Appl. Opt. **24**, 3860–3865 (1988).
10. L. Solymar, D.J. Webb and A. Grunnet-Jepsen. *The Physics and Applications of Photorefractive Materials.* Clarendon Press, Oxford, 1996.
11. P. Yeh. *Introduction to photorefractive nonlinear optics.* Wiley Series in Pure and Applied Optics. Wiley, New York, 1993.
12. P. Günther and J.-P. Huignard. *Photorefractive Materials and their Applications I.* Springer,Berlin, 1988.
13. P. Günther and J.-P. Huignard. *Photorefractive Materials and their Applications II.* Springer, Berlin, 1989.
14. K. Buse, *Light-induced charge transport processes in photorefractive crystals I: models and experimental methods*, Appl. Phys. B **64**, 273–291 (1997).
15. K. Buse, *Light-induced charge transport processes in photorefractive crystals II: materials*, Appl. Phys. B **64**, 391–407 (1997).
16. J. Feinberg, *Asymmetric self-defocusing of an optical beam from the photorefractive effect*, J. Opt. Soc. Am. **72**, 46 (1982).
17. A.A. Zozulya, M. Saffman and D.Z. Anderson, *Propagation of light beams in photorefractive media: fanning, self-bending, and the formation of self-pumped four-wave mixing phase conjugation geometries*, Phys. Rev. Lett. **73**, 818 (1994).
18. O.V. Lyubomudrov and V.V. Shkunov, *Self-bending specklon in photorefractive crystals*, J. Opt. Soc. Am. B **11**, 1403 (1994).
19. O.V. Lyubomudrov and V.V. Shkunov, *Effect of limiting space-charge field on a rate of self-bending trajectory of speckle beam in photorefractive crystals*, Kvant. Elektron. **21**, 863 (1994). (Quantum Electron. **24**, 805 (1994)).
20. M.D. Ewbank, R.R. Neirgaonkar, W.K. Cory and J. Feinberg, *Photorefractive properties of strontium-barium-niobate*, J. Appl. Phys. **62**, 374 (1987).
21. N.V. Kuktharev, V.B. Markov, S.G. Odulov, M.S. Soskin and L. Vinetskii, *Holographic storage in electrooptic crystals I*, Ferroelectrics **22**, 949 (1979).
22. R. Jaura, T.J. Hall and P.D. Foote, *Simplified band transport model of the photorefractive effect*, Opt. Eng. **25**, 1068 (1986).
23. T.J. Hall, R. Jaura, L.M. Connors and P.D. Foote, *The photorefractive effect - a review*, Prog. Quantum Electron. **10**, 77 (1985).
24. J. Feinberg, D. Heiman, A.R. Tanguay and R.W. Hellwarth, *Photorefractive effects and light-induced charge migration in $BaTiO_3$*, J. Appl. Phys. **51**, 1297 (1980).
25. N.V. Kukhtarev, G.E. Dovgalenko and V.N. Starkow, *Influence of optical activity on hologram formation in photorefractive crystals*, Appl. Phys. A **33**, 227 (1984).
26. R. Orlowski and E. Krätzig, *Hologram storage in electrooptic crystals*, Solid State Commun. **27**, 1351 (1978).
27. M.B. Klein and G.C. Valley, *Beam coupling in $BaTiO_3$ at 442 nm*, J. Appl. Phys **57**, 4901 (1985).

28. G.C. Valley, *Erasure rates in photorefractive materials with two photoactive species*, Appl. Opt. **22**, 3160 (1983).
29. A.A. Zozulya and D.Z. Anderson, *Propagation of an optical beam in a photorefractive medium in the presence of a photogalvanic nonlinearity or an externally applied electric field*, Phys. Rev. A **51**, 1520 (1995).
30. D.N. Christodoulides and M.I. Carvalho, *Bright, dark, and gray spatial soliton states in photorefractive media*, J. Opt. Soc. Am. B **12**, 1628 (1995).
31. P. Refregier, L. Solymar, H. Rajbenbach and J.-P. Huignard, *Two-beam coupling in photorefractive $Bi_{12}SiO_{20}$ crystals with moving grating: Theory and experiments*, J. Appl. Phys. **58**, 45 (1985).
32. L.B. Au and L. Solymar, *Space charge field in photorefractive materials at large modulation*, Opt. Lett. **13**, 660 (1988).
33. C.H. Kwak, S.Y. Park and E.H. Lee, *Intensity dependent two-wave mixing at large modulation depth in photorefractive $BaTiO_3$ crystal*, Opt. Commun. **115**, 315–322 (1995).
34. M.R. Belić, D. Timotijević, M. Petrović and M.V. Jarić, *Exact solution of photorefractive two-wave mixing with arbitrary modulation depth*, Opt. Commun. **123**, 201–206 (1996).
35. L.B. Au and L. Solymar, *Higher harmonic gratings in photorefractive materials at large modulation with moving fringes*, J. Opt. Soc. Am. A **7**, 1554 (1990).
36. S.I. Stepanov and S.L. Sochova, *Effective energy transfer in a two-wave interaction in $Bi_{12}SiO_{20}$*, Sov. Phys. Tech. Phys. **32**, 1054 (1987).
37. G. Brost, *Photorefractive grating formations at large modulation with alternating electric fields*, J. Opt. Soc. Am. B **9**, 1454 (1990).
38. N.V. Kuktharev, V.B. Markov, S.G. Odulov, M.S. Soskin and L. Vinetskii, *Holographic storage in electrooptic crystals II*, Ferroelectrics **22**, 961 (1979).
39. H. Kogelnik, *Coupled wave theory for thick hologram gratings*, Bell Syst. Tech. J. **48**, 2909 (1969).
40. I. McMichael and P. Yeh, *Phase shifts of photorefractive gratings and phase-conjugate waves*, Opt. Lett. **12**, 48–50 (1987).
41. C. Denz, J. Goltz and T. Tschudi, *Enhanced four-wave mixing in photorefractive $BaTiO_3$ by use of tilted pump waves*, Opt. Commun. **72**, 129–134 (1989).

4 Spatial Photorefractive Solitons

From the preceding chapter, it is clear that a photorefractive nonlinear medium provides an intensity-dependent, saturable nonlinear response which could be exploited for the purpose of spatial-soliton generation. In this chapter, we explain how this distinct nonlinear response can be utilized to compensate the diffraction of a light beam, allowing the generation of a spatial soliton. In order to compensate diffraction, the self-induced refractive index structure should resemble a focusing lens, and its transverse profile should therefore display a bell-shaped structure similar to the case depicted in Fig. 3.1g. From the quite simplified sketch in Fig. 3.1g, we can conclude that a focusing refractive index profile forms predominately when the light-induced charge carriers become subject to a drift effect.

The existence of photorefractive solitons was first predicted in 1992 [1], and was demonstrated a year later [2]. The first type of photorefractive soliton that was studied, originates from the nonlocal nature of the photorefractive effect and is transient by nature [3]. It was named the *quasi-steady-state soliton*. A second type of photorefractive soliton can be realized in photovoltaic materials, where the internal photovoltaic field provides the necessary drift effect of the optically induced free charge carriers. In this case, a *photovoltaic soliton* [4, 5] evolves and the free charge carriers compensate for the photovoltaic field of the crystal, which leads to a decrease of the electric field in the illuminated area. The third and most attractive type of photorefractive soliton is referred to as the *screening soliton*, which can be generated in a stationary state, where it displays an astonishing robustness. In contrast to the case of the photovoltaic soliton, an external field is applied in the transverse direction in order to provide drift of the free charge carriers [6, 7, 8, 9, 10]. Because of the screening of the externally applied electric field by photoinduced free electrons, the effective refractive index becomes in this case a nonlinear function of the light intensity I, of a type that is analogous to a saturable nonlinearity. The most attractive property of these spatial solitons in photorefractive materials is the very low laser power in the range of milliwatts that is necessary for their generation. This shifts the investigation of spatial solitons to a power region that is accessible with conventional continuous-wave laser sources, and allows us, for the first time, to think about realistic applications to waveguiding and switching owing to the easy availability, compactness and relatively low cost of the experimental devices. Consequently, photorefractive solitons are the most attractive candidates for investigating experimentally the conclusions drawn from theoretical investigations of possible ways to realize stable solitary solutions in (2+1) dimensions, and are the driving motor of recent theoretical attempts to describe (2+1)-dimensional spatial-soliton formation in saturable media.

So far photorefractive materials seem to be the perfect candidate of investigations in nonlinear waveguiding. Nevertheless they display some distinct properties that should always be kept in mind:

- Photorefractive materials exhibit a nonlocal light–matter interaction, leading to a response to the optical beam that is broader than the tails of the beam profile. This nonlocality leads to unexpected effects in soliton interaction.
- The photorefractive nonlinearity exhibits additional effects that are due to the inherent diffusion of the photoinduced charge carriers. This effect, which is commonly referred to as beam bending (see Sect. 3.4.2), comes into play in the self-focusing regime when the transverse size of the light beam is in the range of the diffusion length of the charge carriers.
- The propagation of an axially symmetric beam in a photorefractive medium can, under certain conditions, lead to self-focusing that corresponds to an effective optical lens with astigmatism. This is mainly due to the fact that the external electric field induces an anisotropy in the process. The deviation from the isotropic model therefore leads to astonishing anisotropic effects, for the case of two transverse dimensions.

An immediate consequence of these facts is that solitary solutions in a photorefractive medium have exponential asymptotics for the electromagnetic field, but algebraic asymptotics for the nonlinear refractive index of the medium. Therefore, these solutions can be called semialgebraic in a certain sense. The reason is that the electrostatic potential here is that of a two-dimensional distributed induced dipole, and has algebraic tails even for an exponentially localized beam. In contrast, typical soliton solutions of the nonlinear propagation equations have exponentially decaying asymptotics at infinity as described in Chap. 2. A few interesting exceptions from that rule include the solutions of the Benjamin–Ohno equation (see [11] for a recent review) and of the Kadomtsev–Petviashvili equation [12] that are algebraic in nature.

These differences result in instability phenomena that are not observed in Kerr media and have no direct analogue in nonlinear optics. In particular, we shall mention here the pronounced striped filamentation seen at intermediate stages of the nonlinear decay, which is unique to anisotropic media. Their closest structurally similar counterparts are probably the solutions of the Davey–Stewartson equations [13], describing nonlinear dispersive waves in fluid dynamics, and of the Zakharov equations for parametrically coupled electromagnetic and Langmuir waves in plasmas [14].

These features indicate clearly that investigation of photorefractive spatial solitons is essential from two perspectives. On the one hand, knowledge is required in order to obtain insight into the detailed soliton formation mechanisms in the case of the photorefractive nonlinearity. On the other hand, the general characteristics of the photorefractive nonlinearity provide an ideal means to investigate open questions and theoretical predictions of the general theory of solitons. Therefore, a very important aspect of the soliton investigations presented here is the continual comparison of the experimental photorefractive system with two theoretical models, one of which describes an anisotropic photorefractive system, and the other one which describes an ideal isotropic saturable nonlinear medium, as described in Sect. 2.3.3.

In order to understand the special features of soliton formation in photorefractive nonlinear media, it is important to analyze the steady-state solutions for the space charge field derived in Sect. 3.4.

4.1 Beam Propagation in Photorefractive Media

In general, the propagation of an optical beam $A(\boldsymbol{r})$ in a nonlinear medium can be described by the wave equation in the standard paraxial approximation, as introduced in Sect. 2.1:

$$\mathrm{i}\frac{\partial}{\partial z}A + \frac{1}{2k}\nabla_\perp^2 A + \frac{k\,\Delta n_{\mathrm{pr}}}{n_0}A = 0\,. \tag{4.1}$$

Here, n_0 represents the linear part of the refractive index and Δn_{pr} is a saturating function of the beam intensity.

4.1.1 Beam Propagation in the Isotropic Approximation

As already mentioned in Sects. 3.4.1 and 3.4.2, two different models are suitable for describing beam propagation in nonlinear self-focusing media. In the isotropic approximation [15], which is also applicable to all (1+1)-dimensional configurations, the refractive index modulation is given by (3.30), or by the substitution

$$|\boldsymbol{E}_{\mathrm{sc}}| = |\boldsymbol{E}_{\mathrm{ext}}|\frac{1}{1+I/I_{\mathrm{sat}}}\,. \tag{4.2}$$

If we insert this expression into (3.10), the propagation equation (4.1) becomes

$$\mathrm{i}\frac{\partial \tilde{A}}{\partial \xi} + \frac{1}{2}\nabla_{\perp'}^2 \tilde{A} - \gamma\frac{\tilde{A}}{1+|\tilde{A}|^2} = 0\,. \tag{4.3}$$

In this notation, we have introduced a new set of coordinates where $\xi = z/(kx_0^2)$, $x' = x/x_0$, $y' = y/x_0$; we have also a normalized amplitude $\tilde{A} = A/\sqrt{I_{\mathrm{sat}}}$ and the photorefractive saturation parameter

$$\gamma = \frac{1}{2} k^2 x_0^2 n_0^4 r_{\text{eff}} |\boldsymbol{E}_{\text{ext}}| \,. \tag{4.4}$$

In principle, the spatial parameter x_0 can be chosen arbitrarily, but it is usually set equal to the initial diameter of the beam, which is typically in the range of $10 - 12\ \mu\text{m}$. In this case, the longitudinal scaling constant kx_0^2 equals one diffraction length L_d and varies from 3 to 4 mm for typical experimental conditions assuming an optical wavelength of 532 nm and an extraordinary (unperturbed) refractive index of SBN of $n_0 = 2.3$.

Equation (4.3) has a striking similarity to the propagation equation for a generic saturable nonlinearity (2.48). Therefore, in the isotropic approximation, the photorefractive nonlinearity represents an ideal saturable nonlinear material. The results of the theoretical and numerical investigations based on (2.48) that are presented in Sect. 2.3.4 are directly applicable to the photorefractive system. Parameter regions of stability, as well as oscillations of the beam diameter combined with instabilities, should therefore also be observable in a real experimental system, as long as the isotropic approximation is valid.

Strictly speaking however, the isotropic approximation fails when we are dealing with (2+1)-dimensional systems. Since we have to apply an electric field in one direction to invoke a drift-dominated charge carrier transport, we immediately destroy the lateral symmetry of the system. As a consequence, all photorefractive solitons become elliptical rather than circularly symmetric, a fact which is not covered by the isotropic model. If we disregard this ellipticity, however, the model still gives a sufficiently good description of the main features of photorefractive solitons. Since the entire propagation can be described by only one equation (4.3), numerical beam propagation studies can be performed directly and do not consume a large amount of computational resources.

4.1.2 Beam Propagation in the Anisotropic Approximation

For a more realistic description of the physical processes taking place in the crystal during the propagation of an optical beam, one has to apply the anisotropic model described in Sect. 3.4.2. In the following, we shall assume that the beam is propagating along z and that the external field is applied along the c axis, which is the x axis in our coordinate system, perpendicular to the propagation direction z. This corresponds to our experimental situation and to the choice of SBN as a photorefractive nonlinear medium. Under these assumptions, the propagation equation simplifies to a scalar partial differential equation. However, in this case, the relation between the beam intensity and the optically induced space charge field cannot be expressed in a simple analytic form. It is therefore most convenient to express the space charge field in terms of an electrostatic potential in the form $\boldsymbol{E}_{\text{sc}} = -\nabla\varphi$ [16], where the potential is related to the optical intensity via (3.35). Assuming that the potential φ consists of a light-induced term φ_0 and an additional term due to the externally applied electric field $-\boldsymbol{E}_{\text{ext}}x$, the space charge field can be represented as

$$\boldsymbol{E}_{\text{sc}} = \boldsymbol{E}_{\text{ext}} - \nabla\varphi_0 \, , \tag{4.5}$$

and then inserted into (3.10) and (4.1). Considerung the large electro-optic coefficient r_{33} and recalling the distinct configuration given by (3.8), we only take into account the spatial derivative with respect to x in (4.5). The propagation is then given by

$$\mathrm{i}\frac{\partial \tilde{A}}{\partial \xi} + \frac{1}{2}\nabla_{\perp}^{\prime 2}\tilde{A} - \gamma\left(1 - \frac{1}{|\boldsymbol{E}_{\text{ext}}|}\cdot\frac{\partial|\varphi_0|}{\partial x}\right)\tilde{A} = 0 \, . \tag{4.6}$$

Here, the beam amplitude is scaled with respect to the dark intensity as $\tilde{A} = A/\sqrt{I_{\text{sat}}}$, which was introduced in Sect. 3.3, and $AA^* = I$. To determine solutions or to be able to perform a numerical beam propagation, we first have to obtain sufficient information about φ_0. It cannot be expressed as an analytic function of the normalized light intensity but it can be derived from (3.35) by applying a relaxation technique [17].

$$\begin{aligned}\boldsymbol{\nabla}^2\varphi_0 &= -\boldsymbol{\nabla}\ln(1+I)\boldsymbol{\nabla}\varphi_0 + |\boldsymbol{E}_{\text{ext}}|\frac{\partial \ln(1+I)}{\partial x} \\ &\quad - \frac{k_{\text{B}}T}{e}\{\boldsymbol{\nabla}^2\ln(1+I) + [\boldsymbol{\nabla}\ln(1+I)]^2\} \, . \end{aligned} \tag{4.7}$$

Here, the potential has to fulfill the boundary condition $\nabla\varphi_0(\boldsymbol{r} \to \infty) \to 0$. The two coupled equations (4.6) and (4.7) give a complete description of the propagation of an optical beam in a photorefractive medium. It is obvious that besides the Kerr-type part of the material response, the photorefractive nonlinearity also has an additive part that is due to diffusive charge transport (proportional to $k_{\text{B}}T/e$); this part is responsible for asymmetric stimulated scattering or fanning [18, 19].

4.1.3 One-Dimensional Photorefractive Solitons

The structure of the potential φ in the general three-dimensional case is rather complex. However, one can gain some useful insight from the (1+1)-dimensional model, which takes into account only one transverse coordinate x. Here, it is straightforward to apply the isotropic approximation for the space charge field. Therefore, the propagation equation evolves into the nonlinear Schrödinger equation for a saturable nonlinearity (4.3). Solutions of this equation can be found by analogy to our investigations in Sect. 2.3.1.

From the experimental point of view, the generation of one-dimensional solitons is not straightforward. It is well known that a stripe beam can only form a soliton in a Kerr medium if diffraction is allowed only in one transverse direction. In contrast, a one-dimensional stripe beam propagating in a (2+1)-dimensional bulk medium does not form a self-consistent stripe solitary wave but becomes subject to strong instabilities. This type of symmetry-breaking instability, which is due to the growth of perturbations along the initially homogeneous coordinate parallel to the stripe, is known as a transverse modulation instability [20, 21]. This type of instability exists also for Kerr-like but saturable nonlinear materials, such as photorefractive crystals.

Therefore, optical spatial solitons in the (1+1)-dimensional geometry have so far been observed only in slab waveguide structures [22, 23], and not in a bulk material. For further details of (1+1)-dimensional solitons in a biased photorefractive medium, we refer to [16].

4.1.4 Filamentation of One-Dimensional Gaussian Stripe Beams

To study the stability of the (1+1)- or (2+1)-dimensional soliton solutions in a three-dimensional medium, the tool of choice is linear stability analysis. This analysis can be performed by considering an electromagnetic field with a superimposed periodic modulation in the y direction,

$$A(x, y, z) = a(x) \left[a(x) + \delta a(x, z) \sin(k_y y)\right] \exp(\Gamma_0 z) , \quad (4.8)$$

where $a(x)$ is the soliton solution, k_y is an arbitrary transverse wave number and δa is a small complex-valued perturbation function. Analyses performed by this method (see e.g. [24]) show that the ground-state homogeneous solution is unstable in a certain range of transverse wave numbers k_y. The characteristic spatial scale of exponentially growing perturbations along the y axis is comparable to the characteristic size of the ground-state soliton along the x direction. When the perturbation eigenmode grows, the initially homogeneous solution breaks down into a set of spatial pulses along the y axis.

What does this instability mean for the formation of solitary beams in a real three-dimensional medium? Provided the beam diameter exceeds a certain value indicated by the linear stability analysis, the beam will undergo a modulation instability and break up into smaller entities, which may propagate as stable solitary objects. This effect of filamentation can also be observed when the initial conditions do not match a solitary solution. In this case, the beam cannot propagate in the nonlinear medium without being strongly affected. In a focusing medium, Gaussian beams as well as bright stripes decay first into lines of bright filaments. Wider stripe beams radiate and decay into multiple stripes before the onset of instability for solitary stripes can be seen. For beams that are not too wide the stripes interact with each other and decay into a partially ordered pattern of filamentation. In the (2+1)-dimensional geometry, Gaussian beams of circular symmetry decay into a spatially disordered pattern. This phenomenon has been known since the first investigations of self-focusing in nonlinear media. For a review of self-focusing of large circular beams, see [25].

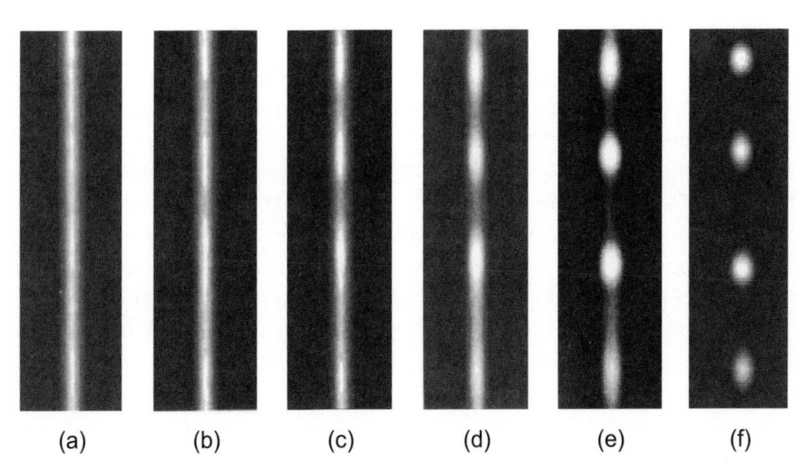

Fig. 4.1. Evolution of a narrow Gaussian stripe beam in a numerical simulation, for normalized propagation distances $z = 5$ (**a**), 10 (**b**), 15 (**c**), 20 (**d**), 25 (**e**) and 30 (**f**). The x coordinate is horizontal and the y coordinate is vertical (after [27], printed by kind permission of Mark Saffman)

The results of the nonlinear evolution can be seen impressively when the region of the initial conditions of a (1+1)-dimensional stripe is subdivided into a region covering narrow beams, with characteristic widths of the order of the width of the solitary solution, and the region of much wider beams. Narrow beams evolve in a fashion that is very similar to that of the solitary solution. They are structurally stable along the inhomogeneous coordinate x and experience some radiative decay. However, larger beams break up owing to the transverse modulation instability mentioned above.

Figure 4.1 shows the results of a numerical simulation of the nonlinear evolution of a narrow Gaussian stripe beam of the form $A_0(x, y) = \sqrt{I} \exp(-4x^2/d^2)$ with initial normalized parameters $d = 4.5$, $I = 3$, $l_y = 60$, $l_x = 20$ and $\varepsilon = 3 \times 10^{-2}$ [26]. The parameters l_y and l_x denote the height and width of the computational window, and l_y was chosen to be about four periods of modulation for the fastest-growing mode for a normalized intensity $I = 3$. The parameter ε characterizes the relative magnitude of the noise embedded in the system, it is defined in [26]. Figure 4.1 shows the intensity distribution of the beam as it propagates in the nonlinear medium for longitudinal propagation distances from $z = 5$ (Fig. 4.1a) to $z = 30$ (Fig. 4.1f). The initial diameter of the Gaussian beam was chosen to be close to that of a solitary solution. For small propagation distances, the beam remains almost in its original shape and shows soliton-like behavior. At this initial stage of the breakup all spatial harmonics of the noise are small, and each is amplified exponentially with its own growth rate. However, at intermediate propagation stages (Figs. 4.1c,d), the fastest-growing modes become noticeable and determine the characteristic spatial scale of the breakup. At larger propagation distances (or, equivalently, a larger nonlinearity), the beam breaks up completely into a quasi-periodic series of (2+1)-dimensional filaments, as shown in Figs. 4.1e,f.

In general, there is no strict selection of a single transverse wave number and even at the nonlinear stage several of them (those possessing the larger growth rates) are present with comparable amplitudes. As a result, the distances between adjacent filaments are not exactly the same, and depend on the particular distribution of the noise and inhomogeneities that are present in the system, although the distance averaged over several filaments is in good correspondence with that determined by the fastest-growing modes of the transverse modulation instability. The intensities of different filaments are also different. Each of the filaments "breathes" and changes its shape and amplitude with the longitudinal propagation distance.

The same behavior was found in experiments where a laser beam derived from a frequency-doubled Nd:YAG laser operating at $\lambda = 532$ nm was focused onto the entrance face of a photorefractive SBN crystal doped with Ce in such a way that the beam had a stripe form similar to that considered in the numerical simulations. The beam propagated in the horizontal plane perpendicular to the crystal c axis and was polarized in the horizontal plane along the c axis to take advantage of the largest component of the electro-optic tensor of SBN. The crystal was biased with an external DC electric field and an incoherent light source in order to enable the formation of a self-focusing nonlinearity. The field controlled the nonlinear coupling strength, whereas the incoherent light allowed us to vary the effective dark intensity in the material. Images of the beam at the input and output faces of the crystal were recorded with a charge-coupled device camera under steady-state conditions. The elliptical beam waist measured 15 μm in the horizontal direction and about 2 mm in the vertical direction. The focal plane was placed slightly in front of the crystal so that the beam diverged significantly in the medium.

The experimental details can be found in Sect. 4.3. Here, we focus only on the results that show filamentation as predicted by theory. Although experiments on planar self-focusing were reported some years ago [28, 29], the instability and the spatial evolution of the beam in this situation were observed only recently [27]. Figure 4.2 shows a situation where the nonlinearity was implemented by raising an externally applied electric field. Here, a fixed level of incoherent illumination several times weaker than the beam intensity of about 25 μW was derived from a white light source and served as a means to adjust the level of the dark conductivity of the material.

If no external field was applied, the beam spreads owing to diffraction during propagation, and a spatially extended beam could be observed at the exit face of the crystal in the near-field distribution (Fig. 4.2a). When the field was applied and increased, the self-focusing process was actuated, forming a self-trapped channel of light. This channel converges to the solitary solution

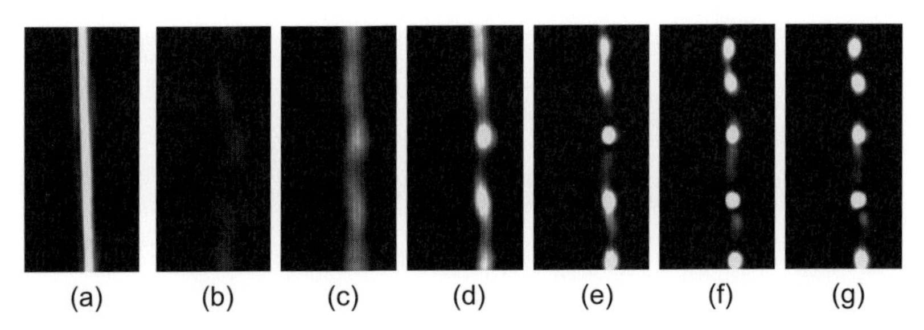

Fig. 4.2. Evolution of a narrow bright stripe in a photorefractive SBN crystal biased with an externally applied electric field: input (**a**) and output (**b**) near-field intensity distributions without external electric field, and development of the filaments with time (c–g). The time sequence starts 5.1 s after the field was applied at the time image (**b**) was recorded; the subsequent time steps Δt are 3.4 s

in the direction in which the input beam diameter is close to the value corresponding to the solitry solution. In the y direction, however, increasing the nonlinearity unavoidably turns on the modulation instability, and the beam breaks up into a sequence of filaments. At higher nonlinearities, it breaks up into several entities. When the field is increased to even higher values, these spots become dynamically unstable and wander around. In Figs. 4.2b–g, the temporal development of the filamentation process for an external electric field produced by applying a voltage of 600 V is depicted.

It is important to note that no artificial seeding was added to the input beam. The instability develops from the natural level of noise present in the beam and/or the crystal. Out of all possible harmonics, the system chooses the one with the largest growth rate. The corresponding value of the wave number determines the spatial period of the modulation (see Fig. 4.2g). This period is equal to be about 40 μm, i.e. $k_y/k \approx 0.015$; this result can also be found from a linear stability analysis (see e.g. [27]) as the fastest-growing mode. As the voltage is increased, the center of the focused channel is displaced along the c axis owing to photorefractive beam bending.

4.1.5 Filamentation of Circular Gaussian Beams

What happens if the diameter of the input beam along the x direction exceeds several times the width of the solitary solution? In this case, shown in a numerical simulation in Fig. 4.3, the loss of structural stability along the inhomogeneous coordinate x is clearly visible. The wide incident beam (Fig. 4.3a) first breaks down into a series of intertwined stripes (Fig. 4.3b) and subsequently evolves into a random pattern of (2+1)-dimensional filaments (Fig. 4.3c) in a manner similar to that for the narrow beam.

The initial breakup into stripes is due to two reasons. First, even in the pure (1+1)-dimensional case, where the homogeneous coordinate y is ex-

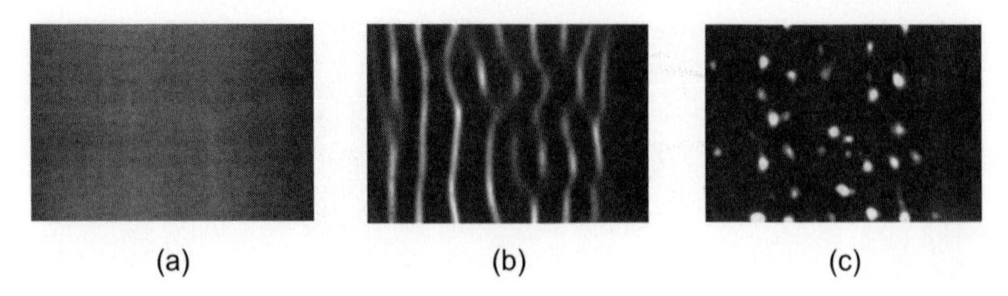

Fig. 4.3. Evolution of a wide super-Gaussian stripe beam with $d = 4.5$ and $I = 1$, for propagation distances $z = 0$ (**a**), 15 (**b**) and 39 (**c**) (after [27], printed by kind permission of Mark Saffman)

cluded from the analysis and considered to be infinitely extended, and when the photorefractive nonlinearity is identical to a saturable Kerr nonlinearity, any wide stripe beam breaks down into narrower (1+1)-dimensional stripes. This phenomenon is analogous to the decay from an arbitrary initial condition into a set of solitons in the case of a (1+1)-dimensional nonlinear Schrödinger equation with a nonsaturable, conventional Kerr nonlinearity. This case has been extensively investigated using the formalism of the inverse scattering method (see Sect. 2.2.2). The second reason for the initial breakup of a wide beam into stripes in the case of a photorefractive nonlinearity is due to the strongly anisotropic response of the material. This is manifested in the structure of the refractive index modulation for photorefractive nonlinear effects, as discussed in Chap. 3.

Even in the case of circularly symmetric initial conditions, this anisotropy favors the decay along the coordinate x. This means that even initially circularly symmetric beams will evolve in an anisotropic manner in a photorefractive medium. To illustrate this point, we show in Fig. 4.4 a comparison of the spatial evolution of a wide, initially circularly symmetric beam for an isotropic saturable Kerr nonlinearity with that for a photorefractive nonlinearity, obtained by numerical simulation. The initial condition in both cases corresponds to a Gaussian beam $A(x, y) = \sqrt{I} \exp(-4x^2/d^2 - 4y^2/d^2)$, with $d = 100$ and $I = 1$. Despite its initial radial symmetry (Fig. 4.4a), the light beam in the photorefractive medium first breaks up into a series of intertwined stripes aligned along the y axis (Fig. 4.4b–d). Each of those stripes subsequently decays into (2+1)-dimensional filaments in a manner similar to the breakup of a narrow stripe, resulting in the appearance of randomly located (2+1)-dimensional filaments. In contrast, a radially symmetric beam in a saturable Kerr medium (Fig. 4.4e–h) for the same conditions, breaks up into a set of filaments without going through the stage of stripes. Since both the nonlinearity and the initial ground state are isotropic, there is no preferred direction for the formation of stripes. When the calculation is extended to longer propagation distances, the array of filaments is in both cases observed to dance and breathe in an apparently turbulent fashion [27]. The

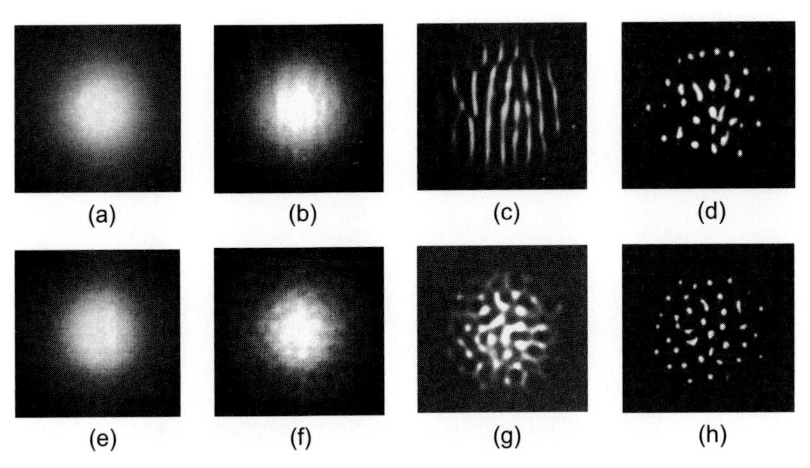

Fig. 4.4. Evolution of a radially symmetric Gaussian beam with the normalized parameters $d = 100$ and $I = 1$ in a photorefractive medium for normalized propagation distances $z = 0$ (**a**), 5 (**b**), 15 (**c**) and 25 (**d**), and in a saturable Kerr medium for propagation distances $z = 0$ (**e**), 5 (**f**), 10 (**g**) and 15 (**h**) (after [27], printed by kind permission of Mark Saffman)

behavior of broad input beams can also be seen clearly in our experiments and is shown in Fig. 4.5. Self-focusing along the x axis results in an anisotropy that results in a initial filamentation along y, followed by a breakup into a series of bright spots due to the development of the transverse modulation instability. An experimental situation of dynamically moving filaments that seem to dance and breathe, indicating a rather turbulent state, is depicted in Fig. 4.5e). It is worth noting that the filaments look approximately circular, despite the anisotropy described by the equations. This is an indication of the global nature of the potential equation (3.35) derived above.

These examples give insight into the effects governing the propagation of a beam in a photorefractive medium in the presence of an external or photovoltaic field. In order to provide a feeling for these effects, the saturation intensity I_{sat} has been assumed to be relatively high. Quantitative results in a more general case can only be obtained numerically, by integrating the propagation equation and the photorefractive material–light interaction equation. These simulations are beyond the scope of the present book, which emphasizes the experimental investigation of saturable, anisotropic soliton behavior. Therefore, in the following, only the main results of the numerical simulations are summarized, in order to give insight into the complexity of the behavior that can be expected in the experimental situation. The publications cited in the following section describe the numerical simulations in more detail.

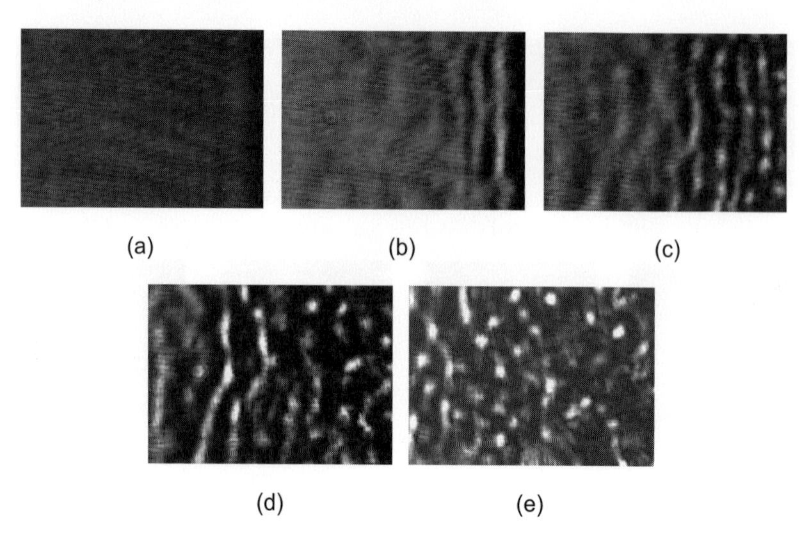

Fig. 4.5. Evolution of a wide radially symmetric Gaussian beam. Near-field intensity distribution output for zero applied field (**a**). From (**b**) to (**e**) the applied field rises to 5 kV/cm. The field is slightly inhomogeneous, leading to an earlier onset of filamentation on the *right* side of the images. Frame (**e**) shows a situation of dynamically unstable filaments

4.2 Two-Dimensional Photorefractive Solitons

Owing to the spontaneous buildup of filamentation in a photorefractive material, the only way to realize a two-dimensional photorefractive soliton is to prevent an incident beam from undergoing filamentation by adjusting the size and shape of its transverse profile to match the expected soliton solution. In order to provide an understanding of the mechanisms of soliton generation in more detail, this section is dedicated to theoretical investigations of the governing conditions. Starting from the generic equations for beam propagation in a photorefractive medium (4.6), (4.7), we shall present a numerical procedure for deriving stationary and stable soliton solutions in an anisotropic photorefractive crystal. We shall make use of the numerical beam propagation method, starting from circular symmetric Gaussian beams and various initial parameters. Thereby, we shall be able to make a direct comparison between numerical simulations and experimental observations. One of the most important results of these investigations is that photorefractive screening solitons can be generated for a wide range of their governing parameters. The anisotropic nature of the nonlinearity is reflected in the transverse shape of the solitons which can be controlled by adjusting the external parameters. We focus on the impact of the diffusion contribution to the charge carrier transport mechanism which leads to a distinct lateral displacement of the self-focused light structure.

4.2.1 Theoretical Approach

If we express the optically induced space charge field in terms of $\boldsymbol{E}_{\text{sc}} = -\boldsymbol{\nabla}\varphi$ only, we can rewrite (4.6) in the following way:

$$\mathrm{i}\frac{\partial A}{\partial \xi} + \frac{1}{2}\nabla_{\perp}^{2} A = -\gamma' \frac{\partial \varphi}{\partial x} A \,, \tag{4.9}$$

assuming that $\gamma' = 1/2k^2 x_0^2 n_0^4 r_{\text{eff}}$. Neglecting diffusion in the potential equation (4.7), which is justified under strong-external-bias conditions, we obtain a reduced form of the electrostatic potential equation,

$$\boldsymbol{\nabla}^2\varphi + \boldsymbol{\nabla}\varphi\, \boldsymbol{\nabla} \ln(1+I) = |\boldsymbol{E}_{\text{ext}}|(\partial/\partial x)\ln(1+I) \,. \tag{4.10}$$

We chose the free parameters to correspond to the experimental conditions, namely $|\boldsymbol{E}_{\text{ext}}| = 3.6$ kV/cm and $r_{\text{eff}} = 180$ pm/V. From our previous investigations, it is clear that the system of equations (4.9), (4.10) is not integrable and that an analytic solution to the problem does not exist. Fortunately, we can make use of a numerical procedure introduced by Petviashvili [30], first applied to seek exact soliton solutions in photorefractive media by Zozulya et al. [31].

Since we are interested in solitary solutions that propagate along ξ without changing their transverse shape we use the ansatz

$$A(x, y, \xi) = a(x, y)\exp(\mathrm{i}\beta\xi) \,, \tag{4.11}$$

where β is an independent propagation constant. This constant changes only the phase of the optical beam, and not its amplitude, which is independent of the propagation coordinate ξ. Inserting (4.11) into (4.9), we obtain a new equation

$$\left(\beta - \frac{1}{2}\nabla_{\perp}^{2}\right) a(x,y) = \gamma'\, \partial_x\varphi(x,y)a(x,y) \,, \tag{4.12}$$

which represents an eigenvalue problem with an eigenvalue β and an eigenfunction a; here, $\partial_x\varphi(x,y) = (\partial/\partial x)\ \varphi(x,y)$. In the following, we Fourier transform (4.12) in order to transform the problem into a pair of fixed-point equations

$$a(\boldsymbol{k}) = \frac{\hat{F}[\gamma'\, \partial_x\varphi a(x,y)](\boldsymbol{k})}{\beta + \boldsymbol{k}^2/2} \,, \tag{4.13}$$

where $\hat{F}(\boldsymbol{k})$ is the two-dimensional Fourier transform in the transverse plane, with a transverse wave vector $\boldsymbol{k}$. Simple iterations of (4.13) do not converge in general. Instead, we use the procedure proposed by Petviashvili, which defines a functional

$$M = \frac{\int \mathrm{d}\boldsymbol{k}\, \hat{F}[\gamma'\, \partial_x\varphi a(x,y)](\boldsymbol{k}) a^*(\boldsymbol{k})}{\int \mathrm{d}\boldsymbol{k}\, (\beta + \boldsymbol{k}^2/2)|a(\boldsymbol{k})|^2} \,, \tag{4.14}$$

which enables us to seek solutions of (4.13). Suppose $a(x, y)$ is a solution to (4.12); then, $M[a(x, y)] = 1$, and multiplying the right-hand side of (4.13) by $|M|^{-3/2}$ leaves the fix-points unaltered, which finally leads to a convergence of the iteration of (4.13). Our numerical procedure consists of the following. Choosing an arbitrary initial function a_0 similar to the expected solution for a, we iterate

$$a_{n+1}(\boldsymbol{k}) = |M|^{-3/2} a_n(\boldsymbol{k}) , \tag{4.15}$$

calculating after each step the new potential φ from (4.10), where $a_n(\boldsymbol{k})$ is given by (4.13). The iteration stops when the relative error becomes smaller than 10^{-5}. By using this numerical procedure, an exact soliton solution can be calculated. For an external field $|\boldsymbol{E}_{\text{ext}}| = 2\text{kV/cm}$ the solution is depicted in Fig. 4.6a by a contour plot. This plot reveals the remarkable feature that the soliton shape is elliptical rather than circularly symmetric. Although the simulation neglects diffusion, the beam is narrower in the horizontal x direction (parallel to the external electric field) than along the vertical y direction. Fig. 4.6b shows the corresponding horizontal and vertical cross sections for $x = \text{const.} = 0$ and $y = \text{const.} = 0$ by the two dashed curves. The solid curve in Fig. 4.6b represents a fitted circularly symmetric Gaussian beam of equal total power. As required by the ansatz (4.11), the elliptically shaped soliton solution depicted in Fig. 4.6a remains entirely unaltered when it propagates through the nonlinear medium. In contrast, a similar Gaussian beam shows strong oscillations in both of its transverse diameters, as depicted in Fig. 4.6c. As a consequence, the beam shape varies from almost circularly

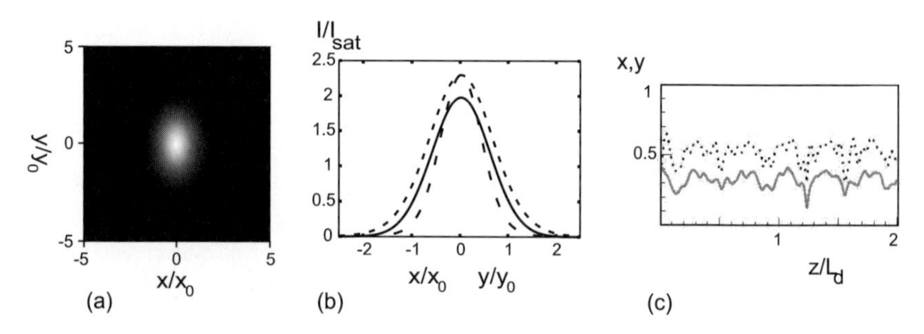

Fig. 4.6. (**a**) Two-dimensional contour plot of a soliton in an anisotropic photorefractive medium, derived by the numerical procedure of Petviashvili. (**b**) Cross sections along the horizontal x axis (*thin dashed line*) and vertical y axis (*thick dashed line*) in comparison with a Gaussian beam of equal total power (*solid line*). (**c**) Spatial evolution of a Gaussian beam with an intensity ratio of $I/I_{\text{sat}} = 1.5$, which is accompanied by oscillations of its diameter in the x direction (*solid line*) and in the y direction (*dashed line*). (Simulations contributed by K. Motzek, Institut für Angewandte Physik, TU Darmstadt)

symmetric to strongly elliptical as it propagates over a distance of 7.2 cm in the nonlinear material.

4.2.2 Anisotropic Refractive Index Modulation

Since the experimental realization of the elliptically shaped solution depicted in Fig. 4.6 is a rather involved task, which is beyond present-day technical feasibility for beam shaping, we now focus on optical beams with a Gaussian profile, which is the typical profile of beams emitted from laser light sources. To understand the physical processes taking place in the photorefractive crystal, we first discuss the properties of the nonlinear refractive index modulation that is caused by illumination with a Gaussian beam of circular symmetry. In this case, the potential equation (4.10) can be solved numerically, and the resulting refractive index modulation is illustrated by a three-dimensional mesh plot in Fig. 4.7. The spatial distribution of the refractive index may be regarded as a strongly nonparabolic lens, induced by the beam in the medium. The central part of this lens corresponds to a positive change in the refractive index, whereas the peripheral regions along the x axis produce negative contributions to the refractive index. In contrast, the cross section along the y axis shows only positive changes. This fact is very important and is unique to the photorefractive nonlinear response. Therefore,

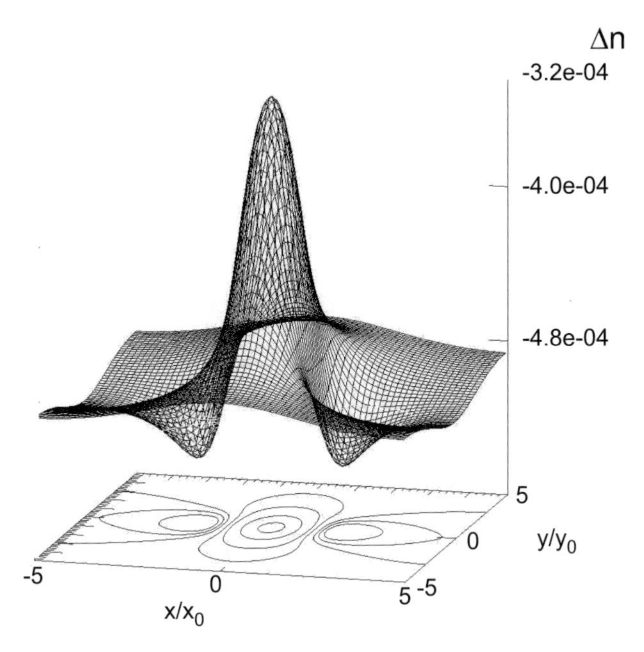

Fig. 4.7. Three-dimensional mesh plot of the refractive index for an initial Gaussian intensity distribution, an external field of 3 kV/cm and $r_{\text{eff}} = 235$ pm/V

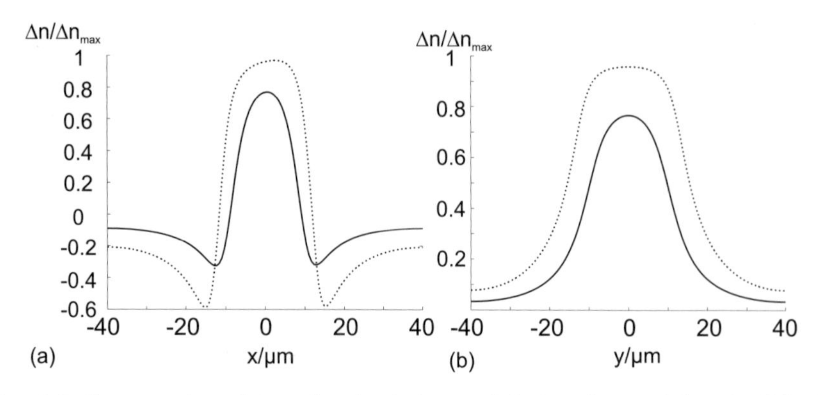

Fig. 4.8. Cross section of the refractive index modulation along x (**a**) and y (**b**) taking account of the impact of diffusion effects at $T = 300$ K, for $I = 10$ (*solid lines*) and $I = 100$ (*dotted lines*), (contributed by K. Motzek, Institut für Angewandte Physik, TU Darmstadt)

two transverse cross sections of the refractive index change are illustrated in Fig. 4.8; Fig. 4.8a shows the change in the direction parallel and Fig. 4.8b the change perpendicular to the applied electric field.

Far away from the center of the beam where the intensity of the light is negligible, the electric field is equal to the external field, and, according to (3.10), the net refractive index change is negative. In the central region, which is illuminated by the beam, the external field is screened by the redistribution of charges, and the net electrostatic field is close to zero. As a consequence, the refractive index increases and has a somewhat parabolic shape that can be interpreted as a nonlinear focusing lens. If we compare the cross sections in Figs. 4.8a,b, the anisotropic nature of the induced refractive index change is apparent. Whereas in the vertical y direction the net refractive index change is bell-shaped (Fig. 4.8b), it shows peculiar characteristics along the axis parallel to the electric field (Fig. 4.8a). Around the circumference of the illuminated area, the net refractive index change is even negative, which has a defocusing effect on the light beam. The solid curve was calculated for $T = 300$ K and $I = 10$, and displays an almost symmetric modulation. If the intensity is increased to $I = 100$, diffusion effects become important. Apart from the fact that the net refractive index change increases, the shape of the refractive index modulation becomes slightly asymmetric in the central region. This asymmetry is only present in the horizontal x direction, parallel to the external field, and is responsible for a lateral deflection of the light beam towards the crystallographic c axis. This effect – commonly referred to as *beam bending* – is due only to the diffusion of excited charge carriers and is always present in an experimental system. In principle, it has no influence on the shape and the general formation process of the self-focused light structures themselves. As will be shown later, the strength

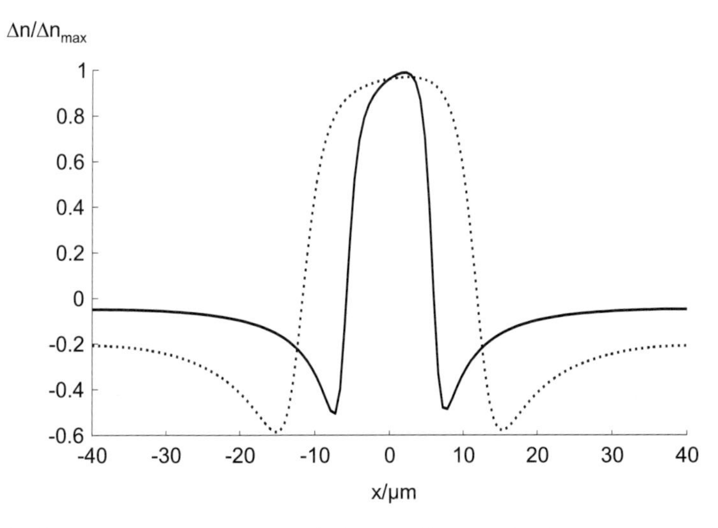

Fig. 4.9. Horizontal cross section of the nonlinear refractive index created by a Gaussian beam with a transverse diameter (FWHM) of 5 μm (*solid line*) and 10 μm (*dotted line*) (contributed by K. Motzek, Institut für Angewandte Physik, TU Darmstadt)

of the bending is strongly dependent on the external parameters intensity and external voltage, and can therefore be exploited for the steering and control of self-focused light structures. Another important parameter that induces a bending effect depends on the transverse size of the light beam. In Fig. 4.9, we illustrate the transverse shape of the refractive index modulation for $I = 100$, $T = 300$ K, and a transverse diameter of the light beam of 5 μm (solid curve) and 10 μm (dotted curve). It is obvious that the asymmetry of the refractive index modulation increases as the transverse diameter of the light beam decreases. This can be explained by the diffusion length of the excited charge carriers, which is typically in the range of a few micrometers in photorefractive crystals. When the beam radius is of the same order of magnitude, the diffusion effect becomes more dominant and the asymmetry in the refractive index modulation becomes more pronounced, which leads to stronger beam bending.

Consequences of the Anisotropic Nonlinearity

Having calculated the nonlinear refractive index, that the beam induces in the nonlinear medium, we can investigate the profile of the beam propagating in its own refractive index channel. Figure 4.10a shows a cross section of the output intensity profile along the x axis after a propagation distance of 2 mm. The dotted curve shows the output intensity of the beam, the solid curve its incident intensity, and the dash–dotted curve the output intensity

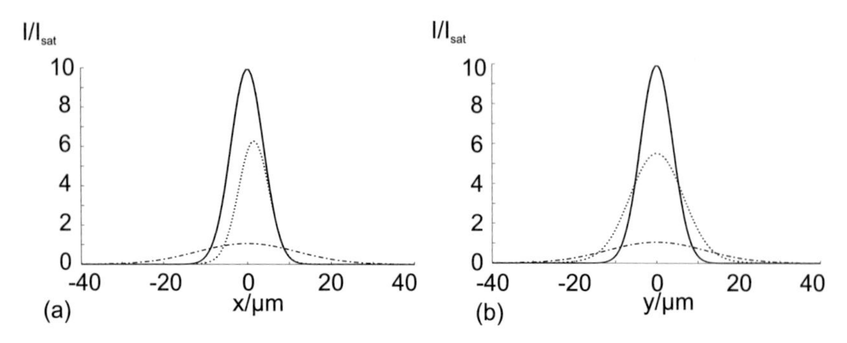

Fig. 4.10. Output intensity distribution *solid lines* of a beam after passing through a nonlinear photorefractive medium with an applied electric field, for a large value of the normalized intensity $I = 10$ (see text for the meanings of the *dotted* and *dash–dotted lines*). (**a**) Cross section along the x axis, which corresponds to the direction of the applied electric field, and (**b**) along the y axis (contributed by K. Motzek, Institut für Angewandte Physik, TU Darmstadt)

of the beam without nonlinearity. When no nonlinearity is present, the beam diffracts linearly and broadens remarkably. This is prevented when the nonlinear effect is present. The shape of the output beam is now comparable to the incident beam width. Further, it is obvious from Fig. 4.10a that the center of the output beam is displaced from its input position. Figure 4.10b shows a cross section of the output intensity profile along the vertical y direction. The solid, dashed and dash–dotted curves have the same meanings as in Fig. 4.10a. The sections are taken through the maxima of the intensity distributions. Note that for this value of I, the photorefractive nonlinearity leads to self-focusing, but the characteristic width of the output beam along the x direction is less than along y, i.e. the beam is focused preferentially along the axis of spontaneous polarization (c axis). Figure 4.10 indicates clearly that in propagating through the photorefractive medium, the initially radially symmetric beam becomes elliptical and can only be fully characterized by two principal diameters d_x and d_y. Thus, in the presence of the nonlinearity the beam experiences *anisotropic self-focusing.* This affects both transverse coordinates, but the magnitude of the effect is different in the two transverse directions x and y.

4.2.3 Evolution of the Solitary Wave

The shape of the amplitude is a crucial factor for the stability and generation of a stable self-focused light structure in a photorefractive material. This fact has severe implications for the form of the solitary-beam solution. In particular, self-channeled propagation with $d_x, d_y = \text{const.}$ is possible if the value of the diameter ratio $F = d_y/d_x$ obeys the relation $F^3 = F + 2$, or $F \approx 1.5$ [26]. Thus, the self-channeled beam is narrower in the direction of the applied

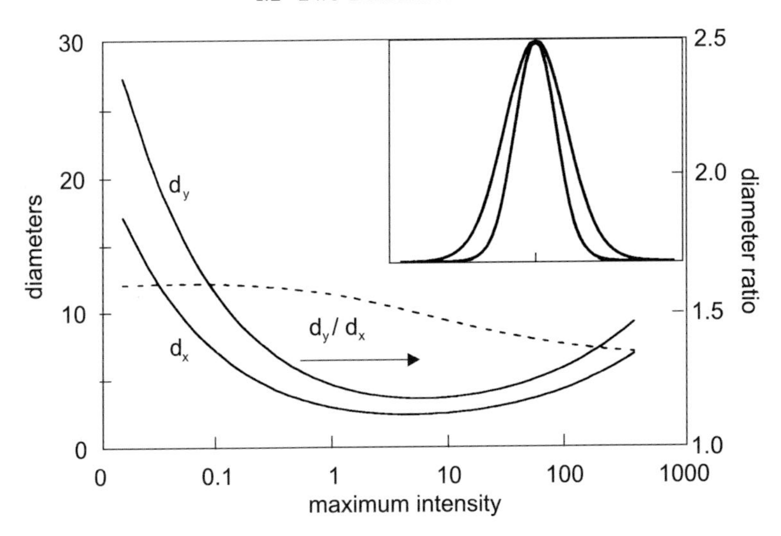

Fig. 4.11. Diameters of the soliton solution and their ratio versus the maximum intensity of the soliton solution. The *inset* shows cross sections of the soliton intensity for $I_{\mathrm{m}} = 5$ (after [26], printed by kind permission of Mark Saffman)

electric field and wider in the perpendicular direction. It can be shown by stability analysis [26] that the evolution of perturbations around this solution results in an oscillatory character, in agreement with the behavior observed for the stability of a soliton in a saturable Kerr medium.

To illustrate the consequences of this anisotropy, we briefly review several important features of the anisotropic photorefractive soliton as derived by Zozulya et al. [26, 31] by numerical simulations. Figure 4.11 shows the diameters d_x and d_y (at the level of half of the maximum intensity) of the soliton solution and their ratio d_y/d_x, as a function of the soliton maximum intensity $I_{\mathrm{m}} = [|A(0,0)|^2/I_{\mathrm{sat}}][d_x(0)d_y(0)/(d_x d_y)]$. The inset shows the cross sections of the soliton intensity for $I_{\mathrm{m}} = 5$ along the x (narrower curve) and the y axis (wider curve). The values of the diameters are inversely proportional to the square root of the maximum intensity of the soliton for $I_{\mathrm{m}} \to 0$ and logarithmically proportional to it in the limit of $I_{\mathrm{m}} \to \infty$. In between, the diameters pass through a shallow minimum. The diameter ratio is close to the value estimated from the parabolic approximation given in [26] (dashed line) and can be roughly approximated by $F \approx 1.5$.

The most important parameter for the rate of convergence to these solutions is the normalized intensity I_{m}. It can be shown that the evolution of an input beam, in general, is characterized by oscillations of both diameters, in agreement with the behavior of solitons in an isotropic saturable system. In the unsaturated limit ($I_{\mathrm{m}} \ll 1$), these oscillations are strongly damped and their spatial period scales as $1/I_{\mathrm{m}}$. The initially circular beam becomes

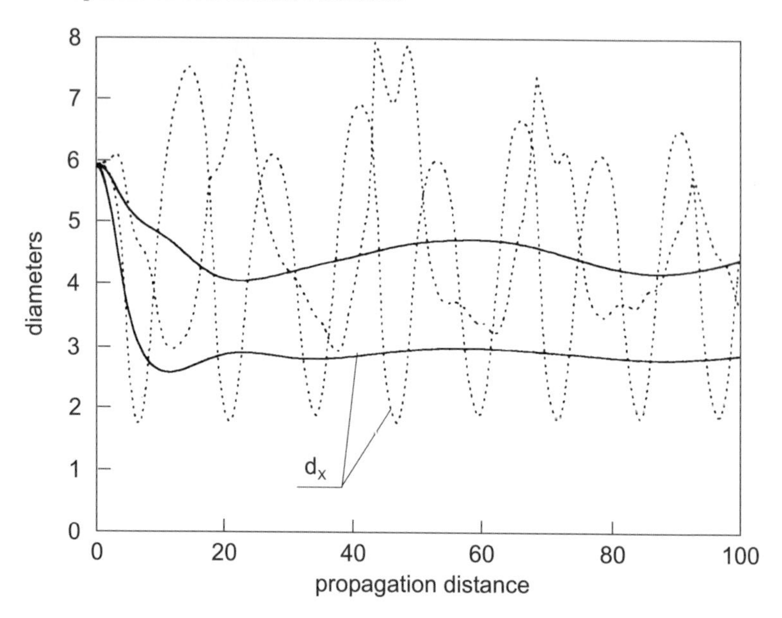

Fig. 4.12. Evolution of the diameters of an initially round Gaussian beam for different incoming relative intensity values: $I = 0.5$ (*solid lines*) and $I = 6$ (*dashed lines*) (after [26], printed by kind permission of Mark Saffman)

elliptical and converges to the typical solitary shape. An increase in I_{m} up to about unity decreases the period of the oscillations, while they still remain heavily damped, thereby decreasing the length of the spatial transient. A further increase in I_{m} up to several times unity continues to decrease the spatial period of the oscillations, but also sharply decreases the relaxation rate and increases the length of spatial transient. In the very high-saturation regime ($I_{\mathrm{m}} > 50$), the oscillation period starts to grow and the relaxation rate remains small, so that the spatial transient remains large. In this case, the characteristic period of spatial oscillations may exceed the length of the medium.

By adjusting the input parameters in an appropriate way, the output profile of the beam in the high-saturation transient regime can be made to have a broad range of shapes, from a beam that is more elongated along the y axis to one elongated along the x axis. The arrangement can therefore also allow the formation of a circularly symmetric output beam, which may give rise to the assumption that circular photorefractive screening solitons represent the general case of the soliton solution [32, 33]. However, the derivations above show that the anisotropy plays an important role and induces a much more complex solitary-beam profile than for standard saturable nonlinear media.

The consequences of this behavior become significant when a circularly symmetric input beam is used. Figure 4.12 shows the spatial evolution of

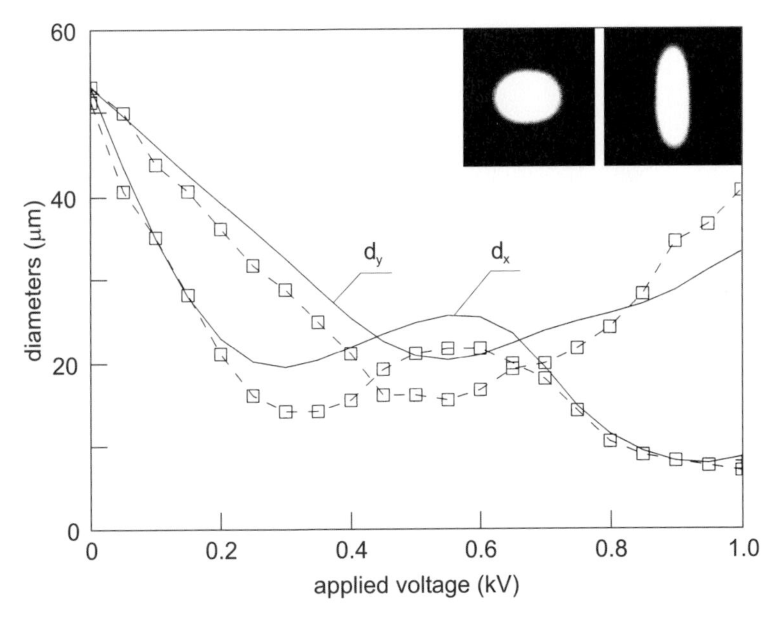

Fig. 4.13. Output diameters as a function of the applied voltage in the high-saturation regime. The power of the input beam was 50 μW (after [26], printed by kind permission of Mark Saffman)

an incident Gaussian beam $A = \sqrt{I}\exp\left[-2\ln 2(r/d)^2\right]$ for $I = 0.5$ (solid lines) and $I = 6$ (dashed lines). Inside the medium, the beam undergoes self-focusing. The relaxation to the soliton solution is rapid for low saturation and has a long transient in the high-saturation regime. In the unsaturated limit ($I_\mathrm{m} \ll 1$) the oscillations of the input beam are damped but the overall evolution scale increases as $1/I_\mathrm{m}$. In the high-saturation regime ($I_\mathrm{m} \gg 1$), the damping rate is very small and the spatial transients last for a very long distance. From an experimental point of view, it is a very involved task to confirm the behavior depicted in Fig. 4.12, since the propagation distance exceeds by far the experimentally accessible length of a crystal.

Appropriate experiments can show clearly the influence of the saturation on the beam profile. Zozulya et al. performed experiments in the two extremal parts of the saturation curve, in the high-saturation and the low-saturation regime. Experiments in the high-saturation regime $I_\mathrm{m} \gg 1$ show that the shape obtained at the output of the nonlinear medium with a circular input beam changes as a function of the voltage. Zozulya et al. observed ellipticities d_y/d_x ranging from about 0.5 to 5, including round output beams. This is illustrated in Fig. 4.13, which shows the output diameters of a circularly symmetric input beam with a diameter of 26 μm as a function of the applied voltage. The insets show the beam profile at 0.5 and 9.9 kV. Note how the diameter ratio d_y/d_x flips over between 0.4 and 0.7 kV and then reverses to

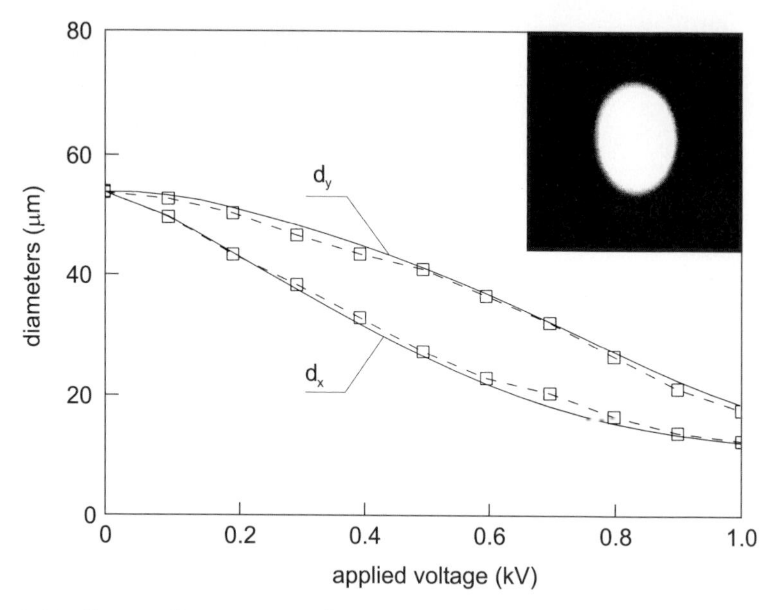

Fig. 4.14. Output diameters as a function of the applied voltage in the moderate saturation regime. The power of the input beam was 4.2 μW (after [26], printed by kind permission of Mark Saffman)

values larger than unity. In the high-saturation regime, the evolution of the light beam is accompanied by long transient oscillations. According to the estimates of Zozulya et al. [26], the spatial period of those oscillations was slightly larger than the size of the crystal. By an appropriate choice of the parameters, the output beam diameter ratio can be made to have many values, including unity. When the saturation is decreased so as to be in the moderate region, the output beam rapidly converges to an elliptical shape elongated along the y axis, which is characteristic of the soliton solution. Figure 4.14 shows the output diameters as a function of the applied electric field for the same beam as in Fig. 4.13, but with a power twelve times smaller($I_{\mathrm{m}} \approx 0.6$).

4.3 Experimental Soliton Realization

Let us now consider an experimental arrangement to generate a photorefractive soliton inside a cerium-doped (0.002% by weight) strontium barium niobate (SBN) crystal, as illustrated in Fig. 4.15. Here, we made use of various crystal samples that measured $5 \times 5\ \mathrm{mm}^2$ in the transverse plane (containing the crystallographic c and b axes) and had lengths of 5, 10, 13.5 and 20 mm. The operating crystal was biased with a DC voltage of $2-3$ kV applied along its polar c axis (x axis of the coordinate system). A circular beam of Gaussian

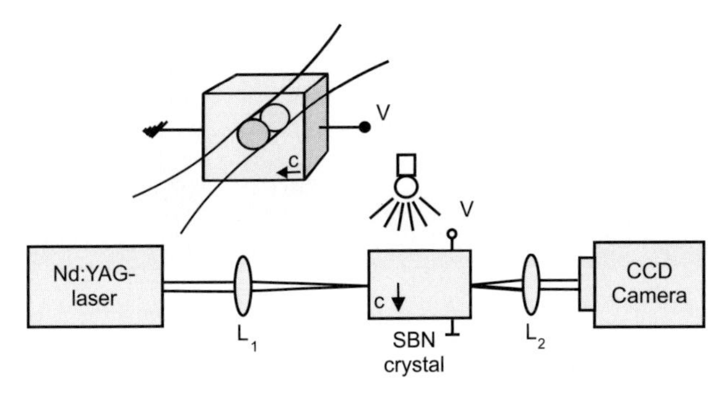

Fig. 4.15. Schematic configuration of soliton formation experiments. L_1, L_2, lenses, V, externally applied electric field

profile was derived from a frequency-doubled Nd:YAG laser ($\lambda = 532$ nm) and focused on the entrance face (x–y plane) of the crystal. The typical beam diameter was around 15 μm. In our experiments, the total power of the incident beam was typically in the range of $1 - 4$ μW. The beam propagated in the horizontal plane perpendicular to the crystal c axis in order to separate beam coupling from nonlinear self-focusing and channeling effects. Therefore, it was polarized in the horizontal plane along the c axis to take advantage of the largest component of the electro-optic tensor of SBN (see Sect. 3.2.2). Together, the above effects compensate spreading due to diffraction and enable the formation of a self-focusing nonlinearity. Images of the beam at the input and output faces of the crystal were recorded with a charge-coupled-device (CCD) camera. All results were recorded under steady-state conditions.

4.3.1 Ellipticity of the Photorefractive Soliton

The process of screening of the space charge field is always influenced by the degree of saturation, which is defined as the ratio of the soliton peak intensity to the background illumination intensity or dark intensity I_d or I_sat (see Sect. 4.2.2). To control the saturation, the crystal was illuminated by a wide beam derived from an incoherent white light source, incident on the SBN crystal from above. The dependence of the photorefractive coupling constant on the dark intensity I_d could be measured in the absence of an applied electric field. In this case, it takes the form

$$\gamma = \gamma_0 \frac{I}{I + I_\mathrm{d}} \,. \tag{4.16}$$

Thus, by measuring the two-beam coupling gain, with the white light source off ($I_\mathrm{d} = 0$) and on ($I_\mathrm{d} \neq 0$), the ratio I/I_d could be determined. However, measuring the intensity of the background white light imposed some difficulties that resulted in an uncertainty in this value. This is due to the

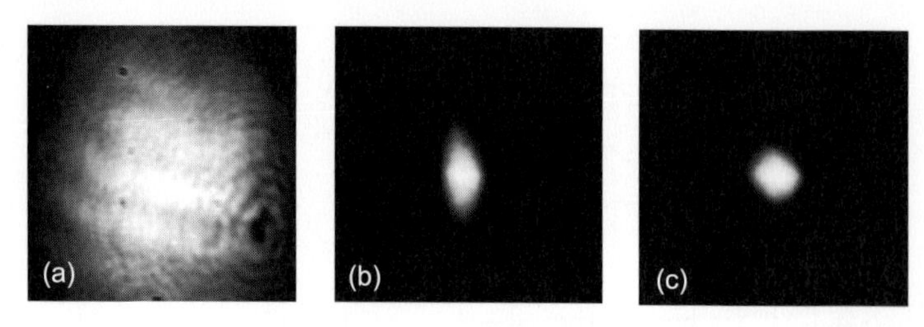

Fig. 4.16. Formation of a soliton beam as a result of focusing a laser beam on the entrance face of an SBN crystal. (**a**) Linearly diffracting beam at the exit face of the crystal without an external electric field. (**b**) Steady state of a soliton in the nonlinear regime for a value of the saturation intensity smaller than unity. (**c**) Steady state for a saturation intensity larger than unity

broad power spectrum of this light source and to the strong divergence of the beam, which gives rise to an inhomogeneity in the transverse distribution of the intensity in this radiation angle. Moreover, the photorefractive crystal has different sensitivities in different regions of the visible spectrum, leading to different weightings of different wavelengths of the white light source in the interaction of the light with the material. This in turn influences the measurement of the ratio of the beam intensity to the background illumination, which as a consequence can only be calculated with an uncertainty of about 30%. This problem has also been considered by other authors [26], who mention the same inaccuracy. In our case, the calculation results in a saturation intensity of $I/I_{\mathrm{sat}} \approx 0.5 \pm 0.15$ nthe typical soliton formation regime. With parameters in this region, the beam would always form an elliptically shaped soliton with a diameter ratio of about 1:1.2 (a width along the x axis of $w_x \approx 10$ μm). Figure 4.16 shows the linear diffraction at the exit face of the crystal and the soliton formation when the external field is applied, in Figs. 4.16a,b, respectively. The figure clearly reveals the expected elliptical shape when the saturation intensity ratio is close to unity. When this condition is varied towards a highly saturated regime ($I_{\mathrm{sat}} \gg I$), the shape transforms into a rather circular one, as can be seen in Fig. 4.16c.

4.3.2 Dynamics of Soliton Formation

In Fig. 4.17, we show the output light intensity distribution of a single soliton at different time steps, as seen at the exit face of the crystal. In these representations, the x axis is horizontal and the external field points to the left.

Without an applied electric field, the beam diffracts to a size of > 100 μm while propagating through the crystal. After the electric field is switched on the beam starts to focus itself within the next 0.8 s to a size of ≈ 10 μm, owing to the drift of the charge carriers and to the corresponding space charge field

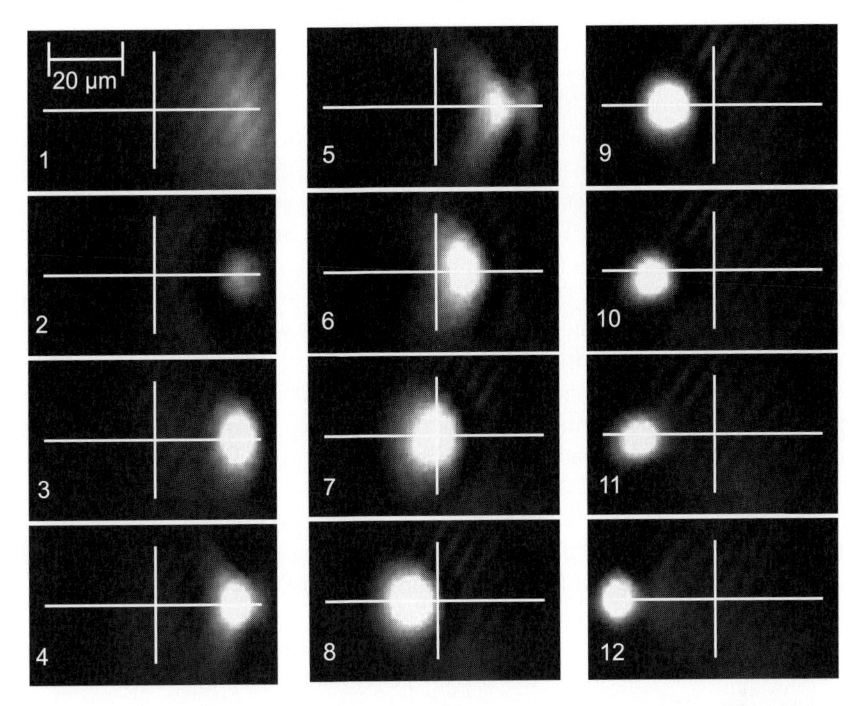

Fig. 4.17. Temporal evolution of a photorefractive screening soliton. The sequence starts about 0.6 s after the external electric field is applied. The time step between each of the frames No. 1–11 is 0.5 s; frame No. 12 illustrates the steady state

inside the crystal. The sequence shown here starts approximately 0.65 s after a beam of 1.85 μW power and an electric field of 3.8 kV/cm were turned on. The time step between the subsequent frames (Fig. 4.17, No. 1–11) is 0.5 s. The last frame (Fig. 4.17, No. 12) illustrates the intensity distribution in the steady state. Initially, the process exhibits strong focusing, predominantly along the axis of the applied electric field. However, as time progresses, the focusing along the other transverse dimension becomes pronounced as well. One can clearly observe the temporal evolution of the beam shape. The size of the beam significantly decreases initially, then slightly increases, decreases again and, finally, after a few seconds, reaches the steady state, when the beam is focused into a slightly elliptical spot. Thus, a closer look at the temporally resolved representation reveals that the beam experiences transient oscillations in its diameter, which may result, for a short time period, in almost a doubling of the beam diameter. It is also apparent from Fig. 4.17 that during the transient phase the beam begins to move along the direction of the applied field. While the soliton is becoming more and more focused, the intensity of the forming soliton becomes much greater than the dark intensity and the diffusion of the charge carriers becomes prominent. Since the charge carriers follow the applied electric field, the resulting space charge field inside the crystal and therefore the refractive index modulation guiding the light beam move in the direction of the field. The bending effect is strongest when the focusing effect and the corresponding beam profile oscillations have settled down almost to the soliton diameter. After some time that depends on the relaxation time of the crystal and the applied electric field, the beam reaches its steady-state position. The situation illustrated in Fig. 4.17 is reached 2.4 s after switching on the electric field, giving a position of the soliton beam that differs considerably from the initial position. The self-bending effect depends on the spatial scale of the beam and can be as high as tens of micrometers over the propagation distance of a few millimeters.

Compared with the theoretical treatment, our experimental realizations impose several constraints that make a complete quantitative comparison of experiment and theory difficult. First, every crystal exhibits absorption, which influences the ratio of the beam intensity to the background intensity throughout the crystal. Therefore, we cannot expect a constant saturation intensity throughout the propagation length. Second, the homogeneity of the crystal itself and the applied electric field are not constant over the whole entrance face of the crystal. Therefore, the nonlinearity across the crystal in the direction transverse to the soliton propagation is not exactly constant, but instead varies with the distorted electric-field lines and with the inhomogeneities of the crystal. Therefore, slight variations in the behavior of the solitons depending on the location of the input beam can be expected. However, when we compare the experimental and theoretical results and consider the complexity of the system, which is often modeled in a simplified way, the agreement of the two is strikingly good. Quantitative differences can be attributed to the effects described above, but they do not disturb the overall good agreement in the cases described in this and in following chapter, which deals with soliton interaction. Therefore, the sources of differences between theory and experiment are not treated in detail for every experimental realization in the following sections.

4.3.3 Beam Bending due to Diffusion Charge Processes

The space charge redistribution in photorefractive crystals due to interaction with an incident light beam is caused by drift of the photoexcited charges in the biasing electric field. This mechanism leads directly to a change of the refractive index and to self-focusing. Therefore, effects attributed to drift are described by local changes of the refractive index. On the other hand, transport of the photoexcited charges also occurs because of diffusion. This process results in a nonlocal contribution to the refractive index change. The strength of the diffusion effect is determined by the width of the beam. In the case of a strong biasing field and a relatively wide beam, the diffusion term is often neglected. However, its contribution can become significant for a very narrow beam (see Fig. 4.9). It is a well-known fact that this diffusion process leads to a strong bending of the trajectories of optical beams in photorefractive crystals. This is also the reason why the beam-fanning effects described in Chap. 3 and illustrated in Fig. 3.2 become so pronounced for diffusion-dominated charge transport effects. Therefore, it is no surprise that diffusion also bends the trajectory of a photorefractive soliton, this was observed quite early on as an unavoidable effect [9, 34, 35, 36].

When this phenomenon is investigated theoretically by variational and numerical analyses, it can be shown that the bending soliton is stable with respect to perturbations of the initial profile. Further, a soliton solution with an elliptical intensity profile propagates along a parabolic trajectory with a constant intensity profile [34, 36]. If the initial beam profile does not match the elliptically shaped exact solution (see. Fig. 4.6), but instead has a Gaussian profile, the beam quickly adapts its shape and transforms itself into the stationary soliton if the saturation parameter is adjusted appropriately. After a transient stage where the beam oscillates and its intensity varies, it propagates in the form of a stationary solution, as a self-bending soliton.

The control parameters that can be used to adjust the amount of bending in soliton experiments are the value of the electric field, the intensity of the incident beam (which is strongly related to the beam diameter) and the saturation intensity, which tunes the dark conductivity of the photorefractive material. By changing these parameters, it is possible to control spatial and temporal development of the bending.

Dependence of the Bending Distance on the Applied Field

In Fig. 4.18, the bending distance (the displacement of the beam due to bending) at the exit face of the crystal is plotted as a function of the externally applied electric field. Figure 4.18a shows experimentally measured data, which are compared with numerically obtained results in Fig. 4.18b. In the experiment, the power of the incident laser was constant at 8.1 μW, and the background illumination was in a range chosen to obtain elliptically shaped solitons. Under these conditions, the soliton formed at a bias electric field of 3.6 kV/cm. When the electric field is increased, the bending-distance rises to 45 μm at 6.0 kV/cm. This distance is more than 2.5 times larger than the diameter of the beam. Above 6.0 kV/cm, the parameter region of the steady-state solution is no longer achieved. At such fields the beam started to experience heavy oscillations and no longer remained in a steady state. Over the whole existence domain of spatial solitons, we found a slightly quadratic relation between the bending distance and the externally applied electric field. Figure 4.18(**b**) shows the corresponding numerical simulation of the bending distance as a function of the applied electric field. The parameters were chosen to be similar to those of the experiment, with an initial beam diameter of 11.25 μm, a propagation length of 12.5 mm and a normalized intensity $I = 0.74$. This plot is in excellent agreement with the experimentally

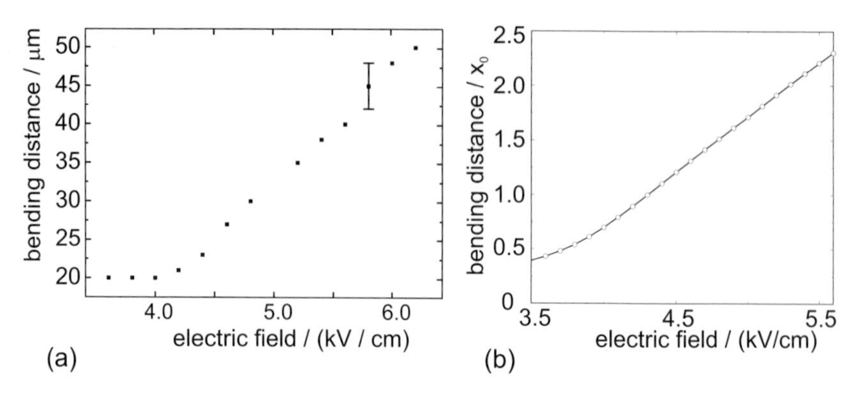

Fig. 4.18. Dependence of the bending distance of a photorefractive soliton on the applied electric field; experimental data (**a**), and numerical simulation (**b**) (contributed by A. Stepken, Institute of Applied Physics, TU Darmstadt)

found dependence of the bending distance on the applied field. In the range of lower saturation $I > 1$, the shape of the curve changes from a quadratic relation to a curve with a square root characteristic. Therefore, an analysis of such curves can be a useful tool for determination of the saturation value.

Dependence of the Bending Distance on the Beam Intensity

An examination of the bending distance as a function of the beam intensity was performed with the same setup, this time maintaining the value of the external electric field and the background illumination fixed, whereas the intensity of the incident beam was varied. The electric-field value was kept constant at 5.4 kV/cm.

When the total power of the beam was varied over the range of 1 µW to 17 µW, the value of saturation varied over a range of about 0.5 to 6, changing also the soliton profile. Figure 4.19 shows that in the low-intensity regime, the bending of the solitons increases strongly with increasing beam intensity, because more charge carriers are subject to diffusion when the gradient of the intensity in the illuminated area is increased. Therefore, the shift of the space charge field in the direction of the bias electric field increases until the beam intensity becomes equal to that of the background illumination. Under these conditions, the beam is focused in the most efficient way and the bending reaches a maximum. Increasing the intensity of the beam further does not allow more charges to be liberated, resulting in a saturation of the effect, and the soliton slowly widens again. In our experiment, this maximum bending distance is reached for a beam power of 7 µW, which corresponds to a normalized beam power of 3.5, assuming a saturation intensity of around 2 µW. If we compare the numerical simulations in Fig. 4.19b with the experimentally obtained data in Fig. 4.19a, both qualitative and quantitative agreement

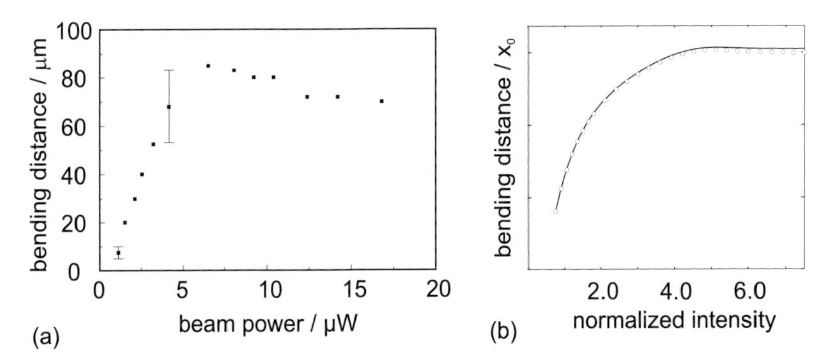

Fig. 4.19. Dependence of beam bending on the incident beam intensity. (**a**) Experimental and (**b**) numerical results (contributed by A. Stepken, Institut für Angewandte Physik, TU Darmstadt)

are obvious. The differences may be attributed to the difficulty of estimating the saturation intensity in the experiments, the slightly different propagation distances in the experiment and the simulations, and to absorption losses, which have been neglected in the numerical calculations.

The possibility of precise control of the bending distance by tuning the applied electric field or the beam intensity allows the realization of a variety of applications in waveguiding, such as adaptive beam-steering devices and optical switches. For all these applications, it is essential to understand the behavior of more than one soliton propagating simultaneously in the material. Investigations concerning this point can be summarized under the heading of "soliton interaction" and are the central topic of the next chapter.

Summary

A space charge field that is induced by a single coherent light beam in a photorefractive crystal provides a focusing refractive index structure, as long as the charge carrier transport process is drift dominated. We showed that the resulting refractive index not only provides a nonparabolic, anisotropic lens, but also is of a nonlocal nature. Despite its anisotropy, however, a photorefractive medium is highly applicable to studies of stable self-focusing of a circularly symmetric light beam in a bulk material. The effects of the anisotropy are mainly reflected in the elliptical shape of the resulting solitons, and have been shown to be present in both experimental and numerical studies. From dynamic studies of soliton formation, we can draw conclusions about the effects of the participating charge carriers inside the crystal. In particular, we explained the typical bending effect, which is due to charge carrier diffusion, and analyzed it by both numerical studies and experiment. The most striking effect about the contribution of the photorefractive anisotropy is the formation of transient states including diameter and beam position oscillations.

References

1. M. Segev, B. Crosignani, A. Yariv and B. Fischer, *Spatial solitons in photorefractive media*, Phys. Rev. Lett. **68**, 923 (1992).
2. G. Duree Jr., J.L. Shultz, G.J. Salamo, M. Segev, A. Yariv, B. Crosignani, P. DiPorto, E.J. Sharp and R.R. Neurgaonkar, *Observation of self-trapping of an optical beam due to the photorefractive effect*, Phys. Rev. Lett. **71**, 533 (1993).
3. M. Segev, *Optical spatial solitons*, Opt. Quantum Electron **30**, 503 (1998).
4. G.C. Valley, M. Segev, B. Crosignani, A. Yariv, M.M. Fejer and M.C. Bashaw, *Dark and bright photovoltaic spatial solitons*, Phys. Rev. A **50**, R4457 (1994).
5. M. Taya, M. Bashaw, M.M. Fejer and G.C. Valley, *Observation of dark photovoltaic spatial solitons*, Phys. Rev. A **52**, 3095 (1995).
6. M.D. Iturbe-Castillo, P.A. Marquez-Aguilar, J.J. Sanchez-Mondragon, S. Stepanov and V. Vysloukh, *Spatial solitons in photorefractive* $Bi_{12}TiO_{20}$ *with drift mechanism of nonlinearity*, Appl. Phys. Lett. **64**, 408 (1994).
7. M. Segev, G.C. Valley, B. Crosignani, P. DiPorto and A. Yariv, *Steady-state spatial screening solitons in photorefractive materials with external applied field*, Phys. Rev. Lett. **73**, 3211 (1994).
8. M. Segev, M.-F. Shih and G.C. Valley, *Photorefractive screening solitions of high and low intensity*, J. Opt. Soc. Am. B **13**, 706 (1996).
9. M. Shih, M. Segev, G.C. Valley, G. Salamo, B. Crosignani and P. DiPorto, *Two-dimensional steady-state photorefractive screening solitons*, Opt. Lett. **21**, 324 (1996).
10. M. Shih, M. Segev, G.C. Valley, G. Salamo, B. Crosignani and P. DiPorto, *Observation of two-dimensional steady-state photorefractive screening solitons*, Electron. Lett. **31**, 826 (1995).
11. Y. Matsuno, *Asymptotic properties of the benjamin-ono equation*, J. Phys. Soc. Jpn. **51**, 667–674 (1982).
12. S.V. Manakov, V.E. Zakharov, L.A. Bordag, A.R. Its and V.B. Matveev, *Two-dimensional solitons of the Kadomtsev–Petviashvili equation and their interaction*, Phys. Lett. **63**, 205 (1977).
13. A. Davey and K. Stewartson, *On three dimensional packets of surface waves*, Proc. R. Soc. A **388**, 191 (1974).
14. V.E. Zakharov, *Collapse of Langmuir waves*, Zh. Eksp. Teor. Fiz. **62**, 1745 (1972). (Sov. Phys. JETP **35**, 908 (1972)).
15. D.N. Christodoulides and M.I. Carvalho, *Bright, dark, and gray spatial soliton states in photorefractive media*, J. Opt. Soc. Am. B **12**, 1628 (1995).
16. A.A. Zozulya and D.Z. Anderson, *Propagation of an optical beam in a photorefractive medium in the presence of a photogalvanic nonlinearity or an externally applied electric field*, Phys. Rev. A **51**, 1520 (1995).
17. W.H. Press, S.A. Teukolsky, W.T. Vetterling and B.P. Flannery. *Numerical Recipes in C++*. Cambridge University Press, Cambridge, MA, 2002.
18. J. Feinberg, *Asymmetric self-defocusing of an optical beam from the photorefractive effect*, J. Opt. Soc. Am. **72**, 46 (1982).

19. V.V. Voronov, I.R. Dorosh, Yu.S. Kuz'minov and N.V. Tkachenko, *Photoinduced light scattering in cerium-doped barium strontium niobate crystals*, Kvant. Elektron. **7**, 2313 (1980). (Sov. J. Quantum Electron. **10**, 1346 (1980)).
20. T.B. Benjamin and J.E. Feir, *The disintegration of wave trains on deep water*, J. Fluid Mech. **27**, 417 (1967).
21. V.I. Bespalov and V.I. Talanov, *Filamentary structure of light beams in nonlinear liquids*, Pis'ma Zh. Éksp. Teor. Fiz. **3**, 471 (1966). (JETP Lett. **3**, 307 (1966)).
22. D. Kip, E. Krätzig, V. Shandarov and P. Moretti, *Thermally-induced self-focusing and optical beam interaction in planar SBN waveguides*, Opt. Lett. **23**, 343 (1998).
23. J.S. Aitchison, A.M. Weiner, Y. Silberberg, M.K. Oliver, J.L. Jackel, D.E. Leaird, E.M. Vogel and P.W.E. Smith, *Observation of spatial optical solitons in a nonlinear glass waveguide*, Opt. Lett. **15**, 471 (1990).
24. A.V. Mamaev, M. Saffman and A.A. Zozulya, *Break-up of two-dimensional bright spatial solitons due to transverse modulation instability*, Europhys. Lett. **35**, 25 (1995).
25. Y.R. Shen. *The Principles of Nonlinear Optics.* Wiley, New York, 1984.
26. A.A. Zozulya, D.Z. Anderson, A.V. Mamaev and M. Saffman, *Self-focusing and soliton formation in media with anisotropic nonlocal material response*, Europhys. Lett. **36**, 419 (1996).
27. A.V. Mamaev, M. Saffman, D. Anderson and A.A. Zozulya, *Propagation of light beams in anisotropic nonlinear media: from symmetry breaking to spatial turbulence*, Phys. Rev. A **54**, 870 (1996).
28. M. Shalaby and A. Barthelemy, *Experimental spatial soliton trapping and switching*, Opt. Lett. **16**, 1472 (1992).
29. A. Barthelemy, S. Maneuf and C. Froehly, *Propagation soliton et auto-confinement de faisceaux laser par non linarit optique de Kerr*, Opt. Commun. **55**, 201 (1985).
30. I.V. Petviashvili, *On the equation of a nonuniform soliton*, Fiz. Plazmy **2**, 469 (1976). (Sov. J. Plasma Phys. **2**, 257 (1976)).
31. A.A. Zozulya, D.Z. Anderson, A.V. Mamaev and M. Saffman, *Solitary attractors and low-order filamentation in anisotropic self-focusing media*, Phys. Rev. A **57**, 522 (1998).
32. H. Meng, G. Salamo and M. Segev, *Primarily isotropic nature of photorefractive screening solitons and the interactions between them*, Opt. Lett. **21**, 1538 (1996).
33. M. Saffman and A.A. Zozulya, *Circular solitons do not exist in photorefractive media*, Opt. Lett. **23**, 1579 (1998).
34. M.I. Carvalho, S.R. Singh and D.N. Christodoulides, *Self-deflection of steady-state bright spatial solitons in biased photorefractive crystals*, Opt. Commun. **120**, 311 (1995).
35. Z. Sheng, Y. Cui, N. Cheng and Y. Wei, *Photorefractive self-trapping and deflection of optical beams*, J. Opt. Soc. Am. B **13**, 584 (1996).
36. W. Krolikowski, N. Akhmediev, B. Luther-Davies and M. Cronin-Golomb, *Self-bending photorefractive solitons*, Phys. Rev. E **54**, 5761 (1996).

5 Interaction of Spatial Solitons in a Saturable Photorefractive Medium

Of all soliton properties, interactions between solitons – commonly referred to as *collisions* – are perhaps their most fascinating feature, since in many aspects, solitons interact like particles.

At a first glance, soliton physics and collisions do not match. From a classical and strictly mathematical point of view, a soliton is a localized nonlinear wave that regains asymptotically (as $\xi \to \infty$) its original shape (the shape in the limit $\xi \to -\infty$) and its original velocity after interacting at $\xi = 0$ with any other localized disturbance. Here ξ is a universal coordinate, which may represent either the transverse direction x in the case of spatial solitons or the time t in the case of temporal solitons. Therefore, it is clear that classical solitons pass through each other without radiation losses [1] – they remain completely unperturbed when they collide, and preserve their identities and form.

This is a consequence of the feature of the nonlinear Schrödinger equation that is has an infinite number of conserved quantities (see Sect. 2.2). Collisions of Kerr solitons are fully elastic, which implies that the number of solitons is always conserved. Furthermore, because the system is integrable, no energy is transferred to radiation waves, but energy is conserved in each soliton. In addition, the trajectories and propagation velocities of the solitons are conserved, so that the solitons recover to their initial values after each collision. The only long-term effect of such an interaction on a soliton that can be expected is a phase shift.

The situation changes slightly for spatial optical Kerr solitons, because in the optical case coherence comes into play as a degree of freedom, which allows one to adjust the phase relationships between the interacting beams. These phase relationships in turn play a crucial role in the general collision characteristics and lead to repulsive as well as attractive effects.

Two mutually coherent beams that are in phase with respect to each other always display an attractive interaction. In this case the collision process is mainly governed by the collision angle of the two beams. When the solitons are separated by any angle different from zero, the solitons simply pass through each other and remain virtually unaffected by the collision process as expected from the mathematical theory of solitons [1]. However, when an attractive collision occurs at zero angular separation between the solitons, they form a bound pair. Because the energy is now twice as high as that for a stable single soliton, the induced focusing lens is much stronger, and causes the beams to move towards each other. The interaction results in an overshoot that leads to an oscillatory motion, where the solitons combine and separate periodically while propagating. Since the solution constantly changes between a single and a double-humped light structure, it is also referred to as a *breather*.

On the other hand, when the two interacting beams are mutually out of phase, the general interaction behavior becomes repulsive, and the two interacting solitons simply move away from each other. All of the above types of coherent collisions of solitons in Kerr media have been demonstrated experimentally [2, 3] and are well understood.

In contrast, self-guided beams in non-Kerr nonlinear materials may experience a much richer wealth of collision scenarios than in the case of Kerr solitons. First, saturable nonlinear media can support solitons in the (2+1)-dimensional geometry, and therefore collisions can occur in the full three dimensions, giving rise to new effects that simply cannot exist in a planar Kerr medium. Moreover, the self-induced waveguide in a saturable medium can guide more than one mode, giving rise to new phenomena, including fusion, birth and annihilation of solitons. Probably the most striking difference from Kerr solitons is that all collisions in non-Kerr media may lead to the emission of radiation and therefore allow a change in the identity of the interacting solitons [4].

This is due to the fact that these non-Kerr solitons typically have only a limited number of conserved quantities, or invariants, and therefore must radiate during collision processes (see Sect. 2.3.4). In a particle-oriented representation, this fact is equivalent to the occurrence of inelastic collisions, where energy can be lost or transferred from one particle to another. This exciting feature gives rise to new phenomena, such as merging and fission of stable solitons. Therefore, the range of possible interactions is broad and gives rise to a variety of new phenomena that are especially attractive for guiding, directing and manipulating light with light without any intervening fabricated structures. This behavior is typical of a broad class of materials that spans those which represent a minor perturbation of a Kerr material and those which are radically different. Among these materials, photorefractive materials provide an ideal nonlinearity for investigations of all the above complex interaction scenarios.

5.1 General Interaction Scenarios

In general, there are two possible scenarios for interaction, since we are dealing with coherent light beams interacting with the nonlinear medium: coherent or incoherent interactions.

For an intuitive understanding of soliton collisions, one may consider the solitons in terms of their self-induced optical waveguides, which are brought into close proximity. The solitons, which typically exist as eigenmodes of these waveguides, overlap primarily in the central region where the evanescent tails of the modes coexist. Various interaction scenarios can then be classified, considering coherent or incoherent solitary beams, and their interaction geometry – whether the two solitons are launched into the medium in parallel directions, collide at a certain angle or cross each other in a complex geometry.

5.1.1 Coherent Interaction

Coherent interactions occur when the nonlinear medium can respond to the interference effects between the overlapping beams (see Fig. 5.1). In order to investigate these coherent scenarios, the relative phase between the interacting beams must be kept stationary on a timescale much longer than the response time of the medium. When the relative phase of the beams is equal to zero ("in phase"), the total intensity in the central region increases (Fig. 5.1a), which in turn leads to an increase of the refractive index. Consequently, more light is attracted into this region and the centroids of the solitons start moving towards each other, which can be observed as an attractive force. When the interacting beams are out of phase by π, they interfere desctructively, and the refractive index in the central region is lower than it would have been if the beams were far away from each other, as depicted by the solid and dotted curves in Fig. 5.1b. As a result, the induced refractive index change acts as a self-defocusing element on the centers of the beams and the solitons appear to repel each other. These phase-sensitive interactions of spatial solitons have been demonstrated in several experiments with various nonlinear media, including liquids [2] and glass waveguides [3]. In the exper-

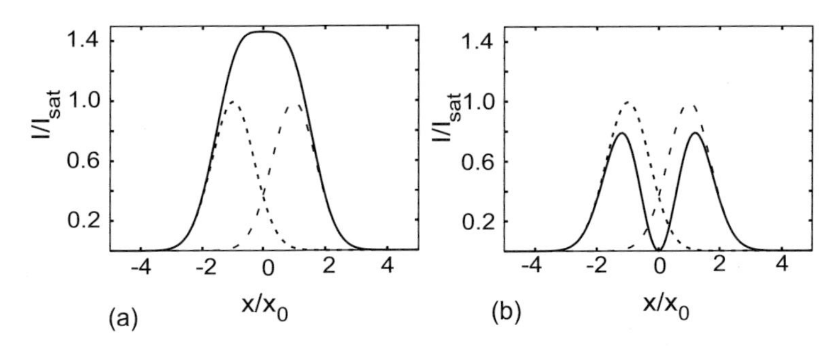

Fig. 5.1. Illustrations of the seperate intensity distributions of two coherent overlapping beams (*dotted curves*), and resulting interference pattern (*solid lines*) for a relative phase difference $\Delta\varphi = 0$ (**a**) and $\Delta\varphi = \pi$ (**b**)

imental part of this chapter, we demonstrate that these interaction effects – attraction and fusion, as well as repulsion and even soliton annihilation and birth – can be realized in photorefractive nonlinear crystals.

5.1.2 Incoherent Interaction

In contrast to coherent interactions, the optical fields of mutually incoherent solitons do not produce any interference pattern. The resulting intensity which affects the material is qualitatively equal to the superposition of the intensities of the two individual components. We would like to emphasize that, in this context, the expression *incoherent interaction* has to be interpreted as the interaction process of two coherent light beams that are mutually incoherent. That is their relative phase varies much faster than the response time of the nonlinear material. Therefore, the material responds only to a time-averaged intensity pattern and the beams appear to be mutually incoherent. Since destructive interference, as depicted in Fig. 5.1b is absent, two mutually incoherent beams always attract and never repel in a saturable nonlinear medium. We shall show below that the anisotropy of photorefractive solitons is the feature that allows one to circumvent this prediction, giving the possibility of achieving both attraction and repulsion in the case of an incoherent interaction also. As in the coherent case, two solitons are able to fuse, but pure phase-dependent effects such as soliton fission or annihilation, which are observed in coherent systems, cannot occur for mutually incoherent beams. From the practical point of view, the interactions of mutually incoherent solitons are very robust, since the effect of attraction is not phase-sensitive. This unique property is already motivating further investigations of interactions. Moreover, several potential applications to reconfigurable waveguide interconnects may be realized on the basis of incoherent interactions of spatial solitons [4, 5, 6, 7, 8, 9].

5.1.3 Waveguide Picture of Soliton Interaction

In a waveguide representation, the phenomenon of soliton fusion and repulsion can be explained in a general way for all non-Kerr nonlinear media by a simple, elegant model. Since both of the solitons induce a waveguide, one needs only to compare the collision angle with the critical angle for guidance in these waveguides, i.e. the angle above which total internal reflection occurs and a beam is guided in the waveguide. In terms of a *potential well*, the complete capturing of a beam depends on the transverse kinetic energy of the colliding wave packets, which has to be smaller than the energy corresponding to the escape velocity. Quantitatively, this picture can be described by two control parameters. The first is the standard waveguide parameter

$$V = k\varrho n_{\mathrm{max}}\Theta_{\mathrm{c}} \, , \tag{5.1}$$

where ϱ is a characteristic half-width of the nonlinear induced waveguide and $\Theta_c \approx (n_{max} - n_{min})/n_{max}$ is the critical angle for total internal reflection. The second control parameter is the ratio of the initial angular separation 2Θ to the critical angle for internal reflection: Θ/Θ_c [4].

All beams with $\Theta/\Theta_c \gg 1$ pass through each other unaffected. They intersect but do not couple light into each other's waveguide. More intuitively, this situation can be compared to the scattering of light from a dielectric slab: at incidence angles well above the critical angle, the light will suffer almost no deviation or attenuation. In contrast, beams with shallow interaction angles, or $\Theta/\Theta_c \ll 1$, stick together when they cross in phase and couple light into each other's induced waveguide. If the waveguide can guide only a single mode (a single bound state), the outcome of the collision will be identical to that of a similar collision in a Kerr medium. However, if the waveguide displays a multimode character, which is the case for a broad range of saturable Kerr-type media, higher modes are excited in each waveguide, and the exchange of radiation energy is generally enabled. In some cases, the waveguides even merge and the solitons fuse to form a single, new soliton.

The effects for both $\Theta/\Theta_c \gg 1$ and $\Theta/\Theta_c \ll 1$ are independent of the waveguide parameter V, provided that V is sufficiently large for the beams to be stable, which holds for $V > 0.86$ for a threshold nonlinearity for example [4]. However, the larger the value of V, the more protracted is the spatial transient required for the soliton "offspring" to reach equilibrium. Collisions for intermediate values of the angular separation Θ/Θ_c depend critically on the waveguide parameter V. In this case, a wealth of phenomena can be expected: stable solitons can annihilate one another as Θ increases from zero. A further increase in Θ can lead to solitons passing through each other. For larger values of V, intermediate values of Θ/Θ_c can lead to the birth of one or more solitons.

When two coherent solitons collide in any medium, pseudolinear interference fringes can be observed. These fringes can in turn induce separate waveguide structures, each of which might evolve into a new stable soliton. This suggests the following necessary conditions for fusion and annihilation of solitons on the one hand, and reproduction or birth of solitons on the other. Fusion can only occur when the first minimum of the interference pattern lies outside the characteristic radius of the beam (i.e. there is minimal power in the side lobes). In contrast, birth may take place if there are at least three maxima inside the beam width, i.e. a significant power level in the side lobes. In terms of the two control parameters defined above, this rough estimate predicts that fusion may be observed for $\Theta/\Theta_c < \pi/(2V)$. Soliton birth may occur for $\Theta/\Theta_c > \pi/V$. Although these estimates are not exact, because the combined beam in the center of the collision is wider than the incident beams, thus changing the parameters for soliton formation, we shall see later that these expressions are able to predict the transition points with surprisingly good accuracy, and that the scaling of the transition is also predicted in the right way.

These examples show, by a waveguide picture, that varying either the phase or the amplitude of the beams allows the output beams to be steered. In particular, in those cases where a central beam emerges (whether through fusion or through birth), that beam can be directed to one side or the other or removed altogether; the latter case occurs at a relative phase $\Delta\varphi \approx \pi$. If the phase difference of the incident beams is varied from 0 to π, the central beam is directed further and further to one side until the pattern becomes the repellor of one of the two beams. This can be explained in terms of the interference fringes mentioned above. When the phase is changed, the middle fringe is no longer central but is offset, and will evolve into a beam steered to one side. As this is a highly phase-sensitive process, it is suggestive of switching applications. Changing the amplitude of one of the interacting beams disturbs slightly the clear concept of relative phase, as the two beams no longer have the same propagation constant, leading to z-dependent fringe patterns. Nevertheless, this provides an alternative way to steer optical beams and enables one to consider these inelastic interactions as a means of optical information processing.

5.1.4 Particle Picture of Soliton Interaction

Another approach to understanding the behavior of non-Kerr solitons during interaction is to consider a particle model, in which the trajectories of interacting self-guided beams can be derived directly from the Lorentz force of classical electromagnetic theory, treating the beams as particles with mass [10]. Because there is a force between any electromagnetic field and a polarizing medium – the Lorentz force of classical electromagnetic theory – parallel linear waveguides exert a force on each other which is mediated by the presence of electromagnetic fields. This force would cause them to attract or repel if they were free to move. Analogously, this should also hold for photorefractive screening solitons.

The arguments used to derive the possible scenarios of interaction and the possible applications are based on the existence of radiation, because they presume inelastic collisions. Therefore, Kerr nonlinearities are excluded from the predictions given above. In contrast, all these types of collisions have been found theoretically for a large class of materials with diverse nonlinearities. It seems that this behavior pattern is universal and is largely independent of the actual physical mechanism that enables the nonlinearity. On the other hand, the characteristics of a pure Kerr material are unique in that they do not display any such interaction scenarios. Hence, Kerr materials are the exception, compared with the general case of soliton interaction considered here.

5.2 Soliton Interaction in Saturable Media

For the case of a general local saturable nonlinear medium, Gatz and Herrmann derived a complete general picture of soliton collisions in the (2+1)-dimensional geometry [11]. Because many of the basic features of the interaction of photorefractive solitons originate from features of this general picture, we shall briefly review the most important results here in order to provide a basis for the understanding of the experimental investigations of interactions of photorefractive solitons described below. Gatz and Herrmann showed that solutions corresponding to the lower solution branch (see Sects. 2.3.2 and 2.3.4) behave quasi-elastically, similarly to Kerr solitons, and retain their shape after collision. In contrast, two solitons corresponding to the upper solution branch may interact in various different ways, depending mainly on the geometry of the interaction [11]. Here, we shall give only a short review of the main results of these derivations.

5.2.1 Interaction of Parallel-Propagating Solitons

In Fig. 5.2, we present the typical propagation dynamics of two coherent solitons launched in parallel in an isotropic saturable medium. The three-dimensional visualization represents an isosurface of the propagating beams at 80% of the normalized beam intensity. The initial parameters are $I(z = 0) = 1$ and $\sigma = 0.2$. Figure 5.2a illustrates a side view when the relative phase between the beams is zero. Although the beams are initially widely separated, they attract each other, and after a certain propagation distance fuse into a single output beam, which is subject to small periodic amplitude oscillations. If the initial separation is increased, the attraction of

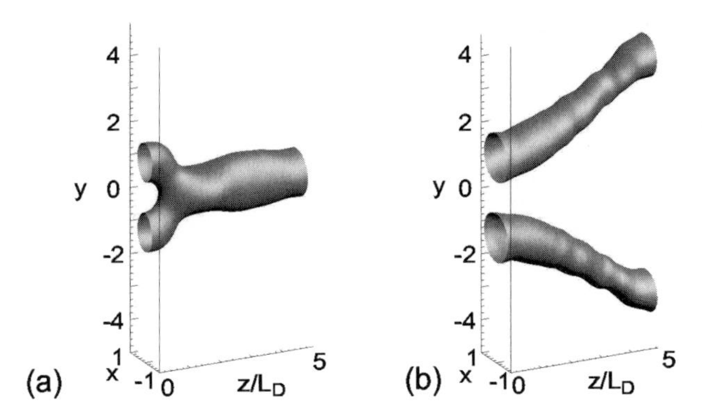

Fig. 5.2. Propagation dynamics of two coherent Gaussian beams in an isotropic saturable nonlinear medium, visualized by an isosurface plot, for $I = 80\%$ of the initial peak intensity of the beams. Depending on their relative phase, the beams attract ($\Delta\varphi = 0$) (**a**), or repel ($\Delta\varphi = \pi$) (**b**)

the solitons is weaker and both solitons propagate unchanged over large distances. The scenario changes dramatically when the beams are out of phase, which is depicted in Fig. 5.2b. Instead of coalescing, the two beams repel strongly, and their initial separation of 2.3 increases to 7.2 after propagation by 5 diffraction lengths. Again, this effect is accompanied by slight beam radius oscillations.

5.2.2 Soliton Collisions

The situation changes if two solitons are allowed to collide with a small crossing angle $\Theta = 2.6°$. In this case, the situation depends strongly on the value of the initial relative phase $\Delta\varphi$ of the two beams. In the case of equal phases ($\Delta\varphi = 0$, see Fig. 5.3a), both solitons fuse into a single beam after collision, but experience periodic oscillations in their amplitude and profile. For a phase difference of $\Delta\varphi = \pi$, which corresponds to an out-of-phase collision, both beams are deflected and retain their soliton properties after collision (Fig. 5.3b). If the solitons collide with a larger crossing angle (e.g. $\Theta = 4°$), the beams do not fuse for a phase difference of $\Delta\varphi = 0$, but lose their soliton property, an indication that a more complex interaction is taking place. For $\Delta\varphi = \pi$, the beams still experience repulsion, but display an oscillatory behavior in both of the beam profiles after the interaction.

For the lower solution branch and the same saturation parameter, two (2+1)-dimensional solitons also fuse into a single output beam with periodically varying beam parameters after collision, for a small crossing angle and equal input phases. This behavior is clearly distinct from the corresponding properties of (1+1)-dimensional solitons of the lower solution branch in materials with a saturable nonlinearity and shows a quasi-elastic collision behavior similar to that of Kerr solitons.

Although these theoretical investigations were performed for a general, local saturable medium, they reveal features that are attractive for possible applications and need to be investigated in an experimental realization.

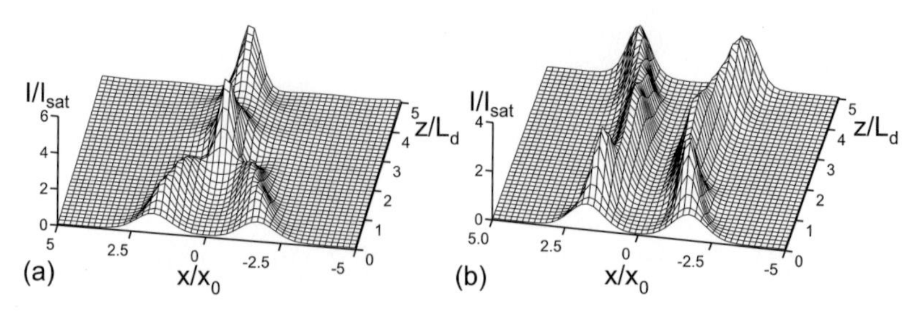

Fig. 5.3. Collision of two solitons for a collision angle $\Theta = 2.6°$ with a mutual phase of (**a**) $\Delta\varphi = 0$, (**b**) $\Delta\varphi = \pi$. Cross sections along the x–z plane in the (2+1)-dimensional geometry

Investigating soliton collisions in photorefractive materials, which are representative of nonlinear materials that are saturable and at the same time anisotropic, allows one to investigate the whole spectrum of scenarios of soliton interaction.

5.3 Soliton Interaction in Photorefractive Media

Spatial solitons in photorefractive media are ideally suited for application in all-optical beam switching and manipulation. This suggestion is based on the possibility of implementing logic operations by allowing solitons to collide in a nonlinear medium [12, 13], as well as on the possibility of soliton-induced waveguides being used to guide and switch additional beams [14]. An efficient implementation of these ideas requires a detailed understanding of the nature of soliton interaction.

Furthermore, photorefractive screening solitons are easy-to-implement representatives of solitons in saturable nonlinear media in general. Therefore, theoretical predictions of soliton interaction in saturable media can be tested experimentally using this class of solitons. However, photorefractive solitons have more features than pure saturable nonlinear media have: owing to their additional anisotropy, unexpected effects can be found. Below, we introduce various interaction scenarios and show how the combination of saturation and anisotropy leads to a variety of interaction phenomena that is much richer than in any other nonlinear material.

As already mentioned in Sect. 5.1, one has to distinguish between coherent and incoherent interaction scenarios. Because the collision processes of mutually coherent solitons are strongly phase-sensitive, they are capable of supporting effects such as soliton fusion, energy exchange, and soliton fission and annihilation. In contrast, in incoherent interaction scenarios, the phase relation between the solitons is destroyed, and phase-sensitive effects can no longer be observed. All interaction processes are purely intensity-dependent. Since a photorefractive crystal has a rather slow response time in the range of milliseconds to seconds, an incoherent interaction scenario can be implemented by imposing a frequency shift on one of the interacting beams. Since the phase control of solitons can be difficult to achieve, incoherent interactions are actually more attractive as far as practical applications are concerned. Incoherent soliton interactions may in general lead to soliton fusion and spiraling, but also to some completely novel, unexpected interaction scenarios that are unique to anisotropic photorefractive materials.

5.3.1 Mathematical Formulation

When two beams propagate simultaneously through a photorefractive nonlinear medium, the optical field can be represented by two equations for A_1 and A_2 and the potential equation:

$$\mathrm{i}\frac{\partial}{\partial z}A_{1,2} + \frac{1}{2}\nabla_{\perp}^2 A_{1,2} = -\gamma\frac{\partial\varphi}{\partial x}A_{1,2} \tag{5.2}$$

$$\boldsymbol{\nabla}^2\varphi + \boldsymbol{\nabla}\varphi\cdot\boldsymbol{\nabla}\ln(1+I) = |\boldsymbol{E}_{\text{ext}}|(\partial/\partial x)\ln(1+I)\,. \tag{5.3}$$

The scaling of the spatial coordinates and the definition of the parameter γ are as given in Sect. 4.2.1. Generally speaking, the light-induced potential φ is due solely to the beam intensity. In the case of a coherent interaction, the intensity is given by $I = |A_1 + A_2|^2$, where a phase-dependent interference is taken into account. In the case of mutually incoherent beams, the total intensity is given by the sum of the intensities of each contributing light beam and equals $|A_1|^2 + |A_2|^2$.

With the help of (5.2) and (5.3), it is possible to describe the propagation of two light beams in a photorefractive medium by means of a split-step Fourier method [15].

5.3.2 Experimental Configuration

All of the above soliton interaction scenarios were investigated experimentally with a configuration shown schematically in Fig. 5.4. Two (or three) circular beams, derived from a frequency-doubled Nd:YAG laser or an argon-ion laser ($\lambda = 532$ nm or 514.5 nm, respectively) were directed by a system of mirrors and beam splitters onto the entrance face of the crystal. At this location, the beams had Gaussian diameters of 15 μm, and were polarized along the x axis (which coincided with the polar axis of the crystal) to make use of the electro-optic coefficient r_{33}, which had a measured value of 180 pm/V (see Sect. 3.2.2 and 3.10) for details of the electro-optic coefficients).

In all of the interaction experiments described below, the power of each input beam did not exceed a few microwatts and the power of the background illumination was set to such a level that the degree of saturation was approximately 2 for all beams, giving a value of $I \approx 0.5$ in normalized units. As in the case of our single-soliton formation experiments, we ensured that these parameters fulfilled the conditions for the generation of a stable elliptically shaped soliton when an individual beam propagated in the crystal. One of the input beams was reflected from a mirror mounted on a piezoelectric transducer. In this way, two operational modes could be realized in the experiment. First, the relative phase of the two input beams could be controlled by a DC voltage, thereby allowing coherent interaction. Second, driving the transducer with an AC signal of several hundred Hz (in our case, 800 Hz) made the beams effectively incoherent, owing to the slow response of the photorefractive medium, allowing incoherent soliton interaction. Thus, coherent and incoherent interaction could be realized in a single configuration. The relative angle of the interacting beams could be adjusted precisely by the mirrors M_1 and M_2. In most experiments this angle did not exceed 1°. The input and output light intensity distributions were recorded with a CCD camera and stored in a computer.

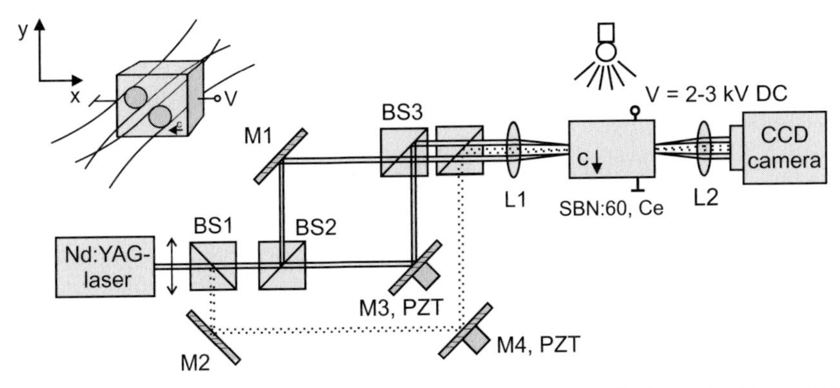

Fig. 5.4. Schematic configuration of the experimental setup for investigations of the interaction of two solitons. BS, beam splitter; M, mirror; L, lens; PZT, piezoelectric transducer; V, externally applied voltage

To detect the position of the solitons without interaction, each beam was first allowed to propagate as a single beam through the crystal. The positions of both beams were registered with the camera and subsequently superimposed in a postprocessing procedure, to indicate how the solitons would behave when interaction was not present.

The process of soliton interaction is, like the process of formation of individual solitons, always accompanied by self-bending of the soliton trajectories (see Sect. 4.3.3). Here, the bending was dependent on the "center of mass" of the soliton and could differ considerably from the bending of a single soliton. In the investigations described below, bending was always present. However, in order to identify the important features of the interaction behavior, it has been removed from the experimental figures. Coherent interaction of solitons propagating in the horizontal plane was always strongly affected by a direct two-wave-mixing process, including phase and amplitude coupling, that involves a phase-independent energy exchange owing to diffraction by the self-induced refractive index grating (see Sect. 3.5 for details of two-beam coupling effects). To suppress this effect, in the case of coherent interaction we used an interaction geometry in which both incident beams propagated in the plane perpendicular to the crystal's c axis.

In this interaction geometry, the results obtained in photorefractive media are qualitatively similar to those obtained in a material with an isotropic saturable Kerr nonlinearity [16]. However, one noticeable difference is expressed in the elliptical shape of the solitons, which is due to the anisotropy of the photorefractive response.

5.4 Interaction of Coherent Photorefractive Solitons

From the above considerations, it is clear that we expect interaction scenarios that depend on the relative phase of the solitons. In particular, two in-phase solitons should attract, while out-of-phase solitons should repel. Moreover, we can expect that photorefractive screening solitons will be able to annihilate each other, fuse or give birth to new solitons when colliding. This kind of behavior is rather generic and is independent of the particular mathematical model for specific nonlinear medium [4]. Consequently, fusion of solitons has already been observed in incoherent soliton collisions in photorefractive crystals [17], as well as during interaction in atomic vapors [18]. However, the special features of the interaction of photorefractive solitons due to the potential for solitons to exchange energy and to the anisotropy of the nonlinearity have not been considered up to now. Therefore, we focus first on the detailed description of phase-controlled fusion, repulsion, energy exchange, birth and annihilation of screening photorefractive solitons.

5.4.1 Fusion, Repulsion and Energy Exchange upon Collision

Our experiments show that collisions that occur at a large angle are basically elastic – the solitons remain unaffected by the interaction. The situation is different for small interaction angles ($\Theta < 1°$). In this case the outcome of the collision depends strongly on the relative phase of the solitons which is a signature of the inelasticity of the collision. In particular, two in-phase beams result in a single bright fringe at the input face of the crystal and for small interaction angles, they tend to collapse into a single solitary beam.

In contrast, out-of-phase beams that have a mutual phase shift of π produce one dark notch in the center. This, in turn, results in a repulsive force that tends to separate the beams significantly (see Fig. 5.1). However, in the general case, the nonlinear attractive and repulsive forces are equally strong and result in the same amount of approach or separation of the two beams.

The degree of fusion observed for incident beams that are in phase depends sensitively on the crossing angle between the beams. If the crossing angle is less than the characteristic diffraction angle of the input beams, a strong attraction can be observed. If the angle is larger, the fusion is less pronounced and eventually the beams begin to pass through each other without fusing – a result that is in full agreement with the general behavior of soliton interactions in saturable media (see Sects. 5.1.3 and 5.2).

For intermediate values of the relative phase and at a slightly larger interaction angle, the soliton interaction leads to energy exchange. After the collision, the initially equal power distribution between the two incident beams becomes asymmetric. This effect is analogous to that found earlier in collisions of solitons described by a perturbed nonlinear Schrödinger equation [19]. In a manner similar to that described in [19] we can invert the direction of the energy transfer by varying the relative phase of the beams.

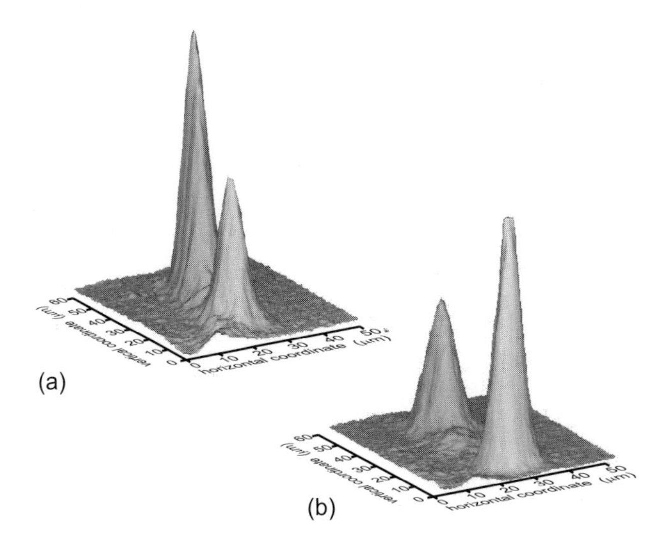

Fig. 5.5. Phase-dependent energy exchange between colliding solitons: (**a**) relative phase close to 90°, (**b**) relative phase close to −90° (after [20], printed by kind permission of Wieslaw Królikowski)

We illustrate this process in Fig. 5.5. The pictures show the output intensity profiles of both beams. In this case, the beams intersect at an angle of about 0.6° and their relative phase is close to either $\pi/2$ (Fig. 5.5a) or $-\pi/2$ (Fig. 5.5b). When a phase shift of π between the input beams was applied, we were able to switch between the two situations. In the case of a shorter crystal or a weaker nonlinearity, the fusion process may not be complete. Instead, it may result in an attractive force for in-phase interaction, that causes the beams to approach each other, but results in only a partial coalescence of the output beams. In this case, switching the phaseshift between the input beams results in a displacement of the output beams by about one beam diameter. This demonstrates the possibility of a high-contrast phase-dependent switch. The reader should also be aware of the fact that the phase relation between two nonparallel propagating beams is not constant over their entire transverse area. When the beam trajectories are directed towards each other, the plane phase fronts of the beams are tilted with respect to each other and the formation of interference fringes is inevitable.

5.4.2 Soliton Birth

It has been predicted that small-angle coherent collisions of two solitons in a saturable medium may give rise to the formation of additional solitons [4]. To observe this *soliton birth*, we chose an interaction angle $\Theta \approx 0.8°$. The relative phase between the two beams was chosen in such a way that without an applied field we could clearly observe three distinct interference fringes at the exit face of the crystal. After the electric field was applied, these fringes evolved into three clearly defined solitary beams, as shown in Fig. 5.6a. Note that the newly formed central soliton does not propagate in the same plane as the two marginal beams. This is due to the higher rate of self-bending experienced by this beam.

5.4.3 Soliton Annihilation

If soliton birth can be realized by proper adjustment of the phases of two solitons, the reverse effect should also be possible – the annihilation of a soliton due to mutual interaction upon collision. Soliton annihilation was predicted theoretically some time ago [4] but has not been observed until now. We have recently presented the first experimental realization of spatial-soliton annihilation in a photorefractive SBN crystal [21].

To investigate this effect, the experimental setup was extended by an additional beam splitter in order to derive a third beam. The beams were then focused into three spots on the entrance face of the photorefractive crystal. We aligned the relative positions and angles of incidence of all three beams in such a way that they intersected in the center of the crystal as depicted in the inset of Fig. 5.6b. In particular, the central beam (C), of power 1.7 μW was approximately four times stronger than the two marginal beams (A and B). We chose the relative phases of the beams in such way that the two outer beams were always in phase, which was easily verified by the observation of their direct collision. For an intersection angle of $\Theta \approx 1°$, they collide without appreciable energy exchange. The relative phase of the central beam with respect to the marginal beams was used as a control parameter; it was adjusted by the piezo-driven mirror. The most interesting results of the three-soliton collision are summarized in Fig. 5.6b,c, which show the light intensity of the outgoing beams after interaction, as seen at the exit face of the crystal. In the example depicted in Fig. 5.6b the phase of C relative to A and B is close to $\pi/2$. The very bright spot corresponds to the central soliton (C), while the two faint spots correspond to the side beams (A and B). A careful examination of this picture reveals that the power of the central beam increases at the expense of A and B. The latter still propagate as well-defined solitary beams, although they are weaker.

The situation changes dramatically when the relative phase of the central beam C is adjusted to $-\pi/2$ (Fig. 5.6c). Clearly, the central beam now virtually disappears, while two new strong, symmetrically located solitons appear. Note that the relative locations of these new beams differ from those in Fig. 5.6a. In fact, it appears as if the soliton C is just split into two separate beams by the simultaneous impact of A and B.

We have also tested the potential use of soliton birth for waveguiding applications [21]. The waveguide structures formed by colliding solitons can in

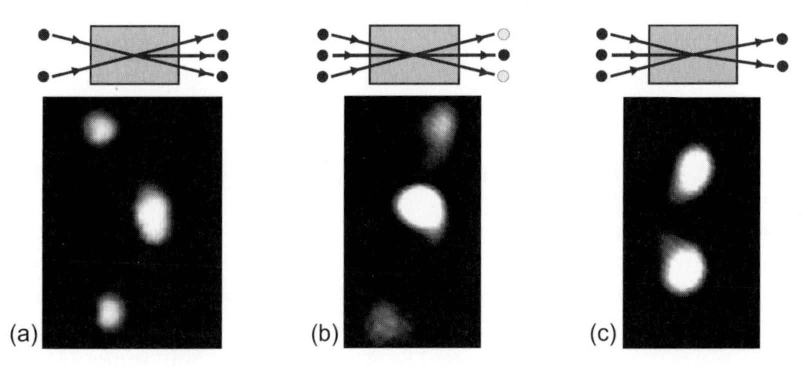

Fig. 5.6. Creation and annihilation of photorefractive screening solitons. Birth of a new beam (**a**), which experiences a stronger lateral shift due to bending. Annihilation of two beams (**b**) and of a single beam (**c**). The sketches on *top* illustrate the y–z propagation plane

principle be utilized to redistribute and to control an external signal beam. We could easily demonstrate this property in our experiments by blocking one of the interacting beams after the steady state was reached. Since the response of the photorefractive effect is slow (in the range of a few seconds for our configuration), the joint refractive index distribution decays slowly after one of the light beams is turned off. During this slow decay, the remaining beam serves as an external signal. Its propagation through the soliton-induced structure results in a redistribution of the beam power among all waveguide channels. Therefore, soliton collision results in the formation of waveguide structures such as crossing junctions (X-junctions) and dividers (Y-junctions), which may be useful for all-optical circuitry [22].

5.4.4 Complex Spiraling Motion

If two coherent beams whose initial trajectories do not intersect propagate in the same plane, they will either attract or repel, depending on their relative phase. A more complex behavior can be obtained if the beams are launched with their trajectories lying in different planes (skewed). It is known that if the mutual force between such solitons is attractive, then they may form a bound state exhibiting a mutual spiral motion as they propagate through the medium [23]. This corresponds to a situation where repulsion due to divergence of trajectories is balanced by attraction. The formation of such a bound state by two coherent solitons is practically impossible because, as we discussed earlier, the mutual force between them will vary with propagation owing to changes of their relative phase. It is therefore much more promising to investigate this problem with mutually incoherent light beams. This will be the topic of Sect. 5.5.5. However, if the input conditions were chosen in such a way that beams with skewed trajectories initially attracted, we were able to observe at least an initial stage of the formation of a bound state [24].

5.5 Interaction of Incoherent Photorefractive Solitons

Although coherent interaction effects are promising for switching applications, their utilization can be troublesome. Phase control of a large number of beams in a switching network may be difficult to achieve. Furthermore, efficient coherent interaction requires that the relative phase between solitons is maintained during propagation. This condition is usually quite difficult to satisfy under experimental conditions, because the soliton phase varies during propagation at a rate determined by the propagation constant β (see (4.11)). Since the propagation constant varies periodically with increasing propagation distance, the mutual phase relation of the two beams varies in the case of nonparallel trajectories. Therefore, the characteristics of the interaction vary periodically between attractive and repulsive. To overcome these limitations, one may consider the use of mutually incoherent solitons, where no type of mutual phase relation exists. Generally, incoherent solitons provide a restricted range of spatial interactions since in typical isotropic self-focusing media, mutually incoherent solitons always attract each other owing to the absence of destructive interference as depicted in Fig. 5.1b. In the waveguide representation introduced earlier, the total light intensity always increases in the region where the beams overlap. This leads to an increase of the refractive index and subsequent attraction of the solitons. Several recent works have studied incoherent soliton interactions in photorefractive media in configurations where the physics is quite similar to that in saturable Kerr-type media [25].

However, this is not the only possible scenario of interaction. In the following sections, we show that it is possible to achieve not only attractive but also repulsive forces between mutually incoherent solitons. This anomalous situation occurs in photorefractive media where the particular anisotropic, nonlocal structure of the nonlinearity results in both attraction and repulsion of parallel beams depending on their relative spatial separation. This effect has severe consequences for various interaction configurations. We shall show that as a result of soliton attraction and repulsion, mutual rotation and complex spiraling of initially parallel and skewed beams becomes possible owing to this anisotropy.

5.5.1 Anisotropic Nonlinear Response

When two solitons are effectively made incoherent by changing their relative phases at a speed that is much larger than the response time of the nonlinear material, the interaction itself becomes dependent only on the intensities and no longer on the relative phase of the two beams.

The unique feature of repulsion of mutually incoherent photorefractive solitons is due solely to the anisotropic nature of the refractive index modulation depicted in Fig. 4.7. This so-called anomalous interaction typically reveals itself when two beams are launched with parallel trajectories in the plane parallel to the external electric field. The initial optical field and the corresponding refractive index modulation are depicted for various lateral distances in Fig. 5.7, in the upper and lower rows, respectively. Owing to the nonlocal response of the material, the refractive index modulation covers a larger area than the optical field itself. Therefore, the individual space charge fields and hence the refractive index modulations of two closely propagating beams are superimposed. In Fig. 5.7a the separation of the beams is in the region of their diameter, the optical fields do not overlap, but the induced space charge fields are superimposed and form a joint refractive index modulation, which is illustrated in Fig. 5.7c. Since the refractive index decreases at the beam's horizontal edges, the net refractive index change between the two solitons becomes negative. Despite the nonzero intensity in the intersection region and the lack of destructive interference, the solitons will repel as they propagate through the crystal. When the separation of the beams is decreased further (Fig. 5.7b), the optical fields start to overlap, and hence the net refractive index change, depicted in Fig. 5.7d, between the two solitons

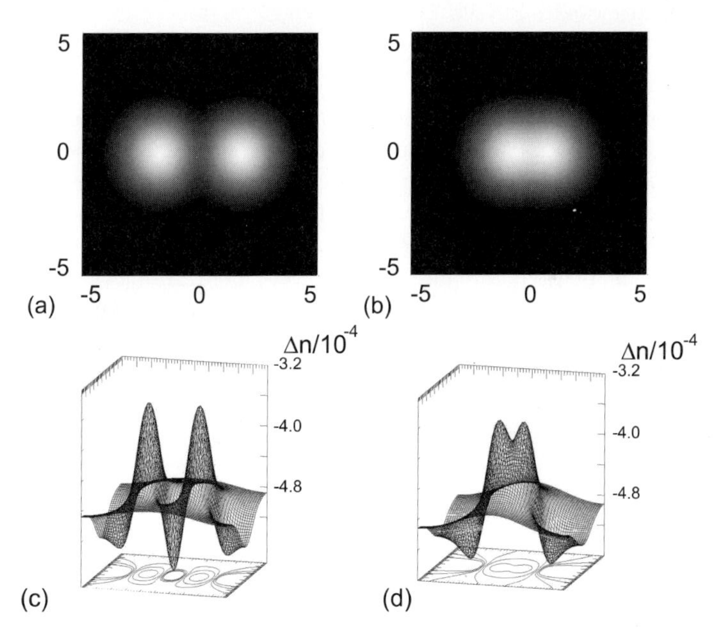

Fig. 5.7. Intensity distribution of two mutually incoherent Gaussian beams (**a**), (**b**) and the corresponding refractive index modulation Δn (**c**), (**d**) for horizontal separations $\Delta x = 1.7$ (**a**), (**c**), and $\Delta x = 3.5$(**b**),(**d**)

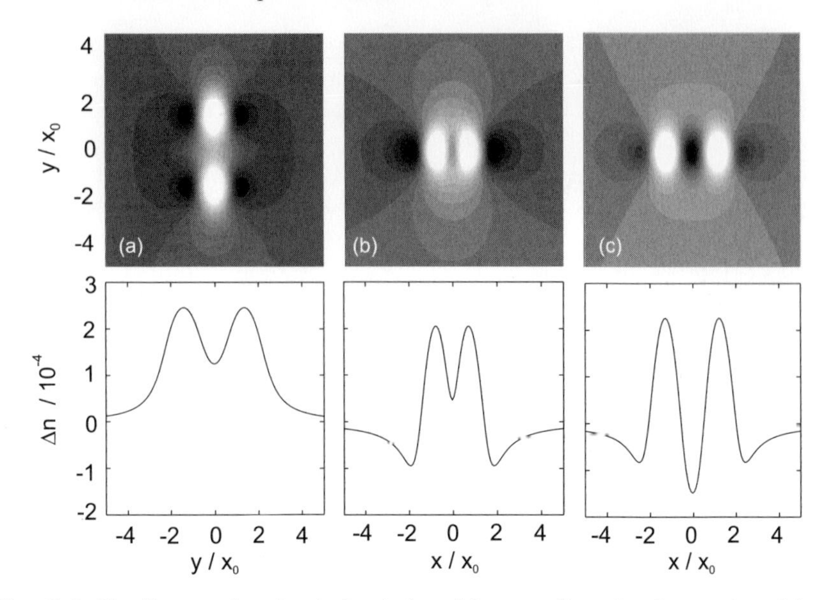

Fig. 5.8. Nonlinear refractive index induced by two Gaussian beams in a biased photorefractive crystal. (**a**) Beams located along the y axis, (**b**) along the x axis, small separation, (**c**) along the x axis, large separation (contributed by A. Stepken, Institut für Angewandte Physik, TU Darmstadt)

becomes positive and the interaction will be attractive. Therefore, we expect a varying interaction scenario depending on the spatial separation of the two beams. For a small separation, the two solitons should attract and eventually merge whereas they will mutually repel if their lateral separation is larger.

Let us now imagine a second configuration, where the two beams are now aligned in the plane perpendicular to the electric field. Along the y axis, where the refractive index decays monotonously to zero, the structure is analogous to that known for an isotropic Kerr-type nonlinearity. Therefore, one may expect that beams separated along the y axis will always attract each other, as is the case for incoherent solitons in isotropic Kerr-type media.

In Fig. 5.8, we illustrate contour plots and profiles of the nonlinear index change in a thin slab of a photorefractive crystal in the presence of two Gaussian beams that are launched in parallel. In the case depicted in Fig. 5.8a, both beams propagate in the z–y plane. In such a configuration, the refractive index in the region between the beams always *increases*, which leads to an attraction of the two evolving solitons. For beams that are propagating in the plane parallel to the applied electric field, the refractive index change depends strongly on their horizontal separation. For beams that propagate in close proximity (Fig. 5.8b) the refractive index increases in the overlap region, which leads to an attractive interaction. On the other hand, for a larger separation (Fig. 5.8c) the refractive index actually decreases in the

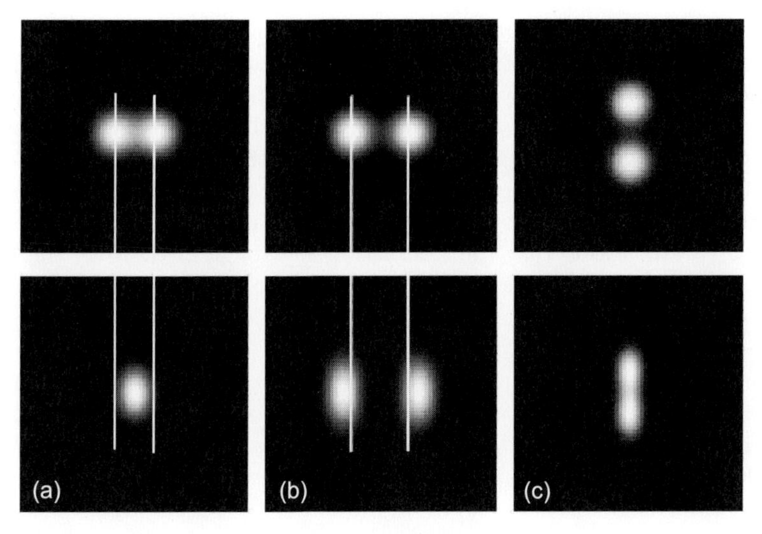

Fig. 5.9. Numerical calculations of propagation over a distance $l_z = 38$ for initial separations $\Delta x = 3.8$ (**a**), $\Delta x = 5.3$ (**b**), and $\Delta y = 5.3$ (**c**). The Gaussian input radius equals 2.5 (all coordinates are dimensionless, see text). The top and bottom row show the input and output beams, respectively

overlap region, which results in a repulsive interaction, which is completely exceptional for mutually incoherent solitons.

5.5.2 Anomalous Interaction

The propagation behavior of two solitons was investigated by means of a numerical beam propagation procedure based on (5.2) and (5.3) for initial separations corresponding to those shown in Fig. 5.8b,c. The closely spaced input beams (Fig. 5.9a, top row) attract each other strongly, and eventually coalesce into a single beam after propagating over a distance of $z = 38L_\mathrm{d}$ (Fig. 5.9b, bottom row). This behavior is generic for solitons colliding at a very small angle in a saturable nonlinear medium. An essentially different behavior can be seen in Fig. 5.9b, where the initial separation Δx is raised to 5.3. The separation between the solitons increases with propagation distance, indicating mutual repulsion. Note also that in this case each beam attains an elliptical shape with a diameter ratio ≈ 1.5, characteristic of photorefractive solitons. If the beams have the same initial separation but are oriented along the y axis (Fig. 5.9c), the beams attract each other and start to fuse, and eventually intersect periodically with increasing propagation distance. This additional and quite remarkable effect of mutual oscillation will be discussed in Sect. 5.5.3.

In order to demonstrate this peculiar behavior experimentally, we aligned two beams in the plane parallel to the electric field at various horizontal distances and recorded the resulting images that appeared at the exit face of the material. The mutual interaction of the two solitons is illustrated in Fig. 5.10. In the upper row, we show the intensity distribution at the exit face of the crystal for noninteracting solitons launched parallel to each other. These images were obtained by superimposing pictures of the propagation of individual solitons. The results of coupled and hence simultaneous propagation are represented in the bottom row of Fig. 5.10. In the case of closely spaced solitons ($\Delta x \approx 15$ μm, Fig. 5.10a) the interaction is strongly attractive, the beams fuse, and emerge from the crystal as a single elliptically shaped solitary beam. The situation changes dramatically when the initial separation Δx between the beams is increased to 25 μm, as shown in Fig. 5.10b. This time both input beams evolve into separate solitary beams, while their separation increases, indicating mutual repulsion. The repulsive force decreases with increasing distance after it has passed through its maximum at about 25 μm separation (see Fig. 5.10c). The repulsion is also clearly detectable by blocking one of the input beams, which induces a lateral motion of the remaining beam towards the center. It is important to stress that the interaction of incoherent solitons separated only along y always results in their attraction, independently of the magnitude of the initial separation. An example of this effect is shown in Fig. 5.10d, for a separation distance, at which repulsion is found in the x direction.

A direct comparison between the experimental observations depicted in Fig. 5.10 and the numerical simulations with corresponding parameters illustrated in Fig. 5.9 shows good quantitative agreement [26]. Both the experiment and the simulation reveal the merging process of two closely spaced beams and their mutual repulsion when the initial spacing is increased. In

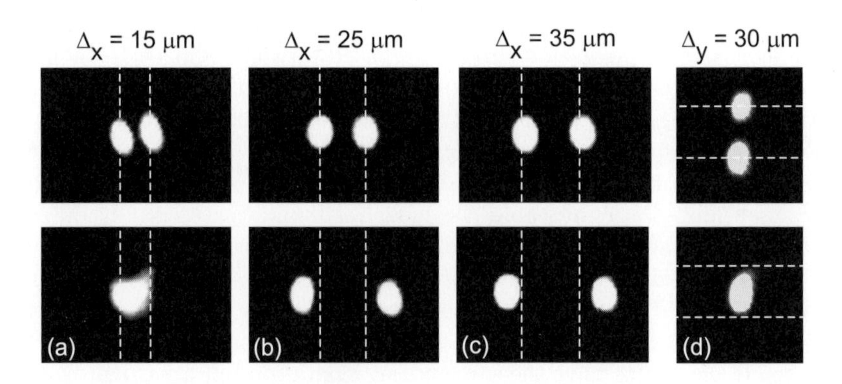

Fig. 5.10. Experimental observation of separation-dependent interaction of incoherent solitons. The x axis is horizontal. *Top row*, noninteracting solitons, and *bottom row*, interacting solitons, for different initial separation distances along the x axis (**a**)–(**c**). (**d**) shows the interaction along the y axis for a comparable separation

the simulation, the final spacing corresponds to 40 μm in Fig. 5.9b, which agrees well with the experimentally determined value of 43 μm in Fig. 5.10b.

5.5.3 Mutual Oscillation

We have already pointed out that for incoherent photorefractive solitons propagating in the plane perpendicular to the applied electric field, the nonlinear response of the medium is analogous to that of an isotropic, saturable Kerr-like nonlinearity.

For such a nonlinearity, solitons propagating in close proximity always attract. Moreover, it is well known that the attraction of these solitons may result in a nontrivial interaction behavior of their trajectories [23]. In particular, beams propagating in the same plane may become bound, with their trajectories periodically intersecting. Therefore, photorefractive spatial solitons should exhibit a similar behavior if arranged properly at the entrance face.

In Fig. 5.11, we show some results of a numerical analysis of the interaction of initially parallel solitary beams propagating in the vertical plane [27]. Figures 5.11a,b depict the incident and output intensity distributions, respectively. For the chosen separation of the beams, the attractive force leads to an interchange of the beam positions. For a better illustration, the two beam trajectories are illustrated in a graph in Fig. 5.11c. Again, the attraction is clearly visible. It leads to a periodic collision of solitons, which is the generic behavior of solitons in a saturable nonlinear medium [4] when they collide at a very small angle.

In Fig. 5.12, we demonstrate experimentally this strong attraction of screening solitons in the direction perpendicular to the electric field. As before, we show in the first row the intensity distribution at the exit face of the crystal for noninteracting solitons launched in parallel to each other, and the results of soliton interaction are displayed in the second row of the figure.

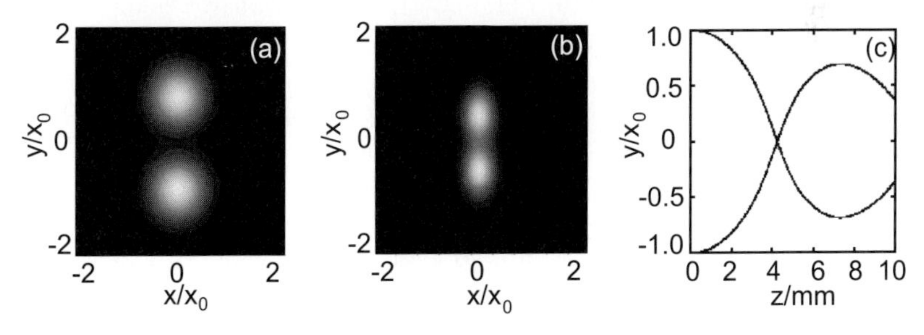

Fig. 5.11. Numerical simulation of the interaction of solitons propagating in the plane perpendicular to the direction of the biasing field. (**a**) Input intensity distribution, (**b**) output intensity distribution, (**c**) soliton trajectories in the y–z plane

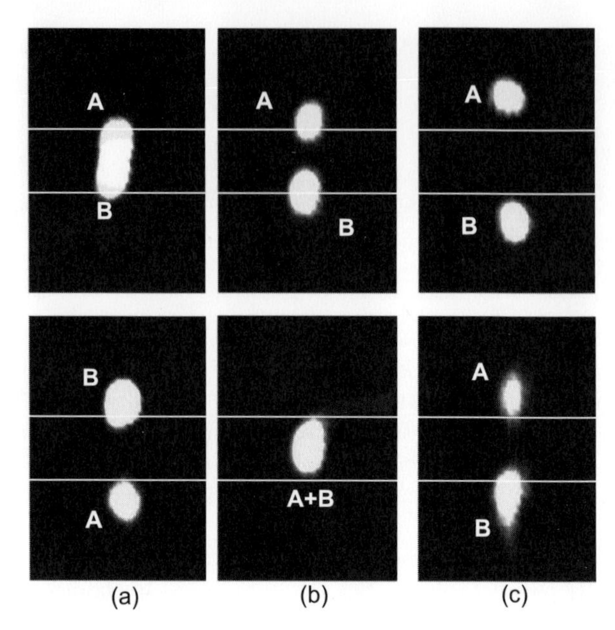

Fig. 5.12. Experimental observation of attraction of incoherent solitons propagating in y–z plane, for initial separations of (**a**) 15 μm (**b**) 30 μm, and (**c**) 50 μm. The *upper row* shows the intensity distribution at the exit face for non-interacting solitons, and the *lower row* the distribution for interacting solitons

For a small separation of the beams (15 μm, Fig. 5.12a) the strong attractive force causes the trajectories of the two solitons to intersect. When the initial separation is increased to 30 μm (Fig. 5.12b) the attraction is obviously weakened, and it takes about the length of the crystal for the solitons to collide. In this particular case the solitons collide very close to the output face of the crystal so the output intensity exhibits a single-peak structure. For a larger initial separation (50 μm, Fig. 5.12c), a significantly weaker attraction results in a decreased separation at the exit face of the crystal.

If the attraction is adjusted in a suitable way, this interaction may lead to a bound state, which in turn represents a stable soliton pair. This is due to the fact that the oscillations are strongly bound, owing to the inelastic nature of the collision. In this case, attraction results in collision and in a change of the positions of both solitons. Subsequently, they stay in their positions, propagating from then on as a bound pair, a phenomenon which has also been investigated for the case of mutually coherent beams that form a bound dipole state [28].

5.5.4 Mutual Rotation

From the preceding sections, it is obvious that the unique anisotropic features of the photorefractive nonlinearity result in a nontrivial interaction behavior of mutually incoherent photorefractive solitons. Considering only the case of two beams that are launched in parallel into the nonlinear medium, it is apparent that various interaction scenarios can occur, depending on the orientation and separation of the two light beams. Two solitons aligned in the vertical direction (perpendicular to the external electric field) experience a strong attractive force, whereas if they are aligned along the horizontal direction parallel to the electric field, they experience both attraction and repulsion. If the two beams are initially separated along both x and y, repulsive forces along x and attractive forces along y may interact. In this case, a more complex behavior featuring mutual rotation occurs. In Fig. 5.13, we show these complex dynamics with transverse sections of interacting solitons which are initially separated along both the x and the y axes (Fig. 5.13a). Here, the propagation of initially parallel beams results in repulsion and a counterclockwise rotating motion around the center of mass of the two beams (Fig. 5.13b).

We confirmed this effect experimentally using the same configuration as described above. We oriented the input beams in such a way, that without interaction, they would propagate parallel to each other in a plane inclined relative to the direction of the horizontal c axis. For this orientation, the strong anisotropy of the electrostatic potential created by the two beams results in an interplay of attraction and repulsion along the two transverse directions. The net result is a rotation of the solitons around their center of mass. Figure 5.14a depicts the intensity distribution of the incident beams. In the upper row, they propagate parallel in a plane which is tilted at $\Theta_x \approx 13°$ with respect to the x axis. Figures 5.14b,c display the output intensity distributions for noninteracting (independent propagation) and interacting solitons, respectively. As is clear from the figure, the mutual interaction results not only in an increased separation between the solitons but also in a counter-

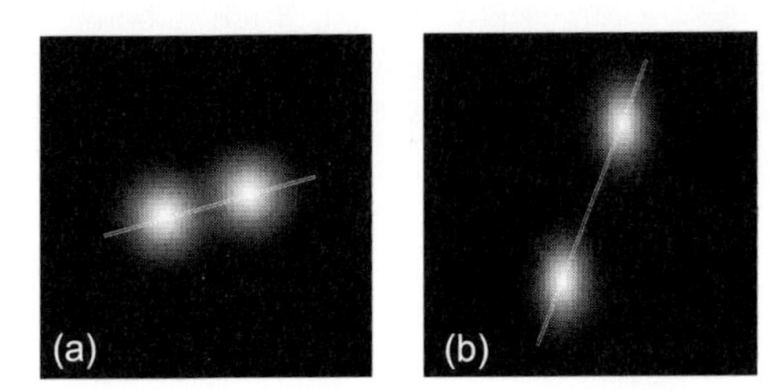

Fig. 5.13. Numerical simulation of the mutual rotation of initially parallel solitons due to the anisotropy of the photorefractive nonlinearity: (**a**) input beams, (**b**) output beams showing rotation due to interaction

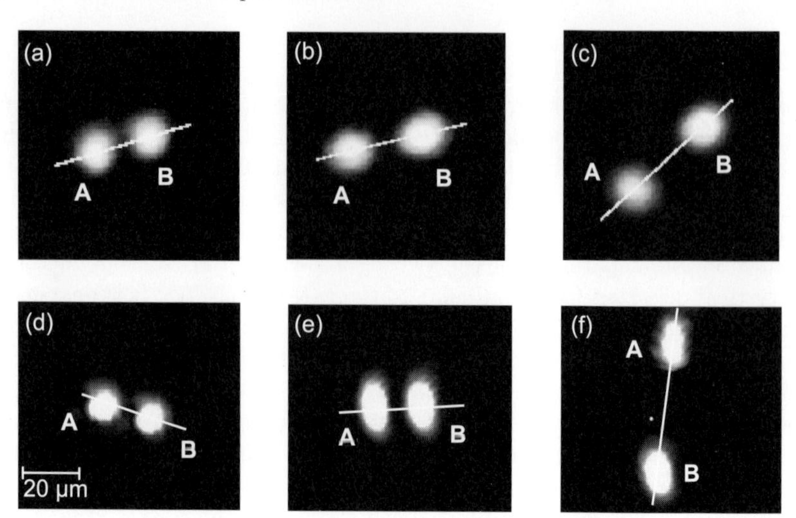

Fig. 5.14. Experimental observation of mutual repulsion and rotation of solitons due to anisotropy of the nonlinearity. (**a**), (**d**) Beams at the plane of incidence on the crystal, (**b**), (**e**) at the exit face without interaction and (**c**), (**f**) when they propagate simultaneously and interact to produce a spiraling motion. *Upper row* (**a**)–(**c**), parallel trajectories, *bottom row* (**d**)–(**f**), slightly tilted trajectories

clockwise spiraling motion about the center of mass of the beams, in exactly the same manner as found numerically in the results shown in Fig. 5.13. It should be emphasized that the spiraling motion is due to the anisotropy of the potential created by the two beams, and occurs despite the fact that they are launched without any tangential velocity. Flipping the initial positions of the beams with respect to the vertical plane (y axis) does not change the qualitative behavior, but changes the sense of rotation. This is a consequence of the fact that the solitons always try to align themselves along the y axis. When one of the solitons has a slight inclination during propagation, the rotation angle may be enhanced, because in this case the anisotropy of the attractive and repulsive forces is changed. Such a behavior is shown in the second row of Fig. 5.14, where beam A has an inclination about $0.1°$ in the direction of the y axis at the entrance face. As a consequence, the attractive force along the y direction is strengthened, leading to a prolonged rotation of the two beams of over $90°$.

5.5.5 Complex Spiraling

Up to now, our configurations for soliton interaction have been restricted to the cases of collision at a certain point and of parallel propagation. The mutual interaction was based on the nonlocal characteristics of the potential of the two solitons. However, it is possible to add another degree of freedom to the interaction – the angular separation of the two beams. If this degree of freedom is introduced, the result is a skewing of the beams, leading to complex interaction scenarios depending on both the relative angle and the relative spatial position of the two solitons throughout the nonlinear material. For closely spaced solitons, a wealth of new phenomena arises: mutual spiraling of two solitons propagating initially in different planes becomes possible.

In order to observe such a spiraling behavior, it is necessary to have an attractive interaction between the solitons. In this case, the initial apparent repulsion of the beams due to the divergence of their trajectories can be compensated for by their mutual attraction. While this situation can certainly be realized in the case of isotropic self-focusing Kerr-type materials, the situation for photorefractive media is more complex, owing to the anisotropy between the two transverse directions. Because two incoherent solitons may experience both attractive and repulsive forces, there exist domains of attraction and repulsion in the transverse plane, which lead to a nontrivial topology of the soliton trajectories. The reason for this behavior is that the beams are now propagating slanted relative to the externally applied electric field. This leads to trajectories that have a more complicated appearance than a simple spiral [29].

As already demonstrated, a pair of solitons launched in a plane tilted at some angle with respect to the direction of the applied field will initially rotate, trying to align itself along the y axis. However, the momentum acquired by the solitons during this movement produces an overshoot. When the beams cross the y axis, the anisotropy of the screening slows down the speed of rotation until the direction of rotation is completely reversed. The distance between the solitons decreases, and the pair twists and turns around the z axis in a damped motion. When viewed along the z direction, the motion initially resembles spiraling followed by oscillations predominantly along the y direction. In general, the whole process ends up in fusion of the two solitons. However, the distance at which the beams fuse might be much larger than the typical length of a crystal, thus revealing this result only for nonlinear materials that provide a sufficient interaction length.

This behavior can again be explained in terms of a waveguide picture in a very general way for solitons in non-Kerr, e.g. saturable, media. To understand the underlying equivalence, let us first consider two one-dimensional solitons which are launched at a certain angle with respect to each other. These solitons act like rays propagating in a graded-profile slab waveguide with an index profile of the kind shown in Fig. 5.8. In this case, the refractive index profile has an exponential falloff at larger distances. If the angle is less than some critical angle, which depends on the refractive index value, then the solitons are bound, i.e. they follow a periodic trajectory. The solitons initially move together and then move apart, in a periodic fashion. If the interaction angle exceeds the critical angle, then the solitons move apart. Since the refractive index profile has an exponential shape, we know that the period of the interacting beams depends exponentially on the beam separation. The period length increases with increasing beam separation. Therefore, the beam interaction is characterized by the critical inclination angle, the period of oscillation of the bound state and the maximum separation for any given trajectory, in analogy to rays in a graded-profile guide.

Consider next two beams of circular cross section. This suggests the analogy of rays propagating in a graded-profile fiber with a circular geometry. Now, the beams are characterized by an angle of inclination to the axial direction (z axis), as in the case of a planar waveguide, and by an azimuthal (or skewness) angle. If the initial directions of the two solitons are in the same plane, the trajectories are like meridional rays and behave exactly like the planar beams described above. However, when the beams are launched initially in a skewed way relative to each other, they undergo a helical trajectory. Planar solitons with a skewness angle larger than the critical angle diverge, but beams of circular cross section, launched sufficiently skewed relative to each other, remain mutually bound within their characteristic double helical orbit – this is called *soliton spiraling*. It is important to note that the two beams must be identical for these effects to be realized – a condition that is difficult to realize in experiments, especially when the anisotropy of the photorefractive nonlinearity comes into play.

In principle, soliton spiraling should be possible with either mutually coherent or mutually incoherent solitons. A necessary condition is, of course, that the mutual repulsion due to the centrifugal force is balanced by attraction. The argument for focusing our theoretical and experimental investigations on incoherent solitons stems mostly from the experimental point of view. The force between mutually coherent solitons depends on the relative phase, which can be manifested in either attraction or repulsion. In this case the mutual force is subject to phase variations, and it is almost impossible to adjust the balance between the attractive and repulsive forces, which needs to be maintained as constant as possible throughout the entire propagation. Using a mutually incoherent pair of optical beams to observe the spiraling process avoids this specific stability problem. The interaction force between the solitons is governed only by their lateral separation and their initial directions relative to the electric field, and not by the relative phase of the beams [29].

Numerical Simulation of Soliton Spiraling

The various possible scenarios that arise when the two trajectories of the solitons converge in a photorefractive nonlinear medium have been the topic of numerous experimental and numerical studies [10, 23, 30, 31, 32, 33, 34]. Belic et al. investigated theoretically various scenarios that are close to a spiraling behavior [34]. The beams always rotate initially. For low intensities above saturation and for parallel launching or insufficient tilt of the two

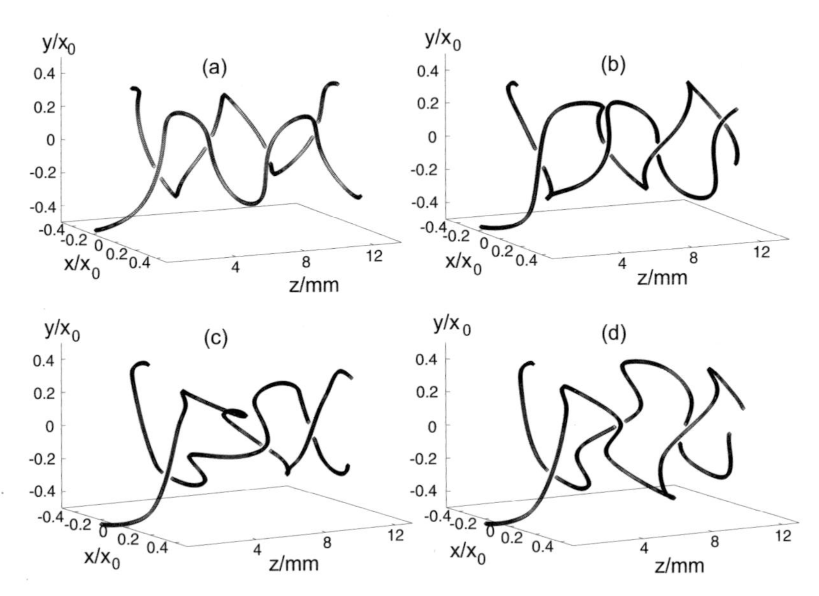

Fig. 5.15. Three-dimensional trajectories of interacting solitons for different input conditions. (**a**) $I = 2$, parallel beams: rotation and damped oscillation. (**b**) $I = 2$, tilted beams: spiraling by more than 2π. (**c**) $I = 5.5$, strongly tilted beams: strong, irregular spiraling of almost 3π. (**d**) $I = 5.5$, less tilted beams: complex oscillation (after [34], printed by kind permission of Friedemann Kaiser)

beams, the initial rotation is followed by oscillation perpendicular to the direction of the applied field, and the solitons eventually fuse. For intensities far above saturation, in contrast, the initial rotation, can lead to long-lasting spiraling when the launching tilt is sufficient. In this case, the influence of the anisotropy is decreased, and a behavior similar to that of an ideal isotropic Kerr nonlinearity is approached. The distance at which the beams stop spiraling and start to oscillate might then be much larger than the typical length of a photorefractive crystal, making the spiraling seemingly indefinite. The various possible scenarios are shown in detail in Fig. 5.15. For $I = 2$ and beams launched in parallel to each other (Fig. 5.15a), the two solitons start to rotate. However, after one twist the beams unwind and each then remains on the same side of the other, producing a damped oscillation. In Fig. 5.15b, the same pair of beams is launched with a tilt in the direction of the applied field only ($\Theta_x = \pm 0.8$; angles are expressed here in multiples of π). The solitons spiral by more than 2π before they are captured along the y axis. When the initial conditions are changed to a higher intensity ($I = 5.5$) and the same initial tilt of $\Theta_x = \pm 0.8$, the beam pair may spiral by about 3π. The beams spiral for the entire length of the numerical simulation, with a swing about the y axis each time it is crossed. In the case of a smaller tilt ($\Theta_x = \pm 0.3$), the same intensity results in an oscillation of the beams. It should be noted that in all cases of spiraling, an extension of the propagation length leads to a finite spiraling behavior, which is followed by oscillations that are damped down to an orientation of the two solitary beams along the y axis. Moreover, the trajectories observed are never smooth spirals. The beams wobble while spiraling, and follow complex trajectories in space in the nonlinear medium. The long attractive well along the y axis and the repulsive areas along the x axis break the symmetry and prevent infinite spiraling. In addition, when the solitons are close to each other and interact strongly – which is a necessary condition when large tilting angles are used – the beams become entangled or partially fuse, and their individual identities can no longer be followed. The light intensity distribution does not show two distinct beams anymore, but a field of beams and radiation energy. As soon as the beams become disentangled again during propagation, two bright spots reappear, displaying solitary properties again.

Experimental Investigations of Soliton Spiraling

In the case of experimental photorefractive screening solitons, it was not clear for a long time whether soliton spiraling could be realized or not. This was mainly due to the fact that, in contrast to the theoretical investigations described above, it is not possible to follow the trajectories of the solitons along their propagation through the nonlinear medium. Therefore, spiraling, oscillating and rotating behaviors cannot be distinguished when the solitons are observed at the exit face of the crystal. This leaves ample space for interpretation and gave rise to controversial results.

On the one hand, the most critical requirement is to find the condition under which the attractive force can compensate exactly the centrifugal force caused by the acquired angular momentum. On the other hand, spiraling needs to be demonstrated by an appropriate visualization method in order to discriminate it from pure rotary or oscillatory behavior.

Segev and coworkers [30] solved these two problems in their experimental realization by using an intensity ratio of the soliton to the background of $I \approx 5$. In the case of a smaller intensity ratio, the attractive force is too weak – the gradient of the refractive index is too small – to compensate the centrifugal force. If the intensity ratio is set to a higher value, the soliton-induced waveguide displays a multimode structure, which means that nonfundamental guided modes can be excited in the collision process. This breaks the two-dimensional symmetry and causes the solitons to deteriorate. Therefore, high intensities are at the border the conditions for formation of stable solitons. In this regime, the solitons become extremely sensitive to perturbations. In order to solve the problem of demonstrating the spiraling behavior, Segev et al. used two different orientations of the same crystal. By rotating the crystal about the c axis, investigation and direct comparison of beam propagation along the crystallographic a and b axes was possible. The generic light propagation characteristics along these two axes can be considered as exactly equal, and a direct comparison is therefore valid. Since the two axes were different in length, two different stages of propagation could be observed.

To investigate these complex interaction scenarios, where two solitons wind in a complex manner, twisting and turning in the complex potential that they have created, it is essential to identify each soliton during its propagation. The standard procedure for solving this problem is to distinguish between the output beams by blocking or modulating one of the beams at the input for a time period that is much shorter than the response time of the photorefractive material. The photorefractive nonlinearity is not affected by such a fast modulation, because the nonlinear index change does not have time to adjust.

In order to observe a spiraling behavior, Segev et al. set the initial separation of the beams to about 15 μm. Otherwise, the interaction would lead to repulsion or merging of the two beams. After a propagation distance of 6.5 mm, the plane of the beams rotated by roughly 270°, whereas propagation along the 13 mm side resultd in a rotation of 540°. In addition, Segev et al. observed an energy exchange of roughly 19% after 6.5 mm of propagation and 30% after 13 mm of propagation. This is due to the fact that the soliton-induced waveguides propagate in close proximity, and therefore that energy from one soliton beam can be coupled into the waveguide induced by the other. Since both solitons induce single-mode waveguides, this energy exchange does not break the symmetry and does not affect the interaction.

However, this one experiment was not sufficient for a complete characterization of the rather complex spiraling characteristics of spatial solitons. Therefore, another technique to detect a spiraling motion inside the nonlinear crystal was introduced [31]. If the lateral separation of the two beams is changed while their relative angle is kept constant, an entangled motion is detectable at the exit face of the crystal. In this way, a series of interaction states resulting in a steady state that one allows to track the action of the interaction force on the final trajectory can be obtained.

We have also applied this technique. By changing the position of one of the solitons relative to the other, we found that when the initial separation d between the two solitons in the x direction was larger than 10 μm, the solitons barely interacted, but instead passed through the crystal almost independently. Interaction effects become pronounced in the region $d < 10$ μm. As d is reduced into this region, the solitons begin to experience the common noncentral potential. The attractive well along the y axis and the repulsive region along the x axis break the symmetry and cause a complex rotational motion, with the same characteristics as described above.

A third technique for gaining insight into the complex interaction process has been introduced by our group [32]. The principle is that one can draw conclusions from the formation process and the dynamic behavior as to

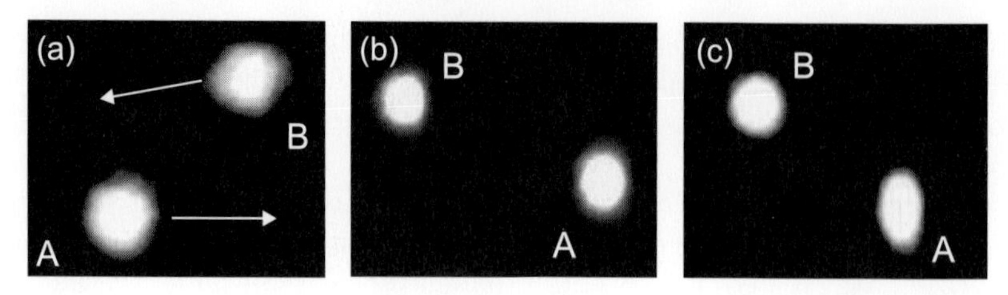

Fig. 5.16. (**a**) Beams at the entrance face of the crystal, indicating the direction of skewing of the two beams. (**b**) The steady-state situation at the exit face without interaction. (**c**) The corresponding situation with interaction

whether the solitons propagate in an entangled way or not. In Fig. 5.16, we illustrate the steady state of two solitons that have been launched in a skewed way into the crystal. Figure 5.16a depicts the plane of incidence and the intensity distribution of the input beams. The beams were separated along an axis tilted by $\Theta_x \approx 44^\circ$ with respect to the x axis, and had a symmetric initial angular tilt of 0.4° in the x direction and an asymmetric angular momentum of 0.1° (beam B) in the y direction, as indicated by the the two arrows. Figures 5.16b,c display how the solitons emerged from the crystal after propagating 13.5 mm, in a noninteracting and an interacting configuration, respectively. In this particular case, the analysis of the exit face of the crystal in the steady state does not give any hints about the question of whether there exists a spiraling at all. This result can be interpreted either as a very small rotation by a few degrees or as a rotation by almost 2π. However, by monitoring the temporal development of soliton formation and analyzing the behavior at subsequent time intervals, we were able to resolve the behavior in time. By doing this, we made the interaction behavior much more transparent.

The sequence depicted in Fig. 5.17 starts when one of the beams (A) has reached its steady state. Launching the second beam (B) leads first to the formation of a soliton beam in its launching position (Fig. 5.17, frame 2). Subsequently, both beams break up into various spatial components (Fig. 5.17, frame 3), and later reappear in a clockwise-rotated position (Fig. 5.17, frame 6). During this stage, a remarkable power exchange takes place between the two beams. Then beam B rotates counterclockwise around beam A, and reaches its steady-state position (Fig. 5.17, frames 7–12). It is interesting to note that the solitons rotate around each other in an elliptical orbit, as was also described in [30]. The reason for the change in the rotation direction is exactly the same as that found in the theoretical investigations and originates from the anisotropic structure of the potential. The pair initially rotates clockwise, trying to stabilize itself along the y direction, which is the direction of strongest attraction. The momentum acquired by the solitons produces an overshoot, and, as the beams cross the y axis, the anisotropy of

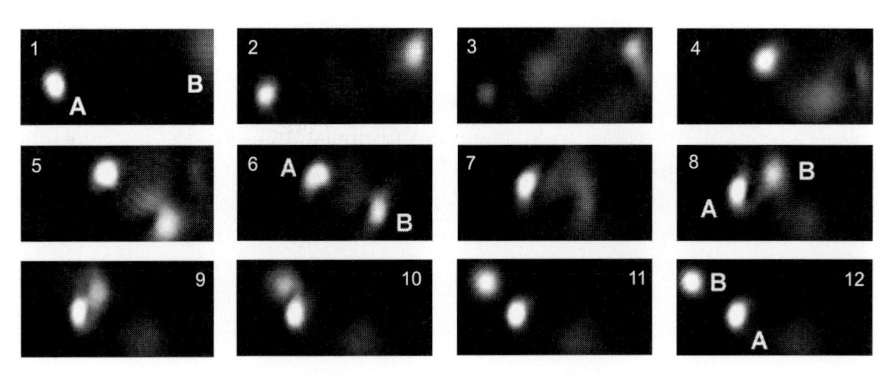

Fig. 5.17. Time-resolved complex spiraling of a soliton pair. The sequence starts when a second beam is launched in the presence of a preexisting steady-state soliton. The time interval between consecutive frames is 0.36 s

the screening slows down the rate of rotation and reverses its direction. The distance between the solitons decreases, and the pair twists and turns around the y axis. This behavior, especially the different rotation speeds at different stages of the development, is typical of situations where the beam intensities are adjusted relative to the background intensity to produce stable spatial solitons [35].

This behavior is in complete agreement with the theoretical predictions presented earlier in this section. Therefore, the launching of skewed beams at higher intensities would offer the possibility of prolonged spiraling. Because the experimental realization of this behavior is beyond the coupling efficiency of standard photorefractive crystals, however this cannot be done at present.

Note that the change in the sense of rotation and the nonuniform rotation speed predicted theoretically are also observed in the experimental results shown in Fig. 5.17. Although the difference between the two cases is that the experiment was performed with less intense beams at a different separation distance, most of the important features of the spiraling interaction are present in both experiment and theory. Therefore, resolving the behavior in time gives insight into the complex interaction scenarios that take place during soliton propagation. In particular, oscillations of the beam diameter and of the beam separation distance, and different speeds of rotational motion can be identified.

These investigations show that temporal and spatial resolution of the behavior of soliton interaction may allow us to view the steady-state results of the interaction in a different light which gives insight into the complex interaction scenarios that take place during soliton propagation. One of the most important phenomena during these scenarios is the occurrence of power exchange between the interacting beams.

All of these interactions – especially the wealth of different incoherent interactions – are suited in general to applications in adaptive waveguiding and switching. It is even possible to realize large, complex arrays of interconnecting configurations exploiting soliton interaction [36]. In the following chapter, we shall discuss the potential of these applications and several complex waveguide configurations, emphasizing the technical features that will be necessary for successfully realizing various complex waveguide configurations using optical photorefractive solitons.

Summary

One of the striking consequences of the special features of photorefractive solitons is their ability to interact during collisions. In this chapter, we described the main mechanisms for the various interaction scenarios, and their experimental verification. The wealth of observed phenomena includes attraction, repulsion, deflection, rotation and even complex spiraling motion. The fact that the nonlinear refractive index change is of anisotropic nature gives rise to the observation of an exceptional type of interaction phenomenon which is manifested as an anomalous interaction for incoherently interacting solitons. Owing to this anisotropy of the refractive index modulation, separation-dependent interaction scenarios occur. By temporally resolving a rather complex spiraling interaction scenario, we were able to gain insight into the characteristics of the entangled soliton trajectories.

References

1. V.E. Zakharov and A.B. Shabat, *Exact theory of two-dimensional self-focusing and one-dimensional self-modulation of waves in nonlinear media*, Zh. Eksp. Teor. Fiz. **61**, 118 (1971). (Sov. Phys. JETP **34**, 62 (1972)).
2. F. Reynaud and A. Barthelemy, *Optically controlled interaction between two fundamental soliton beams*, Europhys. Lett. **12**, 401 (1990).
3. J.S. Aitchison, A.M. Weiner, Y. Silberberg, D.E. Leaird, M.K. Oliver, J.L. Jackel and P.W. Smith, *Experimental observation of spatial soliton interactions*, Opt. Lett. **16**, 15 (1991).
4. A.W. Snyder and A.P. Sheppard, *Collisions, steering and guidance with spatial solitons*, Opt. Lett. **18**, 482 (1993).
5. A. Guo, M. Henry, G. Salamo, M. Segev and G.I. Wood, *Fixing multiple waveguides induced by photorefractive solitons: directional couplers and beam splitters*, Opt. Lett. **26**, 1274 (2001).
6. S. Lan, E. DelRe, Z. Chen, M. Shih and M. Segev, *Directional couplers with soliton-induced waveguides*, Opt. Lett. **24**, 475 (1999).
7. M. Shih, M. Segev and G. Salamo, *Circular waveguides induced by two-dimensional bright steady-state photorefractive spatial screening solitons*, Opt. Lett. **21**, 931 (1996).

8. Z. Chen, M. Mitchell and M. Segev, *Steady-state photorefractive soliton-induced Y-junction waveguides and high-order dark spatial solitons*, Opt. Lett. **21**, 716 (1996).
9. J. Petter and C. Denz, *Guiding and dividing waves with photorefractive solitons*, Opt. Commun. **188**, 55 (2001).
10. D.J. Mitchell, A.W. Snyder and L. Poladian, *Interacting self-guided beams viewed as particles: Lorentz force derivation*, Phys. Rev. Lett. **77**, 271 (1996).
11. S. Gatz and J. Herrmann, *Propagation of optical beams and the properties of two-dimensional spatial solitons in media with a local saturable nonlinear refractive index*, J. Opt. Soc. Am. B **14**, 1795 (1997).
12. W.J. Firth and A.J. Scroggie, *Optical bullet holes: robust controllable localized states of a nonlinear cavity*, Phys. Rev. Lett **76**, 1623 (1996).
13. T.-T. Shi and S. Chi, *Nonlinear photonic switch by using the spatial soliton collision*, Opt. Lett. **15**, 1123 (1990).
14. R. de la Fuente, A. Barthelemy and C. Froehly, *Spatial soliton-induced guided waves in a homogeneous nonlinear Kerr medium*, Opt. Lett. **16**, 793 (1991).
15. T.R. Taha and M.J. Ablowitz, *Analytical and numerical aspects of certain nonlinear evolution equations. II. Numerical, nonlinear Schrödinger equation*, J. Comput. Phys. **55**, 203 (1984).
16. A.V. Mamaev, M. Saffman and A.A. Zozulya, *Phase-dependent collisions of (2+1)-dimensional spatial solitons*, J. Opt. Soc. Am. B **15**, 2079 (1998).
17. M.-F. Shih and M. Segev, *Incoherent collisions between two-dimensional bright steady-state photorefractive spatial screening solitons*, Opt. Lett. **21**, 1538 (1996).
18. V. Tikhonenko, J. Christou and B. Luther-Davies, *Three dimensional bright spatial soliton collision and fusion in a saturable nonlinear medium*, Phys. Rev. Lett. **76**, 2698 (1996).
19. Y. Kodama and K. Nozaki, *Soliton interaction in optical fibers*, Opt. Lett. **12**, 1038 (1987).
20. W. Krolikowski and S. Holmstrom, *Fusion and birth of spatial solitons upon collision*, Opt. Lett. **22**, 369 (1997).
21. W. Krolikowski, B. Luther-Davies, C. Denz and T. Tschudi, *Annihilation of photorefractive solitons*, Opt. Lett. **23**, 97 (1998).
22. A.W. Snyder, D.J. Mitchell, L. Poladian and F. Ladouceur, *Self-induced optical fibers: spatial solitary waves*, Opt. Lett. **16**, 21 (1991).
23. L. Poladian, A.W. Snyder and D.J. Mitchell, *Spiralling spatial solitons*, Opt. Commun., 59 (1991).
24. W. Krolikowski, C. Denz, M. Saffman, B. Luther-Davies and S.H. Holmstrom, *Interaction of coherent and incoherent photorefractive spatial solitons*, Asian J. Phys. **7**, 698 (1998).
25. M.-F. Shih, *Incoherent collisions between one-dimensional steady-state photorefractive screening solitons*, Appl. Phys. Lett. **69**, 4151 (1996).
26. W. Krolikowski, M. Saffman, B. Luther-Davies and C. Denz, *Anomalous interaction of spatial solitons in photorefractive media*, Phys. Rev. Lett. **80**, 3240 (1998).
27. W. Krolikowski, B. Luther-Davies, C. Denz, J. Petter, C. Weilnau, A. Stepken and M. Belic, *Interaction of two-dimensional spatial incoherent solitons in a photorefractive medium*, Appl. Phys. B **68**, 975 (1999).

28. A.V. Mamaev, A.A. Zozulya, V.K. Mezentsev, D.Z. Anderson and M. Saffman, *Bound dipole solitary solutions in anisotropic nonlocal self-focusing media*, Phys. Rev. A **56**, R1110 (1997).
29. W. Krolikowski, C. Denz, A. Stepken, M. Saffman and B. Luther-Davies, *Interaction of spatial photorefractive solitons*, Quantum Semiclass. Opt. **10**, 823 (1998).
30. M. Shih, M. Segev and G. Salamo, *Three-dimensional spiraling of interacting spatial solitons*, Phys. Rev. Lett. **78**, 2551 (1997).
31. A.V. Buryak, Yu.S. Kivshar, M. Shih and M. Segev, *Induced coherence and stable soliton spiraling*, Phys. Rev. Lett. **82**, 81 (1999).
32. C. Denz, J. Petter, C. Weilnau and W. Królikowksi, *Time-resolved formation and incoherent interaction of photorefractive screening solitons*, Phys. Rev. E **60**, 6222 (1999).
33. A. Stepken, M.R. Belić, F. Kaiser, W. Królikowski and B. Luther-Davies, *Three dimensional trajectories of interacting incoherent photorefractive solitons*, Phys. Rev. Lett. **82**, 540 (1998).
34. M. R. Belić, A. Stepken and F. Kaiser, *Spiraling behavior of photorefractive screening solitons*, Phys. Rev. Lett. **82**, 544 (1999).
35. A.A. Zozulya, D.Z. Anderson, A.V. Mamaev and M. Saffman, *Self-focusing and soliton formation in media with anisotropic nonlocal material response*, Europhys. Lett. **36**, 419 (1996).
36. J. Petter, J. Schröder, D. Träger and C. Denz, *Optical control of arrays of photorefractive screening solitons*, Opt. Lett. **00**, 00 (2002).

6 Waveguiding in Photorefractive Solitons

Because every phenomenon of confined light propagation can be interpreted by means of linear waveguide theory [1], the special case of a self-trapped optical beam can be explained by means of the self-consistency principle (see e.g. (Chap. 7) and [2, 3]). In this model, an incident light beam induces a refractive index waveguide, which is capable of guiding the beam itself as an eigenmode. Moreover, (1+1)-dimensional Kerr-type spatial solitons are able to guide not only their basic induced mode but also other beams, owing to cross-phase modulation effects [4, 5]. Therefore, a single spatial soliton basically creates a waveguiding structure in which other beams of different power, wavelength or transverse geometry can be guided. This optical control of one beam by another one has an immense potential for applications in all-optical switching and waveguide coupling. Owing to these generic and obviously promising features of nonlinear spatial solitons in the framework of linear optics, the investigation of the waveguide characteristics of self-trapped optical beams in saturable photorefractive media has become an important branch of nonlinear transverse optics, heading directly towards applications. Combined with the inherent parallelism of conventional optics, solitons therefore open up new attractive ways in which waveguide arrays and lattices, and also waveguide-coupling devices, can be created. Here, the photorefractive effect offers several important advantages, which are due to the ease of creation of solitons at very low laser powers, in the microwatt range, and the tunability of those solitons with artificial dark conductivity and the externally applied electric field. Moreover, a helpful feature that facilitates the realization of waveguide structures is the strong wavelength dependence of the photorefractive effect. Therefore, it is possible to derive a soliton from a weak incident beam and guide a much stronger beam of different wavelength for which the material is much less photosensitive. Consequently, waveguiding was observed soon after the experimental realization of photorefractive solitons in the case of quasi-steady-state solitons [6], and subsequently for stable screening photorefractive solitons [7, 8, 9].

6.1 Photorefractive Waveguides

Exploiting the wavelength selectivity of the photorefractive effect is the simplest approach for demonstrating the waveguiding properties of photorefractive spatial solitons. Shih et al. [7] demonstrated guiding of a 15 μW He–Ne laser ($\lambda = 633$ nm) probe beam in a soliton, which was induced by a writing beam ten times weaker of power 1.5 μW, ($\lambda = 488$ nm), derived from an Ar^+ laser. In the absence of the writing beam, the probe beam diffracts linearly in the biased crystal and does not form a self-confined structure. However, when the writing beam is present, the refractive index of the material changes, the required waveguide is induced and the probe beam is trapped in the soliton. The coupling efficiency of the red beam into the soliton waveguide could be determined to be 85%, normalized to the absorption of the material ($\alpha = 0.21$). If the dark conductivity of the material is low, the refractive index channel formed by the photorefractive soliton can even be persistent enough to guide other beams without the writing beam being present.

In our experimental investigations, we were able to use such a waveguide structure to guide a beam at $\lambda = 633$ nm for several days without significant degradation of the waveguide structure [8]. For details about fixing photorefractive solitons see [10], for example. Figure 6.1 shows the formation of such a waveguide at $\lambda = 532$ nm and the guiding of a probe beam at $\lambda = 633$ nm. Both beams had a total power of about 30 μW and propagated in a 13.5 mm long SBN crystal. During the formation of the soliton waveguide channel, the effects of self-focusing, which is accompanied by beam diameter oscillations and beam-bending effects, can be clearly identified in the soliton G, as well as in the guided beam R. Note that in the steady state (Fig. 6.1f), the typical anisotropic nature of the waveguide also defines the shape of the guided beam.

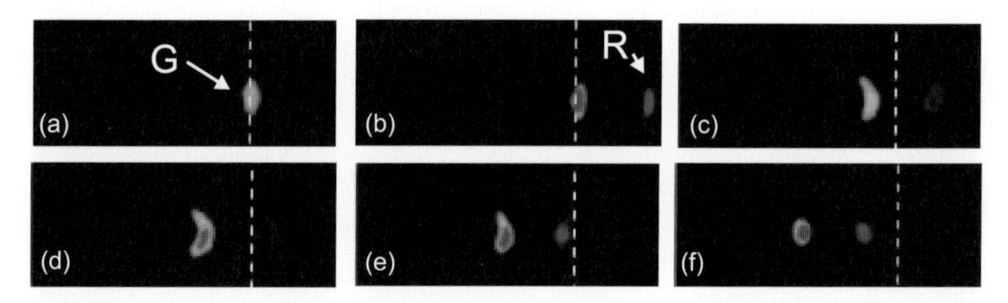

Fig. 6.1. Dynamics of soliton formation by a beam at $\lambda = 532$ nm (G) when it is guiding a probe beam at $\lambda = 633$ nm (R). The probe beam is located to the right of the soliton in these images owing to a spatial-frequency selection obtained by using a prism in front of the recording CCD camera. The time interval Δt between the consecutive frames (**a**)–(**f**) is 1 s

6.1.1 Multimode Waveguides

Generic multimode waveguides are capable of guiding not only modes with a different wavelength, but also modes with a different, more complex geometry. In contrast to Kerr media, where only relatively weak beams can be guided by a soliton, a saturable nonlinearity supports the realization of multimode soliton-induced waveguides, where the probe beam can even be much stronger than the guiding beam. Because the number of waveguide modes supported is directly proportional to the degree of saturation, it increases with increasing ratio of beam to background intensity. In the regime of moderate saturation, which corresponds to the region of the two-dimensional soliton solution, only a limited number of modes can be excited, this reflects the anisotropic nature of the photorefractive soliton.

Therefore, guiding of a double- or triple-humped transverse light structure resembling a TEM_{01} or TEM_{20} transverse laser mode has been demonstrated successfully in both transverse directions in the case of high saturation (where the background intensity was 120 times stronger than the waveguide-inducing structure) [7, 11]. However, in this case, the transverse structure formed by the nonlinear refractive index change was no longer close to the soliton solution, thereby demonstrating that the formation of a stable solitary beam is not obligatory for waveguiding purposes.

In a recent numerical and experimental study [9] that emphasizes the anisotropic nature of the refractive index modulation and hence the self-induced waveguide structure, the waveguiding features have been exploited to visualize the anisotropic refractive index modulation of the photorefractive nonlinearity. Using a soliton-induced waveguide about 15 μm in diameter, a probe beam enlarged to a diameter of 100 μm was used to scan the entire transverse structure of the waveguide. Since the probe beam is much wider than the solitary structure in this situation, only part of the incident light is coupled into the waveguide. The remaining light diffracts and gets scattered at the borders of the waveguide. When the probe beam propagates coaxially with the waveguide, only the TEM_{00} transverse mode is excited, which is illustrated in Fig. 6.2a. Note the elliptical shape of the guided beam and the dark (defocusing) domains at either side, where the refractive index change is negative according to the anisotropic model depicted in Figs. 4.7 and 4.8. Higher-order modes of the waveguide can be excited by shifting the probe beam a few micrometers with respect to the center of the waveguide. In this case, only part of the intensity of the probe beam will be coupled into the waveguide, thereby selecting a certain range of transverse k-vectors. As a result, the light intensity confined by the waveguide propagates nonparaxially as a higher-order mode. This method and the excitation of modes by slightly tilting the probe beam are generic for all kinds of linear waveguides [1]. In the configuration of Fig. 6.2b a vertical shift of the probe beam by 15 μm led to the excitation of a TEM_{01} mode. Shifting the probe beam by another 15 μm excited the TEM_{02} mode, as shown in Fig. 6.2c. In contrast, it is not

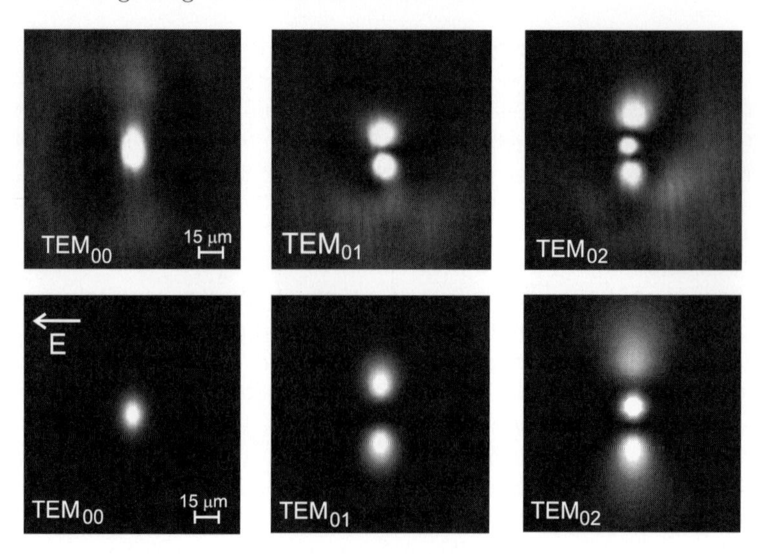

Fig. 6.2. Excitation of ground and higher-order modes of a waveguide written at $\lambda = 532$ nm with a probe beam at $\lambda = 633$ nm: experimental results (*upper row*) and theory (*lower row*, contributed by A. Stepken, Institute of Applied Physics, Darmstadt University of Technology). (**a**) Fundamental mode TEM_{00}, and the higher-order modes (**b**) TEM_{01} and (**c**) TEM_{02}

possible to excite any TEM_{10} or TEM_{20} modes. The inherent anisotropy of the waveguide is reflected in its modal structure, since higher-order modes can be excited predominately in the vertical direction, perpendicular to the external electric field. According to numerical simulations [9], this behavior is a consequence of a general rule that horizontal modes can only be excited when the transverse size of the waveguide exceeds the typical soliton diameter of 15 μm.

6.1.2 Arrays of Photorefractive Soliton Waveguides

The parallel nature of optical information processing suggests the extension of soliton waveguides to two dimensions. If the number of solitons assembled together in parallel is very large, *soliton lattices* can be formed, which allow parallel waveguiding and more complex features similar to those of photonic crystals. However, up to now only a few publications have studied the parallel propagation of several spatial solitons [12, 13]. A crucial point in the parallel propagation of photorefractive spatial solitons is their anisotropic mutual interaction as described in Chap. 5. Because the refractive index modulation induced by each individual soliton reaches beyond its effective waveguide, phase-dependent coherent and separation-dependent incoherent interactions such as repulsion, attraction and fusion naturally appear, as discussed extensively in the last chapter. These interaction effects also affect the waveguiding

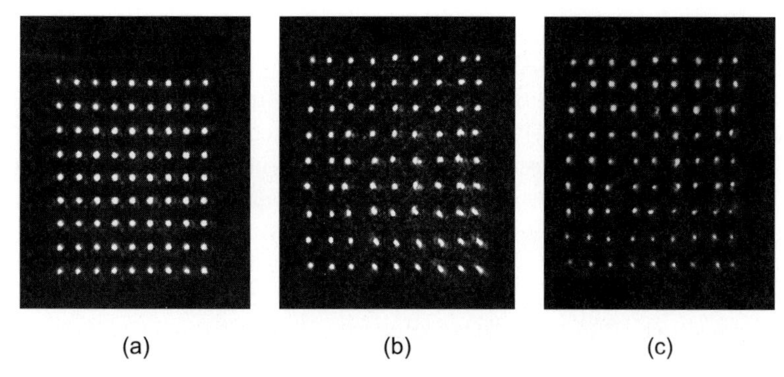

Fig. 6.3. Realization of a 9×9 waveguide array induced by photorefractive solitons. (**a**) Front face of the crystal with the incident spot array (distance between spots $\Delta x = 100$ μm, $\Delta y = 120$ μm; intensity in each channel 110 nW; diameter of each spot 15 μm). (**b**) Array of 81 focused solitons, (**c**) Waveguiding in separate channels of the array, using a probe beam at $\lambda = 633$ nm

properties of a soliton channel. Therefore, the parallel propagation of a multitude of solitons can be achieved only if the separation is carefully chosen in such a way as to prevent all forms of interaction.

To create an array of solitons in a photorefractive crystal, the typical setup for soliton formation (see Fig. 4.15) was modified in such a way that the laser beam derived from a frequency-doubled Nd:YAG laser emitting at $\lambda =$ 532 nm, illuminated a spatial light modulator, which imprinted the image of a spot array onto the beam. The spatial light modulator was imaged, by passing the beam through a set of lenses, onto the front face of a photorefractive SBN60:Ce crystal. For the creation of a soliton array and its waveguiding properties, a regular pattern of 9×9 spots, each with a diameter of about 15 μm and a power of $\approx$ 110 nW, was imaged onto the front face of the crystal (Fig. 6.3a).

In the linear case, without an applied electric field, the beams diffract on their way through the crystal and display a typical interference pattern. When an external electric field is applied, self-focusing in this now nonlinear material forms an array of solitons without interaction (Fig. 6.3b). To obtain propagation without mutual interaction, we took care that the initial distance between individual channels was just large enough to prevent soliton interactions. Therefore, the initial horizontal and vertical separations were chosen to be $\Delta x = 100$ μm and $\Delta y = 120$ μm in order to exceed the range of the various transverse interaction effects.

A separation smaller than the critical distance for coherent interactions between the solitons would cause the solitons to fuse owing to their mutual attractive force. The slight enlargement of the array in Fig. 6.3b compared with the spot array at the front face is due to the imaging optics, while the slight deviation from symmetry is due to inhomogeneity of the crystal.

To test for the waveguide properties of the individual channels of such a soliton-based array, the wavelength selectivity was exploited to scan the array with an intense red probe beam without significantly destroying the individual soliton channels. When we positioned the red probe beam successively on the positions of the previously induced solitons on the front face of the crystal, we found the probe beam to be guided only in the selected channel, for each of the 81 channels. The result of the scan of this array with the red probe beam is shown in Fig. 6.3c. To obtain information about the complete array, every individual channel is scanned separately, and the 81 individual images are superimposed electronically.

6.1.3 Soliton Lattices for Information Processing

Naturally, larger lattices of solitons can also be formed. The number of solitons in such lattices is mainly limited by the aperture of the photorefractive crystal and the resolution of the spatial light modulator that is used for producing the solitons. In larger lattices, numerous applications of parallel soliton and waveguide formation can be imagined. As an example, a digitized image consists of large array of pixel-like spots arranged in a square lattice. In the linear regime, such an image can be reconstructed only within a range that is limited by the depth of focus of the imaging optics, owing to the blurring effects of diffraction. However, if the array of spots propagates in a nonlinear, solitonic regime, we are able to enlarge the range of focus depth to the length of the soliton, i.e. to the length of the nonlinear crystal [12].

The formation of larger soliton patterns is not in principle limited to symmetric arrays. Consequently, reconstructions of selective array configurations that correspond to images can be processed. Figure 6.4a shows a configuration corresponding to the letters "AP" at the front face of the crystal, Fig. 6.4b illustrates the corresponding blurred image resulting from the linear diffraction when the external voltage is switched off, and Fig. 6.4c depicts the reconstructed image when the voltage is applied.

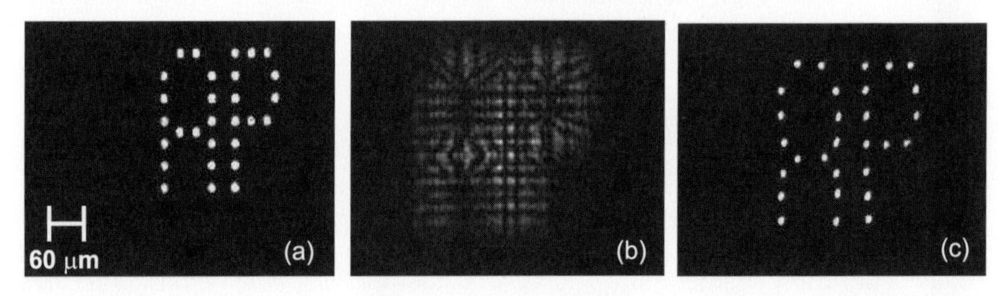

Fig. 6.4. Application of a soliton waveguide array to focus enlargement of the depth in an image. (**a**) Front face of the crystal, where the spot array is imaged; (**b**) interference pattern due to linear propagation of the beams; (**c**) array of solitons

6.2 Soliton-Induced Couplers and Junctions

In the previous sections we demonstrated that the two separate effects of waveguiding and scalar interactions of spatial solitons (see Chap. 5) may lead to fascinating and rather complicated light configurations. Combining these effects leads to a demonstration that waveguiding in interacting solitons can be exploited to design a whole variety of various waveguide couplers and dividers.

Here, we restrict ourselves to a demonstration of the capability to guide two incoherent interacting solitons. Naturally, many other configurations, e.g. coherent soliton birth or annihilation, can be used in a similar way for waveguiding purposes. In this incoherent case, we showed in Sect. 5.5 that two different cases of soliton interaction can be distinguished. Depending on the initial launching conditions (the distance between the solitons and their directions), the two simultaneously propagating solitons can display attraction, separation, rotation or even a spiraling motion, accompanied by an exchange of energy (see e.g. Sect. 5.5.5). In situations in which solitons propagate parallel to each other, where attraction, repulsion or rotation can be expected, a soliton which guides a signal wave can be moved to another position by launching a second soliton, which serves as a controlling beam. This function is the basis of every adaptive all-optical switch. In order to demonstrate such a scenario, two beams G1 and G2 each with an intensity of $I \approx 30$ mW/cm^2, were focused onto the front face of our photorefractive SBN crystal. Their initial diameter was about $d_x \approx 15$ µm in the x direction and $d_y \approx 5$ µm in the y direction. When self-focusing of the beams started, two solitons formed and interacted strongly and thereby their propagation plane rotated through an angle of about 60°.

By coupling a He–Ne laser probe beam R ($\lambda = 633$ nm) into either one of the soliton channels, the probe beam could be shown to be guided only in that particular soliton, without energy coupling into the other soliton. An example of such a situation is shown in Figs. 6.5a,b, where the probe beam is guided in the solitons G1 and G2 respectively. Similarly to Fig. 6.1, the probe beam appears laterally shifted with respect to the writing beams G1 and G2, owing to the wavelength-selective prism in front of the camera. Even though the beams do not propagate in a perfect solitary form, the probe beam R is guided correctly in the structure written by the beam G1 or G2. To use this situation as an all-optical switch, beam G2 could be used as a steering beam for the beam G1, guiding the signal beam R.

In the case of interaction with mutual exchange of energy, waveguiding is expected to result in the splitting of a single beam into several soliton channels. To investigate this possibility, the two interacting beams were launched into the photorefractive crystalin a skewed way , thereby inducing an angular momentum comparable to that in the scenario of spiraling scalar solitons, as described in Sect. 5.5.5. For this purpose, the two writing beams G1 and G2 were launched with a lateral spacing of approximately one spot diam-

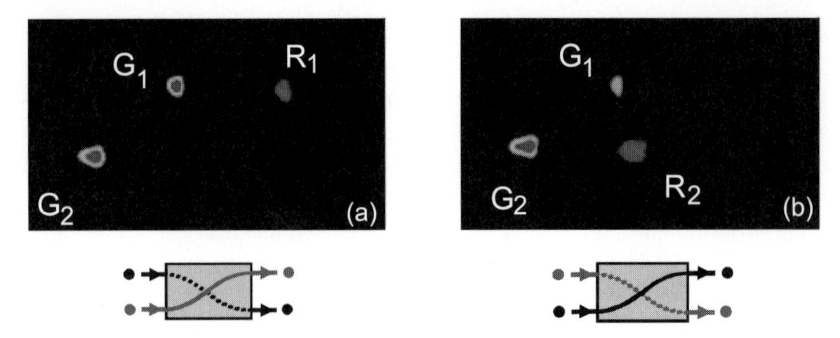

Fig. 6.5. Independent guidance of a beam in two interacting solitons in a configuration that induced a rotation of the solitons relative to each other. The figures show the guidance of the probe beam R in (**a**) soliton G1 and (**b**) soliton G2

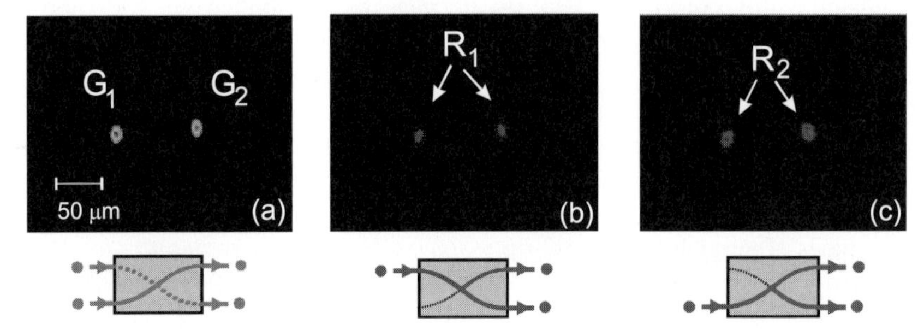

Fig. 6.6. Division of a probe beam R due to the incoherent interaction of two solitons G1, G2. (**a**) exit face of the crystal when the solitons G1 and G2 perform a spiraling motion inside the bulk material. The red probe beam R splits into two channels when it is coupled into the soliton channel G1 (**b**) or G2 (**c**)

eter ($d_x \approx 15$ μm) and at a relative angle of approximately 0.1° towards each other. The mutual exchange of energy of the two solitons could easily be proved by blocking one of the beams, which resulted in a light intensity which was still apparent in both channels at the back surface of the crystal. Owing to the persistence of the waveguides once written in our sample, the probe beam could be guided easily while the writing beams were blocked. We illustrate the back face of the crystal for such a configuration in Fig. 6.6a. When the probe beam R was directed into one of the soliton channels at the front face of the crystal, we found the intensity of the probe beam to be guided in both channels. This situation is shown in Fig. 6.6b,c, where we coupled the probe beam into the soliton channels G1 and G2, respectively. In both cases the principal feature of the division of the probe beam are obvious. Owing to the complex interaction of the beams on their way through the bulk material, the solitons lose their identity in favor of a multisoliton complex, which leads to crosstalk in the waveguiding structure. Therefore, by exploiting the features of such energy-exchanging soliton interaction, an effective Y-coupler can be created.

6.2.1 Interaction in Waveguide Arrays

With the experimental realizations described in the previous sections – namely, the realization of soliton arrays and soliton interactions for waveguiding purposes – we have available all the tools that we need to realize steerable interconnections of waveguide arrays. In order to demonstrate an example of the various types of interaction, we have demonstrated a controlled interaction between two distinct channels of a 3×3 soliton array. To manipulate the mutual attractive interaction between coherent spatial solitons, a third beam positioned between two adjoining channels, was used to steer distinct spots of the soliton array. To do this, a separate beam derived from the Nd:YAG laser was focused appropriately onto the front face of the crystal. While, in this experiment, each soliton of the array had an intensity of $I \approx 55$ mW/cm^2, the additional steering beam was about three times stronger. In Fig. 6.7a, the front face of the crystal is shown with the uncontrolled array, with a lateral distance between the individual spots of $\Delta x = 50$ µm, $\Delta y = 70$ µm and a spot diameter of 12 µm (FWHM). Figure 6.7b displays the undisturbed array of solitons after propagation for 20 mm in the biased crystal. Once the control beam is positioned between the central lower two solitons, as depicted by the sketch in Fig. 6.7c, a new soliton array forms, which leads to the attraction and subsequent fusion of the central and the lower central spot. Figure 6.7d shows the probe beam at $\lambda = 633$ nm, guided in each channel of the controlled array separately (again, the individual snapshots have been superimposed electronically). The fusion of the two lower middle channels, creating a Y-coupler within the waveguide arraym, is obvious. Now coupling the probe beam into either the central or the lower central channel on the front face of the crystal leads to effective guiding into the same output.

In principle, the fusion of neighboring channels in the direction of the electric field is also possible. In this case, however, the distance between the

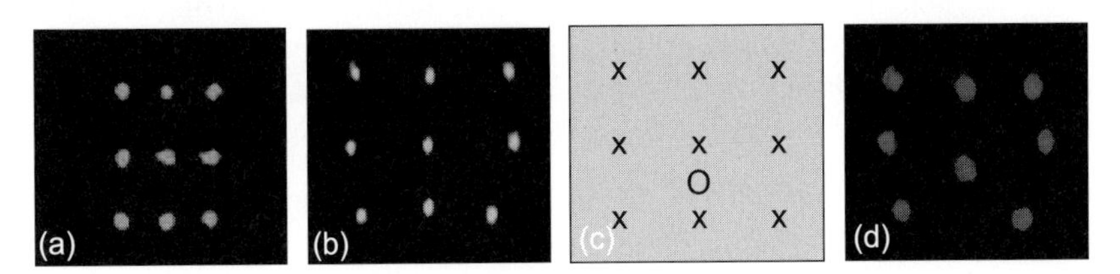

Fig. 6.7. Optical control of a soliton array. (**a**) The spot array at the entrance face of the nonlinear crystal (distance between spots $\Delta x = 50$ µm, $\Delta y = 70$ µm; diameter of each spot 12 µm), (**b**) the uncontrolled soliton array, (**c**) sketch of the position of the steering soliton beam, (**d**) the controlled array probed by a beam at $\lambda = 633$ nm

control beam and each of the channels has to be smaller than that required in the y direction in order to compensate for the different range of the coherent interaction.

These examples show clearly that all-optical control of individual channels in soliton arrays can be achieved by inducing a controlled interaction within the waveguide array. Although these initial experiments focused on coherent interactions, the case of mutually incoherent beams is even more promising, as it allows one to realize rotations and energy exchanges between specific spots of large arrays. With these tools, it is possible to realize all-optical, adaptive processing elements that cover all scenarios that may arise in complex optical interconnection tasks. In order to give some insight into the potential of these ideas, the following sections describe in a general way some of the possible tasks of soliton waveguides in optical information processing.

6.3 Applications of Soliton Interaction

A natural application of self-induced solitary waveguides – as indicated by their name – is the possibility to guide waves. However, the various scenarios of interaction also allow one to realize more complex waveguide structures, which may form combiners, dividers and other functional components for complex waveguide interconnects. In addition to this rather classical exploitation of the waveguiding properties of spatial solitons, it is possible to exploit the interaction for manipulating and controlling solitons. Manipulating a soliton with another one or manipulating two interacting solitons by changing the initial conditions is an ideal field for the possible realization of all-optical beam guiding. If this can be done, the interaction of solitons can be applied for all-optical switching and logic gates. A concise overview of the exploitation of soliton interconnects, all-optical soliton switches and soliton logic can be found in [14].

For simplicity, we restrict ourselves to the potential applications of two interacting photorefractive solitons. This system may serve as the basic unit of a switching or logic gate cell. The extension to much larger lattices of interacting solitons , based on the same basic principles, is straightforward.

Almost every interaction scenario is dominated by the initial conditions and can be controlled by adjusting those initial conditions. Therefore, the experimentalist has to control the following parameters:

- the polarization or degree of incoherence between the two beams,
- the mutual coherence and relative phase,
- the intensity ratio of background to signal beam,
- the positions of the beams, and
- the transverse angular orientation of the two beams.

The first condition, the polarization of the two beams or their relative coherence, strongly influences the generic behavior of the interaction. If the beams are incoherent, only their intensities interact, irrespective of the phase of the two beams. In contrast, the second condition leads us to distinguish between in-phase and out-of-phase interaction. Because we know that in-phase solitons at a certain separation may attract, whereas incoherent solitons may attract at one separation and repel at another, the coherence can determine the direction of the force between two solitons. This can also be achieved by switching the phase between the two beams to an out-of-phase condition.

With the above considerations in mind, a second important initial condition can be exploited to submodulate both cases (coherent and incoherent): the geometry. This describes the relative transverse spacing and the angular orientation; the total power of the two beams is closely connected to it. These are all continuous variables, but we have seen in the preceding sections that distinctively different behaviors occur for different ranges of the initial geometrical parameters. These different behaviors can be used to classify the variety of interactions into a relatively small number of different classes with notably different features.

Of the broad range of possible interaction scenarios, several initial geometries are obviously not useful for switching applications. Geometries in which the solitons are separated and initially move away from each other will lead to only minimal nonlinear forces on the solitons, resulting in essentially no switching. Conversely, if the two solitons are separated and initially propagate towards one another, interaction is most probable. Moreover, with a second component of the angular deviation orthogonal to the line of separation, an attractive nonlinear force may result in one of the complex interaction scenarios of soliton spiraling presented in the previous chapter. Although it might be possible to use the resulting tumbling motions to construct a switch or a logic gate, the complexities of the interaction, the motion and the sensitivity to external parameters would make a reliable implementation quite difficult.

6.3.1 Application Potential of Various Interaction Scenarios

These arguments leave four main types of two-dimensional interaction as candidates for a soliton–soliton switch: attraction and repulsion; collisions involving bound states; rotating or spiraling states, provided they can be handled well enough for practical applications, and guiding and breathing.

- In soliton *attraction and repulsion*, the two solitons are initially separated and propagate in a collinear way, at the same angle. We already know that if the two solitons are coherent and in phase, they attract, and if they are out of phase by π they repel. Incoherent solitons attract or repel depending on their separation in the direction of the applied electric field. The repulsive force can form the basis of a switch or logic gate. In a spatial-soliton repulsion gate, two spatial solitons are launched in the same direction, separated by more than a beam width and out of phase by π. The repulsive force on the two solitons creates a change in their angles of propagation.

After a sufficient distance, the presence or absence of one soliton can be detected by the means of position of the other beam, thus forming the basis for a logic decision.

- If the two solitons are initially separated but directed towards each another, they will collide at some point within the nonlinear medium. In anisotropic crystals such as photorefractive ones, these *collisions* can also be achieved by using the anisotropic interaction. This collision interaction is a result of the fact that the interaction is not perfectly elastic in a saturable nonlinear medium. Therefore, an energy exchange can occur during interaction, which gives rise to a nonideal return to the initial propagation directions after the collision. For large-angle collisions, this deviation may be too small to be detected, resulting in a quasi-elastic interaction. However, for small angles, collision may result in severe changes, such as fusion or repulsion. Therefore, by changing the angle of interaction, it is possible to vary the degree of effectiveness of the interaction.
  For two-dimensional solitons, the two degrees of freedom to be adjusted are given by the incidence angles in the two transverse directions. Therefore, the behavior of the collision in this case depends sensitively on the angles of incidence and may result in complex rotational or spiraling motions. If the angular positions of the incident beams can be adequately controlled, these interactions can be used in any large-scale optical circuit.
  The effects of intensity and phase changes can also be combined to be exploited for soliton guidance applications. Consider two identical in-phase solitons fusing into a single beam. By slightly changing the amplitude of one beam, one can change the relative phase at the collision point, and so the single beam may be replaced by two diverging ones. This situation has been exploited, as described in Sect. 5.4, to realize soliton birth and annihilation. One of the beams may be considered as a control beam, and small changes in this control beam can be used to switch between a single and a dual output. Logical AND gates, OR gates and exclusive OR gates could easily be realized.
- If the two solitons are both coincident and collinear, guidance will become possible as long as the solitons do not interact. This can be accomplished if one beam at a certain wavelength induces the nonlinear waveguide, whereas the second beam has a wavelength at which the nonlinear material is insensitive and does not act in a nonlinear way. A broad range of options is opened up when we consider the possibility of *guidance* of a signal beam in one or both of the input (pump) solitons. This beam may be of similar or completely different wavelength and power, thus allowing to create adaptive waveguides with guiding properties that may be switched by light [8].
- The final type of soliton interaction occurs when the two solitons are initially coincident but propagate with different transverse propagation directions, i.e. they are directed into the material at different spatial angles. This *dragging* interaction – sometimes also referred to as trapping [15]

can be viewed as a modification of the collision interaction: the collision itself takes place in a linear medium where no interaction takes place, and the nonlinear effects occur only from the collision point onwards.

If the initial transverse inclination or the energy difference is so large that the two solitons cannot form a bound state, the asymmetry of the interaction – in comparison with symmetric collision at large angles – will still cause the two solitons to emerge with a permanent angular deviation. This *deflection* operation is similar to the trapping geometry except that the total angular shift is reduced.

The dragging interaction has a number of advantages over the use of attraction–repulsion and symmetric collision devices. First, since the solitons emerge from the initial collision point with a permanent angular shift, it is possible to create much larger spacial shifts than those induced by symmetric collision. Unlike any of the previously described interactions, it is also easy to switch an intense pump soliton with a weaker signal soliton, resulting in logical signal restoration (with gain) – an essential feature for any cascadable logic device. This combination of high contrast and high gain is often difficult to obtain in nonlinear switching devices that utilize diffracting beams, owing to the inherent trade-off between beam size (intensity) and confocal distance (interaction length). Solitons overcome this trade-off and make tightly confined beams, that can interact over long distances possible [14].

6.3.2 Criteria for Soliton Information Processing

The theoretical, experimental and numerical studies on soliton interaction and waveguiding discussed above reveal the wealth of physical interactions that could be used to construct practical devices. In many of the possibilities discussed above, the motivation of the work has been the construction of digital logic. It is well known among designers of electronic digital logic, however, that the simple ability to switch one signal with another is not sufficient to construct large-scale logic circuits, but that other, more application-oriented features are just as important. This is illustrated by the fact that electronic logic is still prevalent, whereas, despite numerous advantages and a significant research effort, optical logic is not. By examining the features of the electronic circuits that have become such a success, one can find the essential properties that a proposed optical logic system must possess if it is to make its way to real applications. The most important of these properties are listed in the following, summarized from a detailed analysis by McLeod et al., which can be found in its entirety in [14].

- **Logical completeness**. To be generally useful, the individual logic gates should be able to be interconnected so as to implement any possible logical function. In their simplest form, most soliton logic gates are inverters – the pump is passed through (output equals one) only if the signal is not present to deflect it (input equals zero). The ability to construct a NOT gate is important because all complete logic families are able to perform this operation. NOR gates form such a logically complete set and can be constructed from soliton interactions by use of a cascade in which multiple, successive signal solitons can interact with a single common pump soliton. In other words, the pump soliton is passed through a series of gates, the presence of a signal in any one of which will deflect the pump and cause it to be blocked, producing the correct low output of a NOR operation.

- **Three-port devices**. Logically complete gates can be combined in a reliable way into circuits only if they are true three-port devices. That is, photons either enter the gate (port one) and exit the output (port two), controlled by the input signal (port three). The inputs must be isolated from the output so that processing proceeds in only one direction, and the output must be standardized to one of two binary levels, independent of all variations of the digital value of the input signal. This can be accomplished in soliton logic gates by supplying a pump soliton from the power-supply laser at each gate, such that the pump soliton is passed on as the output only if no signal soliton is present and is blocked otherwise. Thus the position, direction of propagation, energy, polarization, phase and wavelength of the output soliton are restored by the power supply at each gate. In addition, the output of one gate is used as the input of a subsequent gate where these signal photons drag the pump and are then discarded.
- **Thresholding**. In a digital circuit – as opposed to an analog one – signals entering a gate must be classified into their discrete digital values by thresholding the analog input. The nature of soliton propagation provides a natural thresholding operation – a beam can only propagate as a soliton if it fulfills the criteria for soliton stability, e.g. the critical soliton parameters, otherwise it rapidly diffracts. This digital nature of soliton propagation makes solitons natural carriers of binary information.
- **Cascadability**. Logic devices should be able to use the output of one gate as the input to another one. This is possible as long as the solitons remain in the same state during their interaction. Spatial solitons that change their polarization, however, need a means to alternate the polarizations of the pump and the signal soliton at each level. In general, however, most soliton logic gates are cascadable, but they suffer from a more common limit to cascadability – the lack of sufficient gain for multiple cascaded systems.
- **Gain**. Gain is the ability of a gate to produce outputs that have more energy than the inputs. If a logic device has no gain, the output levels must always be less than those of the inputs. Eventually, the output of a subsequent gate will have insufficient energy to switch the next. In optical systems, losses are mostly due to surface deflections, absorption and diffraction, making this point particularly significant when digital optical circuits are being designed. Gain is also essential for cascaded systems, since the output of any gate is often required to drive the inputs of more than one

subsequent gate – a feature that is known as fan-out. Only soliton-dragging gates or gates that use coherent interaction scenarios with energy exchange can achieve large gain and thus be truly cascadable – a fact that conflicts with the difficulties of controlling the phase in soliton interactions. However, one can still achieve gain, because a small signal can drag a larger pump, and this applies also in the case of incoherent interaction, where beam coupling cannot provide gain.

– **Parallelism**. One of the major achievements of modern electronics was the transition from bulky individual transistors to densely packed very large-scale integrated (VLSI) circuits in which thousands of gates operate simultaneously in a small area. Similarly, any optical logic technology which cannot be fabricated to operate many gates with a high density in a small area will not be competitive. Parallelism can easily be achieved with optical spatial solitons, by exploiting the dimensions transverse to the direction of propagation. We have shown in this chapter that two-dimensional waveguide devices can be realized in a highly parallel way and that they can be efficiently used to switch or combine signals.
– **Speed**. High-speed operation is probably the most important feature of a device technology and is the primary reason for investigating soliton logic. In the case of the spatial screening solitons in photorefractive SBN that we have investigated up to now, this feature is not achieved, except for the case of waveguiding another wavelength in channels that have already been built and fixed. In the latter case, however, the adaptivity of the system is lost. Therefore, for high-speed operation that compares well with VLSI circuits and similar technologies, other materials such as semiconductors are advantageous. Also, spatial solitons in second-order nonlinear materials, which are generated by second-harmonic generation have been predicted and may be wellsuited for applications in terms of their processing speed.
– **Phase insensitivity**. A requirement unique to the design of optical logic is that the operation of the gate must be independent of the phases of its optical inputs. Otherwise, the circuit will not be robust in a large system, which will expand and contract owing to heat and vibration and be subject to production tolerances. In a general isotropic medium, soliton repulsion gates cannot be phase-insensitive in a positive-Δn material, but collision, attraction and dragging gates can be made phase-insensitive by using orthogonal polarizations or incoherent beams. In this case, the anisotropy of photorefractive screening solitons becomes vary important: owing to the anisotropy, the anomalous interaction in the direction of the applied electric field also allows incoherent solitons to repel for certain separation distances. All other features, such as collision, attraction and dragging remain the same as in isotropic media. Therefore, photorefractive solitons are ideal candidates for incoherent operations.
– **Low power consumption**. Realistic logic circuits must contain many individual logic gates, each of which must dissipate a very small amount of

energy, both to reduce the power requirement on the source laser and to limit the heat generation in the circuit. Solitons which are fully confined, or completely formed by nonlinear effects, can be energy efficient. However, filamentary structures and processes of soliton interaction in saturable media which include emission of radiation consume large amounts of energy. Decreasing the energy in a soliton reduces the heat generated when this soliton is dissipated in a logical interaction. Therefore, solitons that may be generated at very low laser power, such as photorefractive solitons, are highly advantageous.

It is obvious from all these requirements, that only very few combinations of solitons and interactions are actually candidates for large-scale digital optical logic. Two-dimensional photorefractive solitons are attractive with respect to almost all of these points, except the question of speed, which is essential for most applications. However, in all applications where parallelism and adaptivity are more important, photorefractive solitons may be competitive alternatives to electronic processing devices. Therefore, these solitons may provide in the near future the first practical proof-of-principle test bed for soliton optical-logic devices – this will form the basis of the new field of *soliton-driven photonics*.

Summary

A light beam incident on a photorefractive material induces a refractive index waveguide which is capable of guiding the beam itself as an eigenmode. Moreover, photorefractive spatial solitons are able to guide not only their basic induced mode but also other beams, owing to cross-phase modulation effects. Therefore, a single beam basically creates a waveguiding structure in which other beams of different power, wavelength or transverse geometry can be guided. By adding interaction between solitons, the optical control of one beam by another one can be made possible, leading to junctions, couplers and dividers. Moreover, if the parallelism of optics is exploited, large configurations of soliton lattices can be realized, which allow one to create complex waveguide patterns. These novel ideas have an immense potential for applications in all-optical switching and waveguide coupling, which we have evaluated by discussing the most important criteria for soliton information processing.

References

1. N.S. Kapany and J.J. Burke. *Optical Waveguides.* Academic Press, New York, 1972.
2. A.W. Snyder and A.P. Sheppard, *Collisions, steering and guidance with spatial solitons*, Opt. Lett. **18**, 482 (1993).

3. D.J. Mitchell, A.W. Snyder and L. Poladian, *Interacting self-guided beams viewed as particles: Lorentz force derivation*, Phys. Rev. Lett. **77**, 271 (1996).
4. R. de la Fuente, A. Barthelemy and C. Froehly, *Spatial soliton-induced guided waves in a homogeneous nonlinear Kerr medium*, Opt. Lett. **16**, 793 (1991).
5. B. Luther-Davies and X. Yang, *Waveguides and Y junctions formed in bulk media by using dark spatial solitons*, Opt. Lett. **17**, 1755 (1992).
6. M. Morin, G. Duree, G. Salamo and M. Segev, *Waveguides formed by quasi-steady-state photorefractive spatial solitons*, Opt. Lett. **20**, 2066 (1995).
7. M. Shih, M. Segev and G. Salamo, *Circular waveguides induced by two-dimensional bright steady-state photorefractive spatial screening solitons*, Opt. Lett. **21**, 931 (1996).
8. J. Petter and C. Denz, *Guiding and dividing waves with photorefractive solitons*, Opt. Commun. **188**, 55 (2001).
9. J. Petter, C. Denz, A. Stepken and F. Kaiser, *Anisotropic waveguides induced by photorefractive (2+1)d solitons*, J. Opt. Soc. Am. B **19**, 1145 (2002).
10. M. Klotz, H. Meng, G.J Salamo, M. Segev and S.R. Montgomery, *Fixing the photorefractive soliton*, Opt. Lett. **24**, 77 (1999).
11. M. Shih, Z. Chen, M. Mitchell, M. Segev, H. Lee, R.S. Feigelson and J.P. Wilde, *Waveguides induced by photorefractive screening solitons*, J. Opt. Soc. Am. B **14**, 3091 (1997).
12. A. Bramati, W. Chinaglia, S. Minardi and P. DiTrapani, *Reconstruction of blurred images by controlled formation of spatial solitons*, Opt. Lett. **26**, 1409 (2002).
13. M. Soljačić, S. Sears and M. Segev, *Self-trapping of "necklace" beams in self-focusing Kerr media*, Phys. Rev. Lett. **81**, 4851 (1998).
14. R. McLeod, K. Wagner and S. Blair, *(3+1)-dimensional optical soliton dragging logic*, Phys. Rev. A **52**, 3254 (1995).
15. M.N. Islam, *Ultrafast all-optical logic gates*, Opt. Lett. **14**, 1257 (1989).

7 Vector Solitons in a Photorefractive Medium

Besides providing a suitable nonlinearity for the generation of (2+1)-dimensional self-focused light beams, a saturable nonlinearity generally displays another remarkable and attractive feature. Since the transverse size and the distinct shape of a solitary beam depend strongly on the magnitude of the saturation parameter, a whole variety of solitons with different beam profiles can be generated. In particular, in the high-saturation regime, the transverse size of the solitons increases, and the self-induced waveguide is no longer necessarily a single-mode waveguide. Under certain circumstances, it has a multimodal structure, which gives rise to the generation of rather complex configurations of all-optical and soliton-based waveguide structures. In principle, we have to distinguish two different approaches.

First, a multimode waveguide induced by a single optical beam can serve as a waveguiding device in the linear sense. This waveguide guides several of its modes simultaneously. The guided modes may be of different wavelengths and/or different transverse geometry (e.g. higher-order laser modes). In this case, the guided modes behave passively, which means they do not significantly modulate the refractive index of the nonlinear medium. We have discussed this situation extensively in the last chapter. Guidance and excitation of double- and triple-humped modes (resembling HG_{01} and HG_{02} laser modes) with a different wavelength from the soliton beam that induces the waveguide have been demonstrated. In such a configuration, it is possible to implement all-optical beam-steering and controlling devices.

Second, a multimode waveguide may be induced not by a single beam, but by several optical beams. In this case, the participating modes are not guided in a passive way, but interact with the medium and contribute equally to the formation of the required refractive index profile. This is a substantially different approach compared with the case described in the last chapter, since two or more beams copropagate in a joint refractive index profile generated by themselves. In this case the principle of self-consistency can be applied [1, 2], which gives a very intuitive insight into the soliton formation process.

Using the linear optical model of waveguiding to describe spatial-soliton formation in two dimensions, a two-step process can be identified. First, a light beam which is incident on a self-focusing saturable nonlinear medium induces a focusing refractive index structure. Second, once the waveguide has been formed, it supports the nondiffracting propagation of its eigenmodes. When the incident light beam coincides with an eigenmode of its induced waveguide, the beam itself will be trapped and propagate with an invariant transverse profile. In Chaps. 4 and 5, we described only a single optical beam, which induces a single-mode waveguide structure in which it propagates as a fundamental mode. This is commonly referred to as a *scalar soliton*. In contrast, if the refractive index waveguide is induced by several simultaneously propagating optical beams and therefore represents a multimode waveguide structure, in which all of the beams that take part in creating it can propagate as eigenmodes, the joint light structure is referred to as a *vector* or *multicomponent* soliton.

This type of soliton was suggested by Manakov [3] in the case of a one-dimensional Kerr system and was first realized experimentally in a semiconductor material [4]. In this chapter, we investigate these structures in the (2+1)-dimensional geometry and demonstrate how such sophisticated light structures can be realized, controlled and stabilized. In particular, the question of stabilization of these structures is a very important one for potential applications and will therefore play an important role throughout this chapter.

The most important prerequisite for the generation of this new and fascinating type of optical solitary wave is the absence of any interference between the individual optical beams. Moreover, the self- and cross-phase modulation terms have to be equal for all contributing components. Thus, the refractive index change which is induced by any one beam affects every propagating light component in exactly the same way. The coupling between the solitary beams is due only to the nonlinear response of the medium to the total intensity distribution.

In general, there exist three ways to achieve these requirements. The original suggestion of Manakov was based on two beams with orthogonal states of polarization, from which the expression *vector* soliton was derived. A second approach uses two beams of different wavelength as in the case of all quadratic solitons [5]. Finally, the use of mutually incoherent beams is a very convenient way to realize multicomponent solitary waves, especially in photorefractive systems. Because photorefractive materials are highly wavelength sensitive and the refractive index change due to the electro-optic effect depends on the plane of polarization, the only effective approach to destroying the mutual coherence between the individual components without changing the electro-optic properties is to make the beams incoherent with respect to each other. Experimentally, this can be realized with the same technique as described in Sect. 5.5 for incoherent interactions of scalar solitons. Additionally, the individual constituent beams of the vector solitons are typically of different geometry; this situation can be realized by imprinting the required phase or amplitude modulation on one of the participating beams.

Vector or multicomponent solitons have an enormous potential for future applications, since the individual contributing light components can display different geometries, which gives rise to rather complex scenarios. Moreover, higher-order multicomponent solitons can serve as a model for describing soliton formation with partially spatially incoherent light. Therefore, the self-focusing properties of light beams with a low degree of spatial coherence should be investigated. This in turn motivates further research towards the use of low-cost light sources, such as light-emitting diodes, for all optical control and steering devices for spatial optical solitons.

For the (1+1)-dimensional geometry, incoherently coupled soliton pairs were predicted to exist in photorefractive nonlinear materials in 1996 by Christodoulides et al. [6] and were subsequently demonstrated experimentally by Chen et al. [7]. Typically, one-dimensional beams are unstable in a two-dimensional medium, being subject to a strong modulation instability (see Sect. 4.1.3). However, by using an approach based on a vector solition, Chen et al. could successfully suppress the modulation instability when two mutually incoherent stripe beams copropagated simultaneously. Each beam itself does not propagate in a stable way, owing to its inherent instability. But when both beams propagate simultaneously, the total intensity increases, whereas the degree of spatial coherence decreases. For this reason, the parameters shift towards a region in which the instability is not apparent for the given propagation distance area. Several configurations, such as pairs of bright (1+1)-dimensional photorefractive solitons and bright–dark pairs of such solitons, have been realized by applying this principle of multicomponent solitary waves [7].

In 1998, the first observation of composite multimode solitons in a photorefractive medium was reported by Mitchell et al. [8], who demonstrated the stable propagation of (1+1)-dimensional higher-order-mode solitary beams. In these experiments, the typical transverse spreading of multihumped optical beams could be prevented in the presence of a fundamental stripe beam of Gaussian shape. Multihumped beams, which represent higher-order transverse modes of a waveguide, typically display one or several phase jumps of π across their transverse plane. This, in turn, leads to a mutual repulsion of the individual stripes when they propagate separately in a self-focusing medium. This behavior is, in principle, directly comparable to the coherent interaction described in Sect. 5.4. The authors of [8] demonstrated various stable combinations of such multihumped, multicomponent light structures, consisting of a fundamental and a double-humped or even triple-humped mode. The stability of these solitons was subsequently analyzed by a linear stability analysis [9, 10], which revealed that only the pair made up of a fundamental and a double-humped mode is modulationally stable; in contrast, all higher-order modes of triple-humped type become subject to instability at larger propagation distances, which could not be detected experimentally, owing to the limited length of the crystal sample. As in the case of scalar beams, the

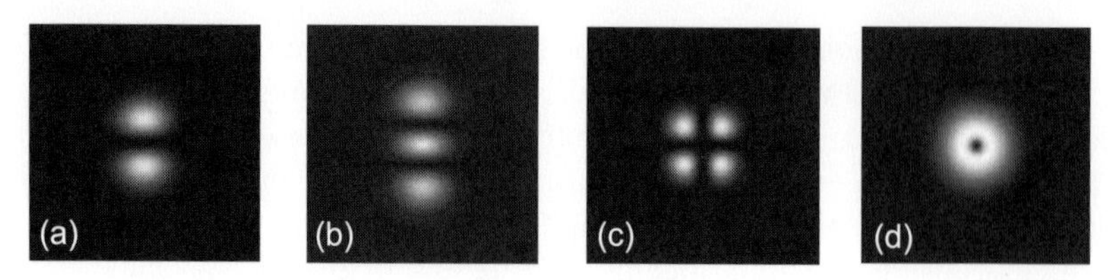

Fig. 7.1. Illustration of some low-order transverse modes. Intensity patterns with rectangular symmetry resembling Hermite-Gaussian HG_{01}(**a**), HG_{02}(**b**), HG_{11}(**c**) modes. Mode of circular symmetry with a doughnut shape(**d**)

expansion from a one-dimensional to a two-dimensional transverse system of vector solitons is a nontrivial task. In contrast to the one-dimensional system, the realization of multihumped beams in two dimensions promises a manifold of possible configurations because of the additional degree of freedom. Higher-order mode beams can display either a rectangular or a circular symmetry, similar to the transverse laser modes of Hermite–Gaussian or doughnut type, respectively. Some of these modes are illustrated in Fig. 7.1. Modes with radial symmetry (Fig. 7.1d) inherently contain a phase singularity, which is analogous to a topological charge. A vector soliton consisting of a fundamental bell-shaped mode and a higher-order mode of doughnut type has been predicted to exist in the case of an isotropic threshold nonlinearity [11, 12], but has been shown to be modulationally unstable in a medium with a saturable nonlinearity [13]. This numerically found instability shows a distinct breakup behavior that leads to the generation of a *dipole-mode vector soliton*, which has been the subject of numerous investigations [13, 14, 15, 16, 17, 18].

7.1 Vector Solitons of Rectangular Symmetry

7.1.1 The Dipole-Mode Vector Soliton

Probably the simplest configuration for generating a vector soliton in both transverse dimensions consists of a dipole-like mode in combination with a fundamental bell-shaped Gaussian beam. The dipole mode depicted in Fig. 7.1a displays a phase shift of π across its transverse plane, and therefore a dark notch that separates the two bright lobes appears in its center. The two bright lobes can in turn be regarded as scalar beams that are phase-shifted by π relative to each other. In a self-focusing medium, they interact in exactly the same way as mutually coherent solitons that are out of phase and propagate in close proximity, and hence they repel strongly (see Sect. 5.4). Now, the question arises of whether the spatial spreading of the two dipole lobes can be prevented in the presence of the second, fundamental mode. In this case both, the Gaussian and the dipole beam would become eigenmodes of their jointly induced multimode waveguide, in which they would both propagate without significantly changing their transverse profile. For a detailed investigation of this question, we have made use of the numerical beam propagation procedure in the isotropic and in the anisotropic approximations.

In order to examine such multicomponent structures experimentally, one can use the principle of the setup used for studies of two-beam interaction shown in Fig. 5.4. One beam serves as the fundamental component, which is effectively made incoherent by the oscillating piezoelectric device. The second beam then represents one of the modes, illustrated in Fig. 7.1. One of the main experimental tasks consists of the generation of such light structures with rather complex geometry. This can be achieved by several different techniques.

Modes with a rectangular symmetry (Fig. 7.1a–c) are typically generated by imprinting one or several phase jumps of π on a Gaussian beam. This phase-imprinting process can be performed most easily by partially inserting a thin glass slide into the optical beam. Owing to the difference between the refractive indices of glass and air, the two parts of the beam experience different optical path lengths leading to a phase shift $\Delta\varphi$, which can be adjusted exactly to π by tilting the glass slide appropriately. Another, but rather challenging technique consists of forming a dipole mode beam from two closely and parallely propagating Gaussian beams, whose relative phase is controlled manually by a piezomechanical device. Here, although the mechanical adjustment requirements present a problem, we have neither absorption losses nor diffraction from the glass plate. A third technique makes use of computer-generated phase holograms. Here, we litographically imprint a computer-calculated interference pattern of any desired phase and amplitude distribution with a plane wave onto a glass substrate. Illumination with a Gaussian beam then leads to the formation of the desired light structure in the first diffraction order. By use of the latter technique, we were finally able to generate various circularly symmetric optical vortices with topological charges $m = 1, 2, 3$ (Fig. 7.1d).

In Fig. 7.2, we present our numerical and our experimental results with respect to the formation process of a multicomponent light structure consisting of a fundamental (Gaussian) and a first-order (dipole) component. Figs. 7.2a,f depict the incident intensity distributions of the dipole and Gaussian beams of equal total power, in an isotropic numerical simulation. When the dipole-mode beam propagates separately in the nonlinear self-focusing medium it behaves as two out-of-phase scalar beams that repel strongly during propagation. Thereby, their separation increases greatly after a propagation distance $z = 6.95L_{\mathrm{d}}$, as depicted in Fig. 7.2b. The Gaussian component. in contrast, forms an ordinary isotropic (and circular) soliton when it propagates separately in the crystal (Fig. 7.2g). The situation changes dramatically when both beams copropagate simultaneously, which is illustrated in Figs. 7.2c,h. The previous natural separation of the dipole lobes is inhibited (Fig. 7.2c), whereas the Gaussian component becomes elongated in the direction of the

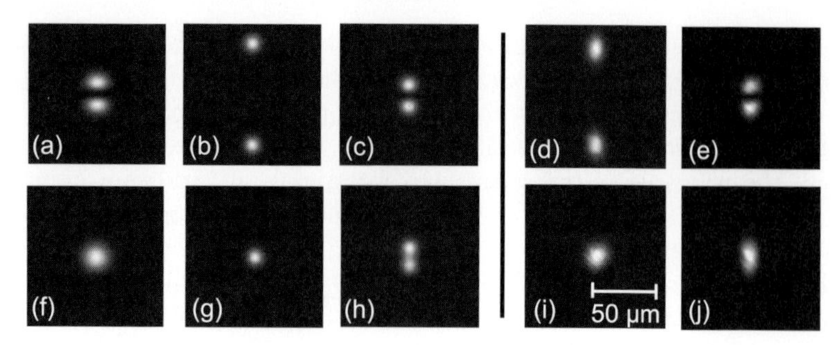

Fig. 7.2. Numerical (*left*) and experimental (*right*) generation of a dipole-mode vector soliton. The dipole and fundamental components are depicted in the *top* and *bottom* rows, respectively. (**a**),(**f**) Incident light intensity. The diverging beamlets are shown for the dipole component after propagation on its own of ($z = 6.95\ L_{\rm d}$(**b**), or $z = 20$ mm)(**d**). Simultaneous propagation leads to a self-confined light structure shown in (**c**), (**h**) and (**e**), (**j**). The repulsion of the dipole lobes (**a**), (**e**) is prevented in the presence of the Gaussian beam

dipole axis and adopts a slightly double-humped shape (Fig. 7.2h). This observation demonstrates a generic feature of unstable structures that copropagate incoherently with a stable soliton beam: both beams are now constituents of a stable, very robust soliton, the *dipole-mode vector soliton* [13]. The experimental results are illustrated on the right-hand side of Fig. 7.2. The frames are arranged to correspond to the numerical simulations. In all four frames, the exit face of the crystal is shown, for the separate propagation of the components (Figs. 7.2d,i) and for simultaneous (Figs. 7.2e,j) propagation. These experimental results are in excellent agreement with the predictions of the isotropic simulations and constitute the first experimental observation of this novel type of multicomponent soliton, i.e. the dipole-mode vector soliton. This type of soliton has been reported independently by two groups [14, 15].

We would like to emphasize that this phenomenon cannot be understood in terms of a linear waveguide theory, since both beams contribute equally to the nonlinear index change. The Gaussian component of the dipole-mode vector soliton not only serves as a stabilizing component but also is affected by the dipole mode. As a consequence, it becomes elongated in the direction of the dipole mode. Note that the numerical data based on the isotropic approach match the experimental situation, which is inherently anisotropic, in a satisfactory way. This is due to the fact that the dipole is oriented along the vertical y axis, in which the anisotropic refractive index modulation is somewhat bell-shaped.

The stability and transverse shape of the dipole-mode vector soliton are extensively investigated in [16], where we have made use of a two-beam expansion based on the Petviashvili approach (see Sect. 4.2.1). Here, we explored the parameter space for numerically exact soliton solutions of the anisotropic photorefractive model. For a given external voltage, the relative beam power

$P_{\mathrm{Gauss}}/P_{\mathrm{Dipole}}$ can range from almost infinity to 0.3, which in turn leads to a Gaussian- or dipole-dominated shape of the resulting vector soliton. In the limit of a vanishing dipole beam, the entire complex can be considered as a scalar soliton that is derived from a Gaussian beam.

In the other limit, the separation of the two dipole lobes increases as the power of the Gaussian beam decreases; at the same time, the Gaussian beam becomes more and more elongated in the vertical direction and even adopts the dipole structure. We were able to demonstrate these specific characteristics experimentally and found excellent agreement with the numerical parameters. In the limit of a vanishing fundamental mode, the specific properties of the photorefractive response lead to the formation of a bound dipole soliton, which has been investigated by Mamaev et al. [19].

7.1.2 Aspects of the Anisotropy

The generation of a dipole-mode vector soliton illustrated above is equivalent to the isotropic case, since the two lobes of the dipole mode are aligned perpendicular to the external electric field. They experience repulsion only because of their relative phase shift, and not because the distinct effects of anisotropy governed by the characteristics of the photorefractive nonlinearity. In this case the isotropic model (see Sect. 3.4.1) is sufficient to describe the soliton formation process in this inherently anisotropic system.

The situation changes dramatically when this vertical orientation of the dipole mode is no longer guaranteed. Suppose the dipole beam is launched into the crystal at a certain angle with respect to the vertical axis. From Sect. 5.5.2, we know that in this case the anisotropy of the refractive index modulation reveals itself and transverse rotational effects can occur, as observed in Sect. 5.5.3 for the case of two scalar beams. In this case, the isotropic nonlinear model fails and we have to apply the anisotropic approach in order to simulate the physical processes inside the crystal. In Fig. 7.3, we illustrate the results of a numerical beam propagation for the simultaneous propagation of a dipole-mode and a Gaussian beam of equal total power. Initially the dipole mode is tilted by 30° with respect to the vertical axis (Fig. 7.3a) and subsequent propagation steps are illustrated in the following frames (Figs. 7.3b–d) for equally spaced stages of propagation ($\Delta z = 0.4 L_{\mathrm{d}}$). It is obvious that the dipole mode does not maintain its initial direction. It starts to rotate clockwise, passes through the vertical position in Fig. 7.3c and then reaches a state where its orientation is opposite to the initial orientation, in Fig. 7.3d after a propagation of $1.2 L_{\mathrm{d}}$. As the propagation continues, the direction of rotation reverses, and the dipole mode displays a rather twisting and wobbling motion through the nonlinear material. This behavior is somewhat similar to the complex spiraling motion of two scalar solitons described in Sect. 5.5.5. In both cases, it turns out that the two lobes of the dipole or the two scalar beams, as appropiate, align preferentially along the direction perpendicular to the external electric field. This twisting and wobbling effect

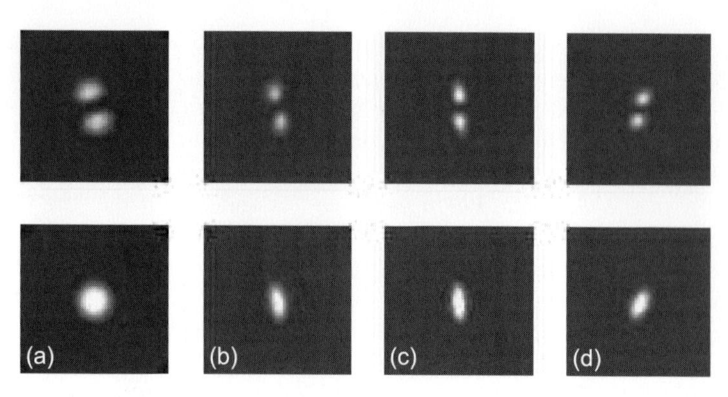

Fig. 7.3. Various stages of propagation of a tilted dipole mode (*top row*) and a Gaussian beam (*bottom row*) for (**a**) $z = 0$, (**b**) $z = 0.4$, (**c**) $z = 0.8$, (**d**) $z = 1.2$

is also accompanied by radiation losses, and therefore the period and the amplitude of the oscillation decrease continously with increasing propagation distance. Additionally, the Gaussian beam does not maintain its circularly symmetric profile, but becomes elliptical and follows the twisting motion of the dipole-mode component.

The experimental verification of this effect is rather complicated, since we cannot follow the beam trajectory inside the crystal. Our approach was to launch the dipole beam at various different angles in the crystal in order to draw some conclusions by monitoring the exit face. In Fig. 7.4, the plane of incidence and the exit face of the crystal are illustrated in the top and bottom rows, respectively. The Gaussian component was always present, but is not illustrated here. We used angles of inclination of 30° (Fig. 7.4a), 45° (Fig. 7.4b), 70° (Fig. 7.4c) and 90° (Fig. 7.4d). Additionally, we made sure that the two lobes propagated in parallel, and that in the linear regime (voltage turned off) the orientation of the diffracting beams corresponded to the initial conditions. By comparing the input and exit faces of the crystal for a fairly small angle (Fig. 7.4a), we can clearly detect a twisting effect. It is remarkable that the initially counterclockwise-tilted dipole rotates clockwise while propagating through the crystal. In Figs. 7.4b,c this effect is even more pronounced, and only when the dipole is aligned exactly horizontally does the rotation not occur (Fig. 7.4d). Thereby, we have qualitatively demonstrated the twisting and oscillating propagation characteristics of a dipole-mode vector soliton, which are due only to the anisotropy of the photorefractive nonlinearity. In terms of a particle model, this represents a state of lowest potential energy for the vector soliton, and therefore can be considered as a global minimum [18]. If the dipole mode is launched at a nonzero azimuthal angle with respect to the vertical axis, it starts to oscillate around its stable vertical orientation. Additionally, there exists a local minimum where the dipole is oriented in the horizontal direction, parallel to the external field. This state is

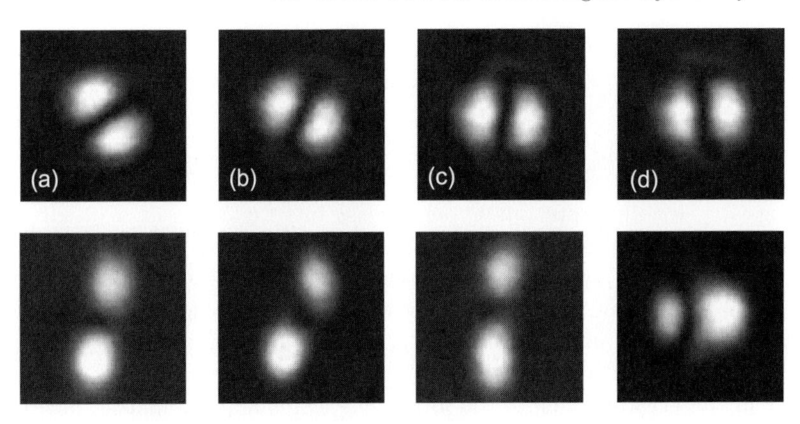

Fig. 7.4. Incident intensity distribution (*top row*) and intensity distribution after copropagating 20 mm (*bottom row*) with a mutually incoherent Gaussian beam (not illustrated), for various initial inclination

not modulationally stable since the double-humped structure tends to align itself in the vertical direction again when it is slightly perturbed. In this way, the anisotropic nature of the refractive index modulation reveals itself not only via the distinct shape and interactions of scalar solitons but also in the formation and stability of vector solitons.

The dipole-mode vector soliton described above is the simplest representative of vector solitons consisting of one fundamental and one higher-order transverse optical mode. It displays a remarkable stability for a large range of its parameters and of external perturbations. Therefore, the question arises of whether optical modes of a more complex geometry may also serve as components of a stable vector soliton.

7.1.3 Vector Solitons of Higher Order

Introducing two phase jumps of π into a Gaussian beam leads either to a triple-humped HG_{02} mode or to a quadrupole-like HG_{11} mode, illustrated in Fig. 7.1b,c. These modes feature a rectangular symmetry and have a vertical and horizontal symmetry axis. Soon after the prediction of the dipole-mode vector soliton and its subsequent experimental observation [14, 15], numerous experimental and numerical investigations dealt with the problem of the stabilization of various multipole structures by copropagation with a fundamental mode [17, 20, 21]. Desyatnikov et al. [21] demonstrated numerically the existence of various types of composite solitary structures in the isotropic regime. According to those numerical investigations, an HG_{11}-like quadrupole mode, as well as a hexapole and even a dodecagon multipole, can indeed serve as a higher-order component of a nondiverging light structure and form a stable vector soliton in combination with a fundamental mode. The experimental verification of these complex light structures is not straightforward, since the anisotropy of the photorefractive crystal will immediately destroy the transverse symmetry of the higher-order light structure.

Quadrupole-Mode Vector Soliton

In Fig. 7.5, we illustrate our experimental investigations of a quadrupole-mode optical beam as the higher-order component of a vector soliton. The top row (Figs. 7.5a–c) depicts the scenario in which the phase jumps of π lie vertically and horizontally in the transverse plane, whereas, in the bottom row (Figs. 7.5d–f), the quadrupole structure is tilted by 45°. In Figs. 7.5b,e the exit face of the crystal is shown when the quadrupole mode propagates separately in the crystal. Each of the four lobes forms a quasi-soliton, but owing to the phase shifts in both transverse directions the individual lobes strongly repel and become separated by a distance that exceeds 100 μm. When a slightly stronger Gaussian beam of power 2.7 μW copropagates with the quadrupole mode, the entire structure becomes self-confined and the spatial spreading of the four lobes is largely prevented. When we resolve the contributions of the individual components to the total light intensity, we observe that the quadrupole structure maintains its profile in the case depicted in Fig. 7.5c, which resembles the incident light intensity shown in Fig. 7.5a. In principle, this is in very good agreement with the isotropic numerical simulations. The situation changes dramatically when the initial orientation of the quadrupole mode is tilted by 45°. In this case the quadrupole mode does not

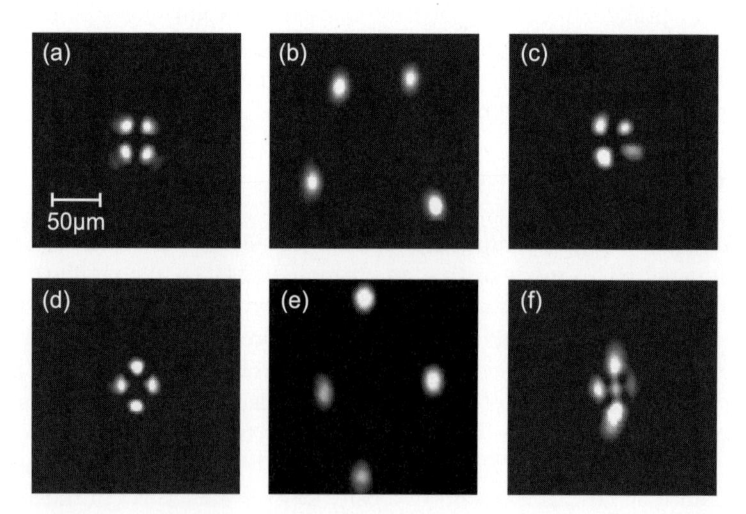

Fig. 7.5. Experimental observation of the generation of a quadrupole-mode vector soliton. Incident HG_{11}-mode beam in two different configurations (**a**), (**d**). Formation of solitary beams after separate propagation of this mode by 20 mm (**b**), (**e**) and after propagation in combination with a mutually incoherent Gaussian beam (**c**), (**f**)

spread either, but its structure changes remarkably. A bright spot becomes clearly visible in the center of the set of beams after propagation by 20 mm (Fig. 7.5f). This remarkable effect is due entirely to the inherent anisotropy of the photorefractive crystal and can also be simulated by means of the anisotropic model [20]. Therefore, we conclude that in an anisotropic material, a stable quadrupole-mode vector soliton can be generated only when the symmetry axes of the light structure coincide with the principal axes of the crystal. When the incident light structure is oriented at an arbitrary angle, the anisotropy of the nonlinearity leads to distortions of the profile of the higher-order mode. So far, a single successful experimental generation of a hexapole vector soliton has been reported [20], but the inherent anisotropy of the photorefractive nonlinearity makes further analysis quite difficult. Highly symmetric transverse optical structures are quite difficult to realize, owing to crystal inhomogeneities and the inherent anisotropy of the photorefractive system. Other light structures with complicated geometries and structures with a higher degree of circular symmetry such as double-ring flower modes [20] have not been demonstrated so far, and are likely to exist only in isotropic nonlinear saturable media.

Triple-Humped Vector Soliton

When two phase jumps are imprinted across a Gaussian beam in the same direction, a triple-humped light structure resembling an HG_{02} mode, as depicted in Fig. 7.1b will evolve. This mode is a quite attractive object, because its structure is similar to that of a dipole-mode beam. When it is aligned in the vertical direction (perpendicular to the external electric field), superposition with a mutually incoherent Gaussian beam gives rise to another stable multicomponent self-focused light structure. The results of our experimental investigations are shown in Fig. 7.6.

Similarly to the case of the dipole mode depicted in Fig. 7.2, the three lobes of the triple-humped mode repel owing to the phase difference between adjacent beamlets (Fig. 7.6b). Under the influence of the copropagating fundamental beam, this repulsion is minimized, and the separation of the three lobes in Fig. 7.6c becomes comparable to the incident intensity distribution. The Gaussian component becomes modulated as well. Its shape becomes rather elliptical and adopts the structure of the triple-humped beam. The entire formation process of this triple-humped light structure is another indication of the nonlocal characteristics of the photorefractive nonlinear response. Since the fundamental mode propagates only in the center of the triple-humped structure, from which the marginal lobes are located at a fairly large distance, the marginal lobes experience a strong attractive force, which leads to the confinement of the whole structure. For the realization of such a nonlocal effect, the photorefractive, anisotropic model is much more suitable than the isotropic model, which is more locally constricted. In our

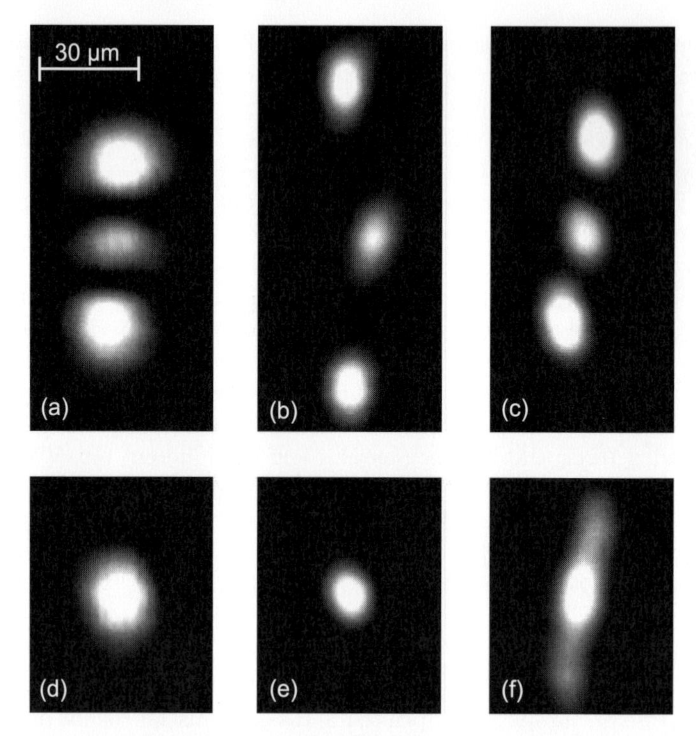

Fig. 7.6. Experimental generation of a triple-humped vector soliton. The triple-humped and the fundamental components as shown in the *top* and *bottom* rows, respectively. Intensity distributions at the front face (**a**), (**d**), at the back face for separate propagation (**b**), (**e**), and at the back face for copropagating components (**c**), (**f**)

numerical simulations based on the anisotropic model, we have observed qualitatively the same behavior as that depicted in Fig. 7.6 [22]. In contrast, the isotropic model predicts a stable self-focusing only for a very restricted parameter range; the anisotropic simulation shows appreciable trapping effects for a large parameter range that corresponds to the experimental values.

7.2 Vector Solitons of Circular Symmetry

After the above investigation of higher-order transverse modes that display either vertical or horizontal symmetry, we now consider modes with circular symmetry that resemble the transverse doughnut modes. The simplest representative of this class of optical modes is depicted in Fig. 7.7. Its intensity distribution displays a doughnut-like pattern (Fig. 7.7a), and its transverse phase distribution is nonuniform. The phase varies azimuthally and covers $\Delta\phi = 2\pi m$ for a rotation of 360°, as depicted in Fig. 7.7b. A phase singularity

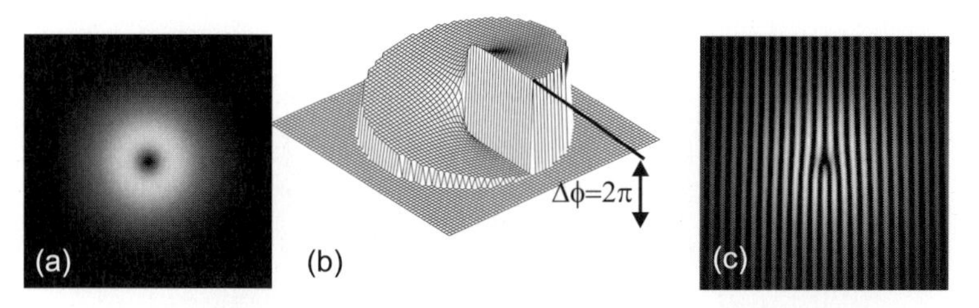

Fig. 7.7. Optical vortex of charge $m = 1$. Numerically obtained intensity pattern (**a**), helical transverse phase distribution (**b**) and interference pattern with a plane wave (**c**)

exists in the central region of the mode, where the intensity drops to zero. The integer m indicates the *topological charge*, which can be determined by interference with a plane or a spherical wave. An interference pattern with the latter displays a helical structure [23], from which the name *optical vortex* was derived. In the case of interference with a plane wave, the phase singularity manifests itself via the splitting of one interference stripe into $m+1$ stripes, as detectable in Fig. 7.7c. Investigations of the stability of optical vortices propagating through a nonlinear self-focusing medium have revealed that they do not propagate self-consistently by themselves [23, 24, 25]. Moreover, multiply charged optical vortex beams disintegrate via a two-step mechanism [26]. A vortex with $m = n$ ($n > 1$) first decays into n singly charged optical vortices (each with $m = 1$), which in turn become subject to further disintegration.

Optical vortices represent the starting point for the generation of dark solitons in the (2+1)-dimensional geometry. They represent a dark spot on a bright background, which can generally be exploited for the generation of a dark soliton in a self-defocusing medium. The self-trapping of an optical vortex in a photorefractive medium has been reported by Chen et al. [27]. Nowadays, optical vortices are commonly used to excite dark solitons in Bose–Einstein condensates, which are governed by the Gross–Pitaevskii equation [28, 29, 30].

We are more interested in optical vortices as the higher-order component of a stable self-focused light structure. Musslimani et al. [11] demonstrated numerically that two-dimensional vector solitons carrying a topological charge do exist in a self-focusing threshold nonlinear medium. The typical breakup of an optical vortex can be prevented in the presence of an additional copropagating beam with a nodeless shape, i.e. a Gaussian beam. In this way, a light beam with a complex internal geometry that does not propagate self-consistently itself in a self-focusing medium can be stabilized when it is combined with a mutually incoherent fundamental beam. After this prediction of multicomponent solitary beams in two transverse dimensions, García-Ripoll et al. [13] showed that the solution of a Gaussian–vortex pair is modulationally unstable and disintegrates into an azimuthally rotating dipole structure after a fairly long propagation distance. The structure has

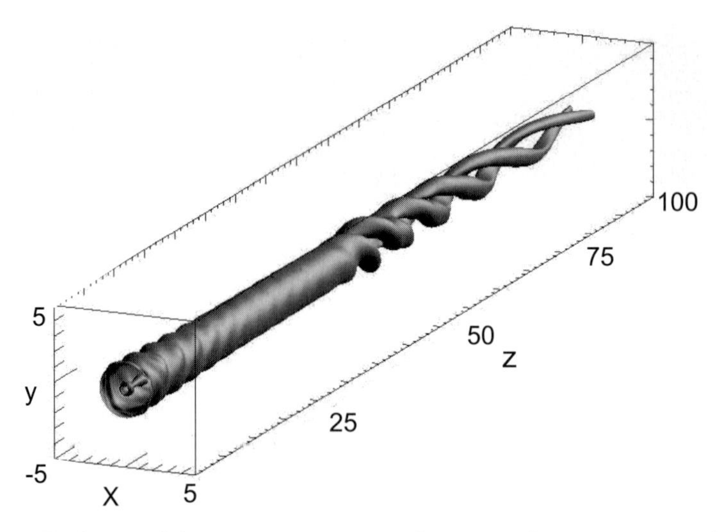

Fig. 7.8. Evolution of the vortex component of a vortex–Gaussian vector soliton on the basis of an isotropic simulation. The presence of the Gaussian beam prevents the transverse spreading of the vortex, which decays after $z = 50L_\mathrm{d}$ into an azimuthally rotating dipole mode

been reported to maintain its shape for some ten diffraction lengths before the inherent numerical noise leads to the instability described. This instability is demonstrated in Fig. 7.8 by an isosurface plot for $I = 0.15I_0$ and $z = 0 - 100L_\mathrm{d}$. The simulation depicted in Fig. 7.8 is based on the isotropic model and shows that the propagation of the vortex component is first accompanied by some minor oscillations of the beam diameter, but that the vortex finally disintegrates into two spiraling lobes that resemble a rotating dipole structure. As the propagation distance increases, it becomes also apparent that the rotation period decreases owing to continuous radiation losses, which can be interpreted as *friction* in terms of the particle model of solitons.

Our further investigations are twofold. First, it is interesting to know if such a breakup that occurs in an isotropic medium can be observed in an anisotropic experimental system. Second, does the evolving dipole-like light structure represent a mode of a stable vector soliton?

To investigate these open questions experimentally, we slightly modified the experimental setup in such a way that one of the two mutually incoherent beams was transmitted through a computer-generated holographic phase mask in order to obtain the desired vortex mode with $m = 1$. Both beams were then superimposed and focused into the crystal. Figure 7.9a depicts the incident light intensity of the vortex component, and Fig. 7.9b shows its rather filamented structure after propagating 10 mm in the crystal separately; this has already been observed in previous studies [25]. When we superimposed an incoherent Gaussian beam of equal total power, the vortex

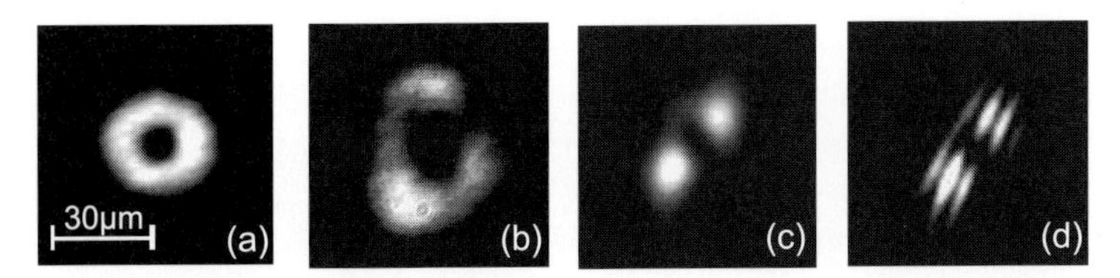

Fig. 7.9. Propagation of an optical vortex ($m = 1$) in a photorefractive crystal. Incident vortex (**a**) and its breakup after separate propagation by 10 mm in the biased crystal (**b**). Simultaneous propagation with a mutually incoherent Gaussian beam leads to a tilted dipole structure (**c**) with a phase shift of π between the constituents, as visualized in the interference pattern obtained when the structure is superimposed on a plane wave (**d**)

component transformed into a tilted double-humped structure, which is depicted in Fig. 7.9c. By visualizing the phase distribution via interference with a plane reference wave (Fig. 7.9d), we were able to prove that the two lobes are out of phase by π. We finally conclude that the vortex mode does not represent a higher-order mode of a stable vector soliton. Instead it decays in a well-defined way and its initial phase singularity transforms into a phase shift of π corresponding to a dipole beam [14].

The general breakup behavior is in good agreement with the isotropic numerical simulations of García-Ripoll et al. [13]. Although the decay into the dipole structure can be observed in the isotropic simulation as well as in the inherently anisotropic experimental system, it occurs at a totally different stage of propagation. Whereas we observe the breakup after about 3.5 diffraction lengths in the experiment, it occurs at a propagation distance roughly 10 times larger in the isotropic simulation. This indicates again that the role of anisotropy is of crucial importance in the propagation of circularly symmetric light structures. As soon as the circularly symmetric vortex beam propagates in the nonlinear medium, it experiences the nonuniform refractive index modulation depicted in Fig. 4.7, which leads to an immediate disintegration and hence to the formation of a dipole structure. Therefore, the more realistic anisotropic model has to be applied to reveal the propagation dynamics of the vortex component. In contrast to the isotropic simulations, numerical beam propagation on the basis of the anisotropic model reveals that decay typically occurs at about one diffraction length, which corresponds to the experimentally observed situation. In a recent publication, Belić et al. [31] demonstrated the difference between the two models and showed that the anisotropic model does not support a homogeneously rotating motion of the resulting dipole component, but a twisting, oscillation motion instead. In analogy to the case of interacting incoherent solitons, the two dipole lobes align themselves preferentially along the vertical y axis, perpendicular to the externally applied electric field, which is a stable orientation in a biased photorefractive crystal [18].

From the above discussion, it is obvious that the photorefractive nonlinearity does not directly support the continuous rotation of light structures in the transverse plane but tends to transform any kind of angular momentum into a twisting or oscillating motion. Nevertheless, a smoothly rotating multicomponent solitary light structure can be realized by imprinting a helical phase distribution onto a dipole beam. In the case of a highly saturated crystal, an additional mutually incoherent Gaussian beam provides a kind of linear waveguide in which a complete rotation might be possible. In this case, the dipole mode performs a continuous spiraling motion while propagating in the biased photorefractive crystal; this has been observed experimentally, and the structure has been called a *propeller soliton* [32]. In a corresponding numerical investigation, Musslimani et al. have reported fascinating, new kinds of interaction of these multicomponent solitary beams – *spin–orbit* coupling upon collision [33, 34] and *delayed-action interaction*. However, even though these intriguing effects reveal themselves clearly in simulations based on the isotropic model, they still lack experimental verification.

7.2.1 Vector Solitons of Higher Order

In analogy to the breakup of a singly charged vortex (see Fig. 7.9), we now describe an investigation of the propagation dynamics of a higher-order optical vortex in the form of a doubly charged ($m = 2$) doughnut-shaped beam in combination with a fundamental Gaussian mode, where we expect characteristics similar to those of the unstable propagation of singly or doubly charged optical vortices themselves [24, 25]. In Fig. 7.10, we show the result of combining an optical vortex of power 1.7 μW with a mutually incoherent fundamental mode of comparable total power of 1.9 μW and proceeding in the same way as described above. Figure 7.10a depicts the incident doughnut mode, which displays a quite disordered disintegration after 13.5 mm of separate propagation in the self-focusing medium (Fig. 7.10b). In the presence of the Gaussian beam, we made use of two different crystal orientations by flipping the sample. In this way, we were able to investigate the propagation along the 5 mm long b axis, and along the 13.5 mm long a axis, which are equivalent in their electro-optic properties. For the shorter propagation

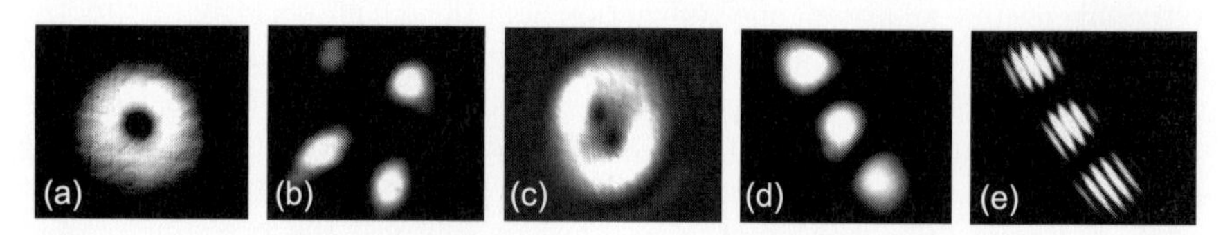

Fig. 7.10. Optical vortex with $m = 2$ (**a**) and its experimentally observed breakup (**b**). Successive decays into a twisting, triple-humped structure (**d**) via an intermediate state of two singly charged vortices (**c**) in the presence of a Gaussian beam. Phase distribution visualized by interference with a plane wave (**e**).

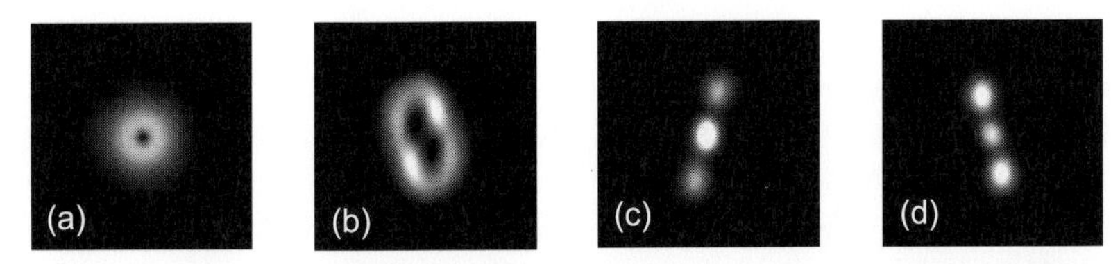

Fig. 7.11. Numerically obtained propagation behavior of an optical vortex ($m = 2$) in the presence of a Gaussian beam for $z = 0$ (**a**), $z = 0.25$ (**b**), $z = 3.5$ (**c**) and $z = 4$ (**d**)

length, we observed the light structure depicted in Fig. 7.10c. A distinct deformation of the initially circular symmetric beam and the evolution of two dark notches in its center are clearly visible. With the help of an interference experiment, we were able to connect each of the dark spots to a singly charged vortex [22]. When we investigated the same beam configuration for the propagation distance of 13.5 mm, keeping the external voltage of 1.8 kV constant, we detected the decay of the vortex into a tilted triple-humped light structure, illustrated in Fig. 7.10d. We visualized the phase distribution of this structure by interference with a mutually coherent plane wave. From Fig. 7.10e it is obvious that neighboring lobes are out of phase by π.

In Fig. 7.11, we illustrate the propagation behavior of the vortex mode as determined numerically in the presence of the stabilizing Gaussian component, for parameters corresponding to the experimental conditions. After propagation by $z = 0.25L_{\mathrm{d}}$, the circular symmetry of the vortex is strongly affected (Fig. 7.11b). At this early propagation stage, the principal features of its shape are directly comparable to the experimental case illustrated in Fig. 7.10c. With increasing propagation distance, the vortex undergoes a strong but transient reshaping process and, after approximately two diffraction lengths, it finally converges into a twisting, triple-humped light structure, as depicted in Figs. 7.11c,d, which illustrate its evolution at $z = 3.5L_{\mathrm{d}}$ and $z = 4L_{\mathrm{d}}$. Again, the numerical results agree qualitatively with the experimentally obtained images.

This twisting behavior is typical of the photorefractive nonlinearity and is analogous to the behavior of a singly charged optical vortex described in the previous section and to the oscillating or spiraling scalar solitons described in Sect. 5.5.5.

According to our experiments, the Gaussian beam has a stabilizing influence on the propagation dynamics of topologically charged doughnut modes, but this influence is not strong enough to compensate for the decay into a well-defined multihumped light structure. This confirms the universal effect whereby all topologically charged light beams transform into a rotating or twisting multihumped object in a self-focusing medium [24, 25]. In the particle model of solitons, the helical phase distribution of an optical vorticex can therefore be interpreted as the mechanical analogue of *spin*. Additionally, we observe the well-known phenomenon that an optical mode of charge m transforms first into m singly charged vortices [26] before it converges into a multihumped light structure with $m+1$ humps. Each initial phase singularity leads to a phase jump of π in the transverse profile of the mode in accordance with previous investigations. Comparing our numerical results [22] with the propagation dynamics in an isotropic atomic-vapor system [24], we conclude that the oscillating transverse motion around the vertical axis stems from the inherent anisotropy of our photorefractive system. Finally, the additional Gaussian beam prevents any further separation of the individual humps with increasing propagation distance. This is underlined by studies of the self-focusing properties of a triple-humped mode combined with a Gaussian beam illustrated in Sect. 7.1.3. Similarly to the dipole-mode vector soliton, this specific triple-humped light structure tends to align itself in the vertical direction perpendicular to the externally applied electric field owing to the anisotropy of the photorefractive medium, as described in Sect. 7.1.3.

7.3 Vector Solitons due to Mutual Stabilization

In the results described up to now, we have achieved nondiverging light structures by using the stabilizing impact of a fundamental mode. The question of whether two or more unstable, diverging light structures can also stabilize each other is therefore a crucial point for the general understanding of vector solitons. The mutual trapping of a double- and a triple-humped mode has been successfully demonstrated in a (1+1)-dimensional photorefractive system [8] but, until the present work, has never been observed in the (2+1)-dimensional configuration. From the scenarios described above, it is obvious that neither the double- nor the triple-humped optical beam can propagate by itself self-consistently in a self-focusing medium. To investigate the propagation behavior of a combination of such beams, we created two mutually incoherent optical fields, which represented a double- and a triple-humped mode respectively (Figs. 7.12a,d). When these two higher-order modes propagate separately for 13.5 mm in the nonlinear crystal, all of their individual beamlets form quasi-solitons. Owing to the relative phase shift of π between adjacent lobes, the beamlets repel strongly as they propagate, which results in an increasing vertical separation illustrated in Figs. 7.12b,e. The simultaneous propagation of the modes, in contrast, leads to the mutual attraction of the individual lobes and the overall spreading of the multicomponent solitary wave is minimized (Figs. 7.12c,e). Comparing the incident light structures with those that evolve from the interaction, shown in Figs. 7.12c,f, it is obvious that the shapes of both higher-order modes are strongly affected. In particular, the two lobes of the dipole and the innermost lobe of the triple-humped beam become elongated in the vertical direction, which is a consequence of the mutual attraction of these neighboring beamlets. Since the resulting light pattern shown in Figs. 7.12c,f deviates clearly from the initial

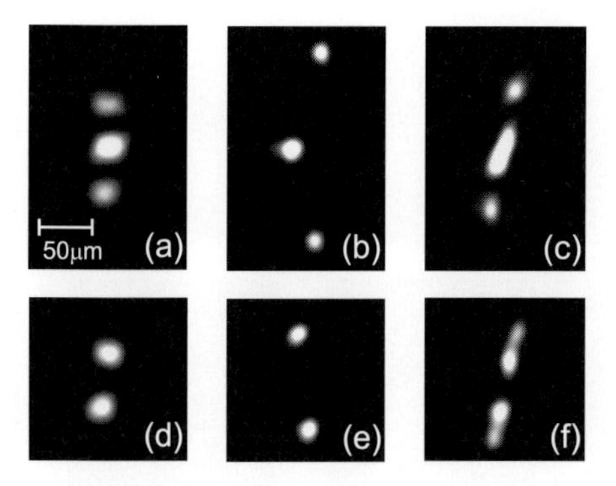

Fig. 7.12. Experimental observation of the mutual stabilization of a double- and a triple-humped mode. Incident light intensity (**a**), (**d**), the spatial spreading after 13.5 mm due to the various phase jumps of π across the beams (**b**), (**e**), and the mutual attraction in the case of simultaneous propagation (**c**), (**f**)

conditions depicted in Figs. 7.12a,d, and because of the limited propagation distance that is experimentally available, we cannot ensure that these light structures represent stable solitary solutions. With increasing propagation distance, they might experience perturbations that lead to disintegration. To gain more insight into the stabilization or any eventual breakup process, we applied the numerical beam propagation routine of anisotropic and nonlocal type with parameters corresponding to the experimental boundary conditions. Starting from circularly symmetric Gaussian beams, the simulations display results comparable to the experiment, as depicted in Fig. 7.13. When each mode propagates separately, the spatial separation due to repulsion is appreciable after two diffraction lengths, as can be seen in Fig. 7.13b. But when the two modes propagate in a coupled way, they mutually stabilize as shown in Figs. 7.13c,d for propagation distances of $z = 3L_{\mathrm{d}}$ and $4L_{\mathrm{d}}$, respectively. Hence, the spatial separation between the lobes reduces to that in the initial situation depicted in Fig. 7.13a. However, this simulation starts from arbitrary initial conditions, and the profiles of the individual lobes do not exactly match an elliptical soliton solution. Comparing Figs. 7.13c,d, we can see that the double- and the triple-humped mode undergo a reshaping process while they mutually trap. From our numerical studies, we could deduce that they typically break up when the propagation distance exceeds $z = 5L_{\mathrm{d}}$, which is beyond the experimentally accessible range. When that happens, the entire structure disintegrates and forms predominately a dipole-mode vector soliton.

In order to seek solitary solutions to the problem, one can make use of the Petviashvili-based numerical procedure, with an extension for two-beam interaction (see Sect. 4.2.1 and [35]). In this way, a self-consistent solution of the double-humped plus triple-humped type that displays stable and hence nondiverging propagation characteristics can be derived. As with all solutions

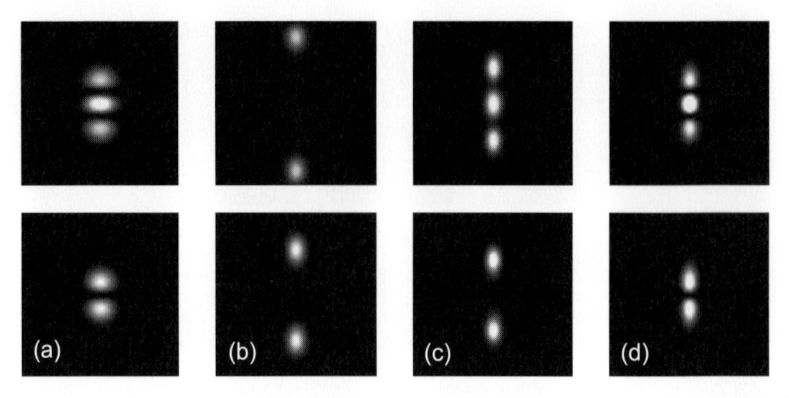

Fig. 7.13. Numerical simulation of the mutual stabilization of a double- and a triple-humped mode. Incident light intensity (**a**), the spatial spreading after $z = 2L_\mathrm{d}$ for separate propagation (**b**), and the copropagating case for $z = 3L_\mathrm{d}$ (**c**) and $z = 4L_\mathrm{d}$ (**d**)

obtained by the Petviashvili method, the transverse shape of the individual lobes is rather elliptical; this shape is a parameter which plays a major role in the stability of these multicomponent structures. The realization of continuously variable beam shapes that would be necessary to perform further experimental studies is an almost inaccessible control parameter. However, taking into account the fact that we have launched beams with a Gaussian shape into the crystal that deviate clearly from the numerically derived profile, which is the desired profile for stable propagation, we are nevertheless able to demonstrate a distinct effect in Fig. 7.12. Generally speaking, we are able to demonstrate the striking effect of the mutual stabilization of inherently diverging higher-order-mode optical beams.

7.4 Controlled Breakup of Multicomponent Light Structures

Recently, Desyatnikov and Kivshar [36] have demonstrated the formation and propagation dynamics of a new type of solitary structure, called a *necklace-ring vector soliton*. Necklace beams have been suggested, in the case of non-saturable Kerr media [37] as beams that display an azimuthally modulated, ring-like intensity profile with phase shifts of π between adjacent lobes. They show remarkable propagation dynamics since they do not disintegrate in a bulk Kerr medium but propagate in a stable way for fairly large propagation distances of up to 50 diffraction lengths. Owing to the inherent phase configuration of the necklace, the structure sooner or later spreads in the transverse dimension in a way that depends on the initial conditions. One approach to compensate for the spatial spreading of these necklace beams consists of applying the principle of vector solitons via the coupling of mutually incoherent necklace beams whose individual lobes attract each other [36].

The simplest ring-like structure that represents such a necklace beam consists of two perpendicularly aligned mutually incoherent dipole modes. In [36], it has been demonstrated that two perpendicularly oriented dipole modes can indeed form a multicomponent solitary ring structure in an isotropic saturable self-focusing material. The repulsion of two oppositely located dipole lobes can be compensated by the mutual attraction of adjacent lobes that are mutually incoherent. A stability analysis via numerical beam propagation reveals an astonishing symmetry-breaking instability. The ring-like structure splits into three individual lobes, positioned at the edges of a triangle, that separate radially with increasing propagation distance.

Here, we want to realize such a scenario in an experimental system, making use of the photorefractive nonlinearity. In order to take account of the experimental conditions of a circularly symmetric incident light structure, we applied the Petviashvili method (see Sect. 4.2.1) to simulate the anisotropic experimental system as realistically as possible [35].

As expected, the simulation routine does not preserve the initial circularly symmetric configuration; instead it transforms the structure consisting of two dipole beams into a squeezed ring, depicted in Fig. 7.14c. The profiles of the incident horizontal and vertical dipole components depicted in Figs. 7.14a,b were obtained from the Petviashvili-based simulation. The separation and the shape of the lobes of the horizontal dipole in Fig. 7.14a vary slightly from those of the vertical component in Fig. 7.14b owing to the inherent anisotropy of the medium. As a result, the overall intensity depicted in Fig. 7.14c resembles a horizontally squeezed ring rather than a circularly symmetric structure. Nevertheless, when the two components copropagate mutually incoherently in the self-focusing material, the repulsion of the oppositely located lobes, which are π out of phase, is prevented and the entire structure remains unaltered. Owing to numerical noise that stems from a quite rough discretization of the numerical array, the self-focused light structure starts to disintegrate at $z = 10L_\mathrm{d}$. The breakup characteristics display a symmetry-breaking instability, since the two components split unequally in, as shown in Figs. 7.14d,e, at $z = 16L_\mathrm{d}$. The initial symmetry of the system is broken, which is revealed in the evolving triangular structure of the total intensity distribution in Fig. 7.14f.

Because the experimental creation of the desired elliptical transverse beam profiles in accordance with the numerical analysis is beyond standard optical configuration capabilities, we approached the problem of creating a configuration of two coupled dipole modes with four Gaussian beams that were pairwise out of phase. The top row of Fig. 7.15 depicts various experimentally observed stages of propagation. The initial total intensity distribution, shown in Fig. 7.15a consists of two perpendicularly oriented dipole modes that do not have exactly the desired shape depicted in Fig. 7.14a–c. After

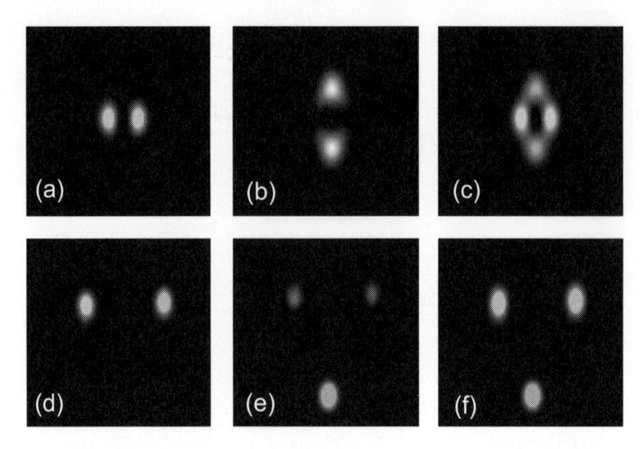

Fig. 7.14. Numerically obtained solution for two perpendicularly oriented dipole beams (**a**)–(**c**). The symmetry-breaking instability after a numerical propagation of $z = 16L_\mathrm{d}$ is depicted in (**d**)–(**f**). The evolution of the horizontal and vertical dipole modes is shown in the first column (**a**), (**d**) and in the second column (**b**),(**e**) respectively. The total intensity is given in (**c**) and (**f**)

the two dipole modes have propagated in the material for 5 mm, we observe in Fig. 7.15b that they are both strongly affected. The two horizontally aligned lobes are clearly focused, whereas the vertically aligned lobes become rather kidney-shaped and torn in the horizontal direction. When we flipped the crystal sample and investigated the propagation along the 13.5 mm long a axis for the same parameters of 2.3 μW beam power for each mode and an external voltage of 3.6 kV/cm, we detected a distinct breakup of the structure into two intensity-modulated stripes, which is illustrated in Fig. 7.15c.

The experimentally observed scenario deviates from the numerical simulation depicted in Fig. 7.14. To account for these deviations, which we believed to be due to different intensities of the vertical and horizontal dipoles, we simulated the experimental propagation behavior by inducing a supplementary initial perturbation, where we changed the intensity of one component relative to the other by 10%. The additionally induced noise leads in fact to a different propagation behavior, which is depicted in Figs. 7.15d–f. The initially perturbed soliton solution is illustrated in Fig. 7.15d and the corresponding spatial evolution after $z = 2.5L_\mathrm{d}$ and $z = 4.5L_\mathrm{d}$, which corresponds roughly to propagation distances of 5 mm and 13.5 mm is depicted in Figs. 7.15e,f, respectively.

These results demonstrate excellent agreement between the numerical and experimental breakup characteristics. In the numerical evolution for larger propagation distances, the two stripes that arise spread further in the horizontal direction, but remain self-confined vertically. This can be interpreted in terms of two dipole-mode vector solitons that separate in the horizontal plane. Owing to the interaction, the initially vertically aligned dipole mode

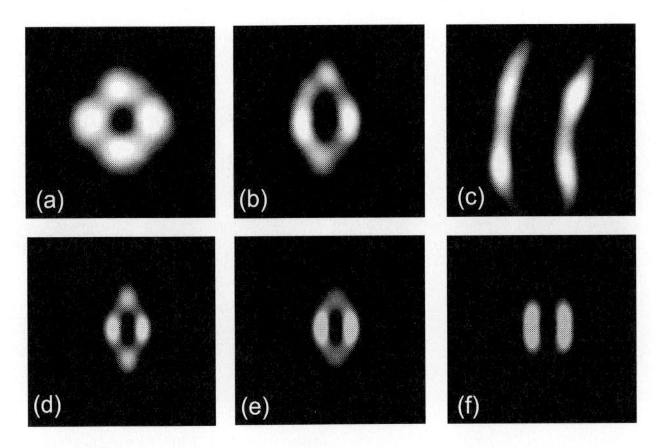

Fig. 7.15. Experimental (*top row*) and numerical (*bottom row*) implementation of the dipole–dipole coupled soliton pair. Total intensity at $z = 0$ mm (**a**), (**d**), $z = 5$ mm (**b**), (**e**) and $z = 13.5$ mm (**c**), (**f**)

splits into four beamlets that are trapped pairwise by the lobes of the horizontal dipole mode.

The simulations and the experimental results demonstrate the mutual stabilization of two perpendicularly oriented coupled dipole-mode beams in an anisotropic self-focusing medium. Both optical modes propagate in a stable and nondiverging way for a few diffraction lengths before they become subject to a noise-induced symmetry-breaking instability. This distinct unstable propagation behavior leads in turn to a conversion of the vertical dipole mode into a horizontally spreading quadrupole-like structure. Therefore, the whole process can be interpreted as a controlled transformation of two dipole-mode beams into two dipole-mode vector solitons.

7.5 Multicomponent Vector Solitons

As illustrated above, the coupling of two perpendicularly oriented dipole modes leads to the formation of a semistable solitary light structure. In view of the well-investigated effect that a fundamental Gaussian beam can trap or stabilize complex transverse light structures so that they become a higher-order component of a stable vector soliton, we now investigated the stabilization of the above-mentioned coupled dipole structure with the help of an additional Gaussian beam. In a recent numerical study [38], we could show that a configuration consisting of two perpendicularly oriented dipole modes and a Gaussian fundamental beam can indeed form a nondiverging entity in the case of an isotropic saturable nonlinearity. These *multicomponent dipole-mode vector solitons* were shown to exist for a wide range of parameters. Since the two dipole modes themselves have a symmetry-breaking instabil-

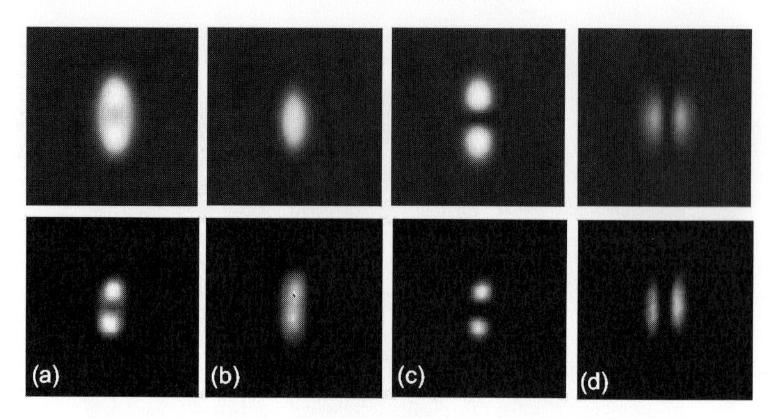

Fig. 7.16. Numerical (*top row*) and experimental (*bottom row*) formation of a multicomponent vector soliton. Total intensity (**a**), and the contributions of the Gaussian component (**b**), the vertical dipole component (**c**) and the horizontal dipole component (**d**)

ity, the power of the stabilizing Gaussian beam relative to the two dipole components is of crucial importance for the dynamical characteristics of this three-component object. We found an instability threshold when the power of the vertical or horizontal dipole mode exceeded a certain value and the beams propagated for an appreciable distance. Near this threshold, an intriguing dynamics accompanied by weak oscillatory instabilities arises. The symmetry is broken only along one axis leading to the formation of a fundamental, a dipole and a triangularly aligned triple-humped structure. The latter evolves from the weaker dipole mode and displays a fascinating *swinging dynamics* as it propagates further. We expect that these vibrational degrees of freedom, which are likely to be associated with long-lived soliton internal modes, should manifest themselves in the rich dynamics of soliton collisions.

Since the generation of a stable three-component solitary structure has been clearly demonstrated in an isotropic simulation, we expect to observe a comparable behavior in an anisotropic or photorefractive medium. In order to compare our experiments with the numerical results, the Petviashvili approach was repeatedly applied to find a numerical solution [39]. The routine indeed supplied a whole family of stable solitary solutions for a vast parameter space and various relative beam powers. From these results, we deduced that the power of the two dipole modes must not exceed a certain threshold value if a stable object is to be generated. Otherwise, the Gaussian component is not capable of trapping the two instability-inducing dipole modes. In contrast to the isotropic system described above, a swinging dynamical behavior near the instability threshold has not been observed. In Fig. 7.16, we illustrate the numerical and experimental generation of such a three-component light object. Figure 7.16a depicts the total light intensity distribution of the numerically calculated (top row) and experimentally ob-

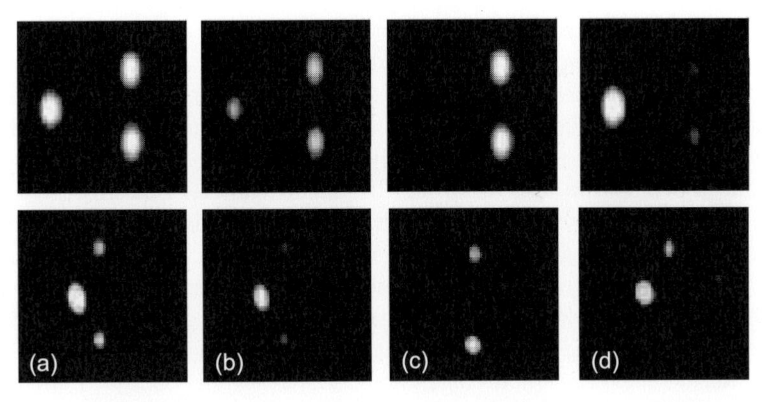

Fig. 7.17. Numerically (*top row*) and experimentally (*bottom row*) observed unstable propagation dynamics of a multicomponent vector soliton. Total intensity (**a**), and the contributions of the Gaussian component (**b**), the vertical dipole component (**c**) and the horizontal dipole component (**d**)

served light structures (bottom row) after 10 mm of propagation in the biased crystal. In Figs. 7.16b–d, the intensity distributions of the Gaussian, vertical dipole and horizontal dipole components are illustrated as they emerge from the crystal.

If we compare the frames in the top and bottom rows, a remarkable agreement concerning the shapes of the individual components is obvious. We can clearly demonstrate experimentally a transformation of the incident beams with Gaussian shape into specific elongated light structures that show a remarkable similarity to the numerically calculated profiles.

When the system is perturbed by varying the initial intensity distributions, the structure breaks down as a whole and displays a symmetry-breaking behavior already known from [36]. Here, the total intensity distribution evolves into a triangular structure, which is asymmetrically distributed among its constituents.

Figure 7.17 illustrates this numerically and experimentally obtained instability in the top and the bottom row, respectively. The total intensity distribution shows three spots of comparable intensity in (Fig. 7.17a) The Gaussian component is distributed almost equally over the three spots (Fig. 7.17b), whereas the vertical dipole merely retains its basic double-humped shape (Fig. 7.17c). In contrast to the latter, the horizontal dipole mode reveals a single, very distinct peak at its left side and two vertically aligned spots around its center. Again, a striking similarity between the numerical simulations and experimental results is obvious. It is clear that the anisotropic nature of the photorefractive nonlinearity leads to a symmetry-breaking instability of a solitary structure when the external conditions do not match the exact solution.

In general, we have demonstrated that the principle of stabilizing optical beams of complex geometry so as to lead to the formation of stable vector solitons is not restricted to a certain number of mutually incoherent constituent beams. We believe that solitary beam configurations of even higher complexity, consisting of several constituents, can in principle be realized [40].

Multimode spatial optical solitary beams have another interesting feature, since they also describe a light wave with partial spatial coherence [41]. The degree of coherence governs the diffraction of a light wave, and in general it is a nontrivial task to trap spatially incoherent light in a self-focusing medium [42]. On the other hand, such trapping offers the possiblity of low-cost implementation of light sources such as light-emitting diodes in all-optical switching devices. Recent investigations of spatially incoherent light in self-focusing media illustrate a whole variety of intriguing effects in nonlinear physics [43, 44], and even the self-focusing of white light from an incandescent light bulb has been demonstrated [45].

Summary

The preceding chapters on scalar-soliton interaction and the description in thos chapter of the creation and investigation of vector or multicomponent solitons demonstrate that the controlled manipulation of nondiverging optical beams by other light beams is generally possible. In the case of saturable nonlinearity, many effects have been demonstated, such as controlled gain and loss, the creation and annihilation of solitons and the generation of (2+1)-dimensional transverse solitary structures themselves. Complex interaction effects that result in three-dimensional soliton trajectories and mutual stabilizing of complex light structures into solitary channels have been realized and can, in principle, be exploited for future applications. In the further investigation of self-focusing of partially spatially incoherent light the question of modulation instabilities, as a precursor of soliton or pattern formation in general arises,. The following chapters are therefore dedicated to the intriguing field of the formation of transverse patterns in a nonlinear photorefractive system.

References

1. A.W. Snyder, D.J. Mitchell, L. Poladian and F. Ladouceur, *Self-induced optical fibers: spatial solitary waves*, Opt. Lett. **16**, 21 (1991).
2. A.W. Snyder, D.J. Mitchell and Yu.S. Kivshar, *Unification of linear and nonlinear wave optics*, Mod. Phys. Lett. B **9**, 1479 (1995).
3. S.V. Manakov, *On the theory of two-dimensional stationary self-focusing of electromagnetic waves*, Sov. Phys. JETP **38**, 248 (1974).

4. J.U. Kang, G.I. Stegeman, J.S. Aitchison and N.N. Akhmediev, *Observation of Manakov spatial solitons in AlGaAs planar waveguides*, Phys. Rev. Lett. **76**, 3699 (1996).
5. G.I. Stegeman W. Torruellas, Yu.S. Kivshar. *In: Spatial Solitons*, chapter Quadratic Solitons, 127–168. Springer, Berlin, Heidelberg, 2001.
6. D.N. Christodoulides, S.R. Singh and M.I. Carvalho, *Incoherently coupled soliton pairs in biased photorefractive crystals*, Appl. Phys. Lett. **68**, 1763 (1996).
7. Z. Chen, M. Segev, T.H. Coskun and D.N. Christodoulides, *Observation of incoherently coupled photorefractive spatial soliton pairs*, Opt. Lett. **21**, 1436 (1996).
8. M. Mitchell, M. Segev and D.N. Christodoulides, *Observation of multihump multimode solitons*, Phys. Rev. Lett. **80**, 4657 (1998).
9. E.A. Ostrovskaya, Yu.S. Kivshar, D.V. Skryabin and W.J. Firth, *Stability of multihump optical solitons*, Phys. Rev. Lett. **83**, 296 (1999).
10. E.A. Ostrovskaya and Yu.S. Kivshar, *Multi-hump optical solitons in a saturable medium*, J. Opt. B **1**, 77 (1999).
11. Z.H. Musslimani, M. Segev, D.N. Christodoulides and M. Soljačić, *Composite multihump vector solitons carrying topological charge*, Phys. Rev. Lett. **84**, 1164 (2000).
12. Z.H. Musslimani, M. Segev and D.N. Christodoulides, *Multicomponent two-dimensional solitons carrying topological charges*, Opt. Lett. **25**, 61 (2000).
13. J.J. García-Ripoll, V.M.Pérez-García, E.A. Ostrovskaya and Yu.S. Kivshar, *Dipole-mode vector solitons*, Phys. Rev. Lett. **85**, 82 (2000).
14. W. Królikowski, E.A. Ostrovskaya, C. Weilnau, M. Geisser, G. McCarthy, Yu.S. Kivshar, C. Denz and B. Luther-Davies, *Observation of dipole-mode vector solitons*, Phys. Rev. Lett. **85**, 1424 (2000).
15. T. Carmon, C. Anastassiou, S. Lan, D. Kip, Z.H. Musslimani and M. Segev, *Observation of two-dimensional multimode solitons*, Opt. Lett. **25**, 1113 (2000).
16. K. Motzek, A. Stepken, F. Kaiser, M.R. Belić, M. Ahles, C. Weilnau and C. Denz, *Dipole-mode vector solitons in anisotropic photorefractive media*, Opt. Commun. **197**, 161 (2001).
17. C. Weilnau, W. Królikowski, E.A. Ostrovskaya, M. Ahles, M. Geisser, G. McCarthy, C. Denz, Yu.S. Kivshar and B. Luther-Davies, *Composite spatial solitons in saturable nonlinear bulk medium*, Appl. Phys. B **72**, 723 (2001).
18. D. Neshev, G. McCarthy, W. Królikowski, E.A. Ostrovskaya, Yu.S. Kivshar, G.F. Calvo and F. Agullo-Lopez, *Dipole-mode vector solitons in anisotropic nonlocal self-focusing media*, Opt. Lett. **26**, 1185 (2001).
19. A.V. Mamaev, A.A. Zozulya, V.K. Mezentsev, D.Z. Anderson and M. Saffman, *Bound dipole solitary solutions in anisotropic nonlocal self-focusing media*, Phys. Rev. A **56**, R1110 (1997).
20. A. S. Desyatnikov, D. Neshev, E.A. Ostrovskaya, Yu.S. Kivshar, G. McCarthy, W. Królikowski and B. Luther-Davies, *Multipole composite spatial solitons:theory and experiment*, J. Opt. Soc. Am. B **19**, 586 (2002).
21. A.S. Desyatnikov, D. Neshev, E.A. Ostrovskaya, Yu.S. Kivshar, W. Królikowski, B. Luther-Davies, J.J. García-Ripoll and V.M. Pérez-García, *Multipole spatial vector solitons*, Opt. Lett. **26**, 435 (2001).
22. C. Weilnau, C. Denz, M. Ahles, A. Stepken, K. Motzek and F. Kaiser, *Generation of higher-order optical (2+1) dimensional spatial vector solitons in a nonlinear anisotropic medium*, Phys. Rev. E **64**, 056601 (2001).

23. M.S. Soskin I.V. Basistiy and M.V. Vasnetsov, *Optical wavefront dislocations and their properties*, Opt. Comm. **119**, 604 (1995).
24. V. Tikhonenko, J. Christou and B. Luther-Davies, *Three dimensional bright spatial soliton collision and fusion in a saturable nonlinear medium*, Phys. Rev. Lett. **76**, 2698 (1996).
25. V. Tikhonenko, J. Christou and B. Luther-Davies, *Spiraling bright spatial solitons formed by the breakup of an optical vortex in a saturable self-focsusing medium*, J. Opt. Soc. Am. B **12**, 2046 (1995).
26. A.V. Mamaev, M. Saffman and A.A. Zozulya, *Decay of high order optical vortices in anisotropic nonlinear optical media*, Phys. Rev. Lett. **78**, 2108 (1997).
27. Z. Chen, M. Segev, D.W. Wilson, R.E. Muller and P.D. Maker, *Self-trapping of an optical vortex by use of the bulk photovoltaic effect*, Phys. Rev. Lett. **78**, 2948 (1997).
28. D.A. Butts and D.S. Rokshar, *Predicted signatures of rotating Bose–Einstein condensates*, Nature **397**, 327 (1999).
29. D.S. Rokshar, *Condensates in a twist*, Nature **401**, 533 (1999).
30. M.R. Matthews, B.P. Anderson, P.C. Haljan, D.S. Hall, C.E. Wiemann and E.A. Cornell, *Vortices in a Bose–Einstein condensate*, Phys. Rev. Lett. **83**, 2498 (1999).
31. M.R. Belić, D. Vujić, A. Stepken, F. Kaiser, G.F. Calvo, F. Agulló-López and M. Carrascosa, *Isotropic versus anisotropic modeling of photorefractive solitons*, Phys. Rev. E **65**, 066610 (2002).
32. T. Carmon, C. Pigier, Z.H. Musslimani, M. Segev and A. Nepomnyashchy, *Rotating propeller solitons*, Phys. Rev. Lett. **87**, 143901 (2001).
33. Z.H. Musslimani, M. Soljačić, M. Segev and D.N. Christodoulides, *Delayed-action interaction and spin–orbit coupling between solitons*, Phys. Rev. Lett. **86**, 799 (2001).
34. Z.H. Musslimani, M. Soljačić, M. Segev and D.N. Christodoulides, *Interactions between two-dimensional composite vector solitons carrying topological charges*, Phys. Rev. E **63**, 066608 (2001).
35. M. Ahles, K. Motzek, A. Stepken, F. Kaiser, C. Weilnau and C. Denz, *Stabilization and breakup of coupled dipole-mode beams in an anisotropic nonlinear medium*, J. Opt. Soc. Am. B **19**, 557 (2002).
36. A.S. Desyatnikov and Yu.S. Kivshar, *Necklace-ring vector solitons*, Phys. Rev. Lett. **87**, 033901 (2001).
37. M. Soljačić, S. Sears and M. Segev, *Self-trapping of "necklace" beams in self-focusing Kerr media*, Phys. Rev. Lett. **81**, 4851 (1998).
38. A. S. Desyatnikov, Yu. S. Kivshar, K. Motzek, F. Kaiser, C. Weilnau and C. Denz, *Multicomponent dipole-mode spatial solitons*, Opt. Lett. **27**, 634 (2002).
39. K. Motzek, F. Kaiser, C. Weilnau, C. Denz, G. McCarthy, W. Królikowski, A. S. Desyatnikov and Yu. S. Kivshar, *Multi-component vector solitons in photorefractive crystals*, Opt. Commun. **209**, 501 (2002).
40. A.S. Desyatnikov and Yu.S. Kivshar, *Rotating optical soliton clusters*, Phys. Rev. Lett. **88**, 053901 (2002).
41. M. Mitchell, M. Segev, T.H. Coskun and D.N. Christodoulides, *Theory of self-trapped spatially incoherent light beams*, Phys. Rev. Lett. **79**, 4990 (1997).
42. M. Mitchell, Z. Chen, M. Shih and M. Segev, *Self-trapping of partially spatially incoherent light*, Phys. Rev. Lett. **77**, 490 (1996).

43. D. Kip, C. Anastassiou, E. Eugenieva, D.N. Christodoulides and M. Segev, *Transmission of images through highly nonlinear media by gradient-index lenses formed by incoherent solitons*, Opt. Lett. **26**, 524 (2001).
44. D. Kip, M. Soljačić, M. Segev, E. Eugenieva and D.N. Christodoulides, *Modulation instability and pattern formation in spatially incoherent light beams*, Science **290**, 495 (2000).
45. M. Mitchell and M. Segev, *Self-trapping of incoherent white light*, Nature **387**, 880 (1997).

8 Self-Organized Pattern Formation in Single-Feedback Systems

Up to now, we have considered the simplest case of light–matter interaction based on the countereffects of diffraction and nonlinearity, namely the passing of a single beam through a nonlinear photorefractive medium. Despite the simplicity of this approach, nonlinear interactions lead to a variety of complex spatial effects, among them the creation of spatial solitons. We have discussed their shape during propagation, their interaction behavior, the formation of complex multisoliton ensembles and the waveguiding properties of all those structures. However, it is not necessary to restrict ourselves to small beams that have a size below the threshold size for filamentation.

We discussed in Sect. 4.1 the appearance of a modulation instability when one is dealing with extended beams in one or two transverse dimensions. A modulation instability corresponds to the appearance of an intensity modulation on the transverse profile with a characteristic modulation period considerably less than the diameter of the beam. As a result of this nonlinear interaction, the beam collapses and filaments into several smaller beams.

In this chapter, we shall consider the full transverse extension of a laser beam or its idealized form, a plane wave. In this case, the transverse degrees of freedom do not appear only in the interaction of different solitary beams. They play a key role in the interaction of the transverse wave-vector components of the beam itself. As in the case of small-scale beams, the beam experiences filamentation under certain circumstances. A key property of wave number interaction in the case of extended beams is a transverse coupling of the filaments, resulting in an ordered, patterned state. This state is reached in a self-organized way by the system itself. The formerly unordered system reaches a completely regular, ordered state. This aspect explaines the fascination and motivation of studying spontaneous pattern formation: a physical system with a large number of degrees of freedom is reduced to the action of a small set of order parameters, with a macroscopic patterned state as the end result.

8.1 Pattern Formation in Nonlinear Optics

Extensive research on pattern formation effects in nonlinear optical systems in general started at the end of the 1980s. Owing to the increasing calculational power of modern computers for theoretical and numerical studies, and the development of stronger lasers and advances in sensitive nonlinear materials for experimental investigations, pattern formation in nonlinear optical materials became a lively field of research. *Transverse nonlinear optics* is not yet a textbook topic, but can be found in a number of review articles and special issues of widely read journals [1, 2, 3, 4, 5, 6, 7, 8, 9, 10, 11]. Contributions cover a great variety of nonlinear optical materials, among them atomic vapors, nonlinear crystals, photorefractive crystals, liquid crystals, liquid crystal light valves, semiconductors, solid-state laser systems and optical fibers. The interaction schemes cover all kinds of nonlinear resonators and cavities, with a wealth of corresponding nonlinear $\chi^{(2)}$ (second-harmonic generation, optical parametric oscillation, etc.) and $\chi^{(3)}$ effects, four and two-wave-mixing geometries, single-mirror feedback systems and cases of simple nonlinear beam propagation of one or more optical beams in a nonlinear medium, including propagation of solitary waves. The diversity of contributions to this research field is as remarkable as the similarities in the experimental and theoretical results among the various interaction schemes and nonlinearities. In particular, analogies with other disciplines have been established, not only in the phenomenology, but also in the governing equations and theoretical models.

In order to understand the principle of the spontaneous formation of symmetric patterns in the transverse profile of a laser beam, let us start with the simplest geometries of nonlinear matter–light interaction. Figure 8.1 depicts two basic interaction schemes capable of exhibiting spontaneous pattern formation: a single primary beam passing through a nonlinear medium and two counterpropagating waves intersecting inside the nonlinear medium. In this representation, a strong plane wave, called F_0 in the following, passing through a nonlinear medium is convectively unstable against the excitation of two spatial sidebands $F_{\pm 1}$ – two plane waves propagating symmetrically at an angle θ to the primary wave (see Fig. 8.1a). Because the satellite and central beams will be coherent, they will interfere, resulting in a transverse intensity modulation with a characteristic spatial scale $l_s = 2\pi/\theta_c k_0$. This transverse modulation is the origin of several more complex transverse modulation effects. The term *convective instability* briefly describes the following situation. When small seed amplitudes of the sideband waves (or Fourier harmonics) are present in the input to the nonlinear medium, they may grow exponentially with distance at the expense of the energy supplied by the strong (main or pumping) wave. In this case, an instability may arise for a range of angles θ between the wave vector k_0 of the strong plane wave and the wave vectors $k_{\pm 1}$ of the spatial sidebands. For a Kerr nonlinearity, the characteristic angle is given by $\theta_c^2 \propto n_2 |F_0|^2$.

The phenomenon of spontaneous pattern formation becomes even more significant when a second beam counterpropagating against the first one, created either independently or by reflection from a mirror, is present. Under certain circumstances, these two strong plane waves (F_0 and B_0) turn out to

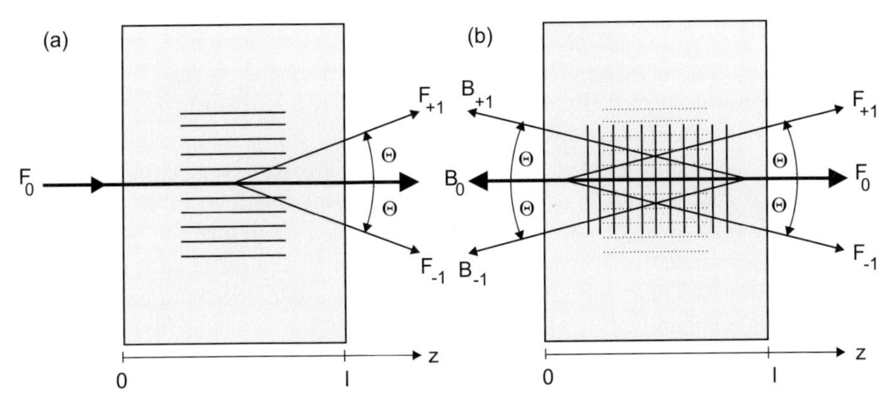

Fig. 8.1. Geometry of interaction for spontaneous pattern formation. (**a**) Excitation of satellite beams by a single primary beam. (**b**) Excitation of pairs of satellite beams by counterpropagating primary beams, The orientation of the resulting interference transmission and reflection gratings is indicated in the figure

be unstable with respect to the excitation of two pairs of waves $F_{\pm 1}$ and $B_{\pm 1}$, as shown in Fig. 8.1b. An important aspect of the case of two counterpropagating beams compared with the case of a single propagating beam is that the system is now absolutely unstable. In contrast to the case of one pumping beam, the instability can start from any arbitrarily small perturbation in the transverse beam profile. In the Fourier plane, this corresponds to infinitesimally small initial amplitudes of the generated sideband waves, which grow exponentially in the framework of the linearized set of equations. Saturation of this growth at a steady-state value is caused by the finite intensity of the pump beam and the saturating character of the medium.

In both cases, the transverse instability of the beam(s) results in the appearance of a characteristic spatial scale in the transverse structure of the beam(s) corresponding to the generation of sidebands with either the highest amplification coefficient in the case of a single pump or the lowest threshold (which in turn specifies the largest growth rate of the corresponding wave number) in the case of counterpropagating pumps. The satellite beams created around the central beam are positioned on a circle in the Fourier plane. Their wave vectors have the same magnitude, but still have a degeneracy concerning their orientation. In the nonlinear stage of the pattern generation process, a finite set of wave vectors is selected by the system, leading to specific geometric patterns. These patterns are stationary or dynamic according to the nature of the instability. Common examples of patterns which may appear are hexagons, stripes (also called rolls owing to their similarity to Rayleigh–Bénard instabilities in fluids), and squares. Each of these pattern configurations may display defects, for example dislocations in roll patterns and penta-hepta defects in hexagonal patterns.

The processes by which spatial patterns arise from the homogeneous state are of fundamental conceptual importance. Spontaneous pattern formation is a direct consequence of the combined action of a nonlinearity and crosstalk mechanisms between the different spatial points, leading to coupling effects and constraints on the system. The latter action can be accomplished by diffusion or diffraction. For the case of diffusion, the mechanism was first elucidated in a pioneering work by Turing [12]. Therefore, it has become established to call those instabilities which lead to the onset of a stationary structure out of a spatially uniform state *Turing instabilities*, independent of the driving mechanism.

In contrast to diffusion or convection, the spatial coupling process of diffraction represents a massless mechanism, which implies considerable advantages when one is observing spatio-temporal structures. The inherent parallelity of optics, the possibility of controling the system parameters and the natural access to the Fourier spectrum render nonlinear optical systems excellent objects of study for pattern formation phenomena and their active control.

In the paraxial approximation, which is currently used in all such studies, diffraction is described by the transverse Laplacian, that is, by the sum of the second derivatives with respect to the two transverse spatial variables. The kind of pattern depends crucially on the type of nonlinearity, which may be, for example, quadratic in the case of parametric down-conversion and second-harmonic generation, cubic in $\chi^{(3)}$ media (four-wave mixing), or a saturable nonlinearity, which governs the single-mode gain in lasers, for example. The nonlinearity determines the allowed and favored combinations of wave vectors in the transverse plane, which in turn produces the macroscopic geometric structure. Despite the fundamental differences in the nonlinearities and the wealth of different patterns, one special orientation of the wave vectors is favored in all of the systems studied: hexagonal patterns, as shown in Fig. 8.2. In the following analysis, we shall focus on systems that display translational symmetry and are driven by plane waves. These systems are ideally suited to studying the spontaneous breaking of the translational symmetry and the onset of patterns from the homogeneous state.

8.2 Pattern Formation in Single-Mirror Feedback Systems

In order to obtain a transverse modulation instability of two counterpropagating beams that results in the formation of spatially symmetric patterns, the presence of a nonlinearity and coupling between neighboring spatial regions by diffraction are necessary ingredients. Consequently, the first spontaneous-pattern-formation experiments were conducted with counterpropagating beams in Kerr media [13, 14, 15].

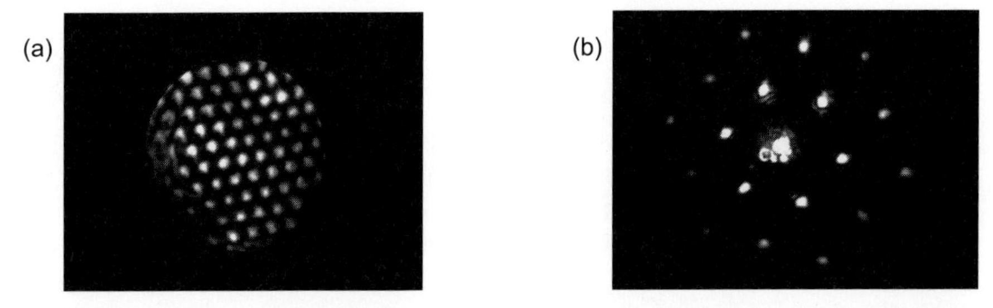

Fig. 8.2. Typical hexagonal pattern for single-mirror feedback systems. (**a**) Filamented laser beams forming a hexagonal spot array (near field); (**b**) spatial Fourier transform of the near-field pattern showing the DC component (central spot) and the six modes with a wave number $|\boldsymbol{k}_\perp|$ related to the spatial periodicity in the near field. Also visible in (**b**) are the second-order terms $\sqrt{3}|\boldsymbol{k}_\perp|$ and third-order terms $2|\boldsymbol{k}_\perp|$

However, optical feedback can also be realized by using a mirror to back-reflect a beam into the nonlinear optical medium. This technique, shown in Fig. 8.3, incorporates all the necessary ingredients for pattern formation, but is at the same time a very simple experimental tool for generating the two counterpropagating beams. Therefore, it is the most studied model system for spatial instabilities in nonlinear optics.

When an incoming laser beam passes through the nonlinear optical material, feedback is provided by a high-reflectivity mirror placed at a distance L from the medium, and a diffraction length of $2L$ results for the reflected beam. The two counterpropagating beams inside the medium have coupled amplitudes and phases (see also Sect. 3.6.1), thereby setting a defined relationship between the two counterpropagating beams. Above a certain threshold value,

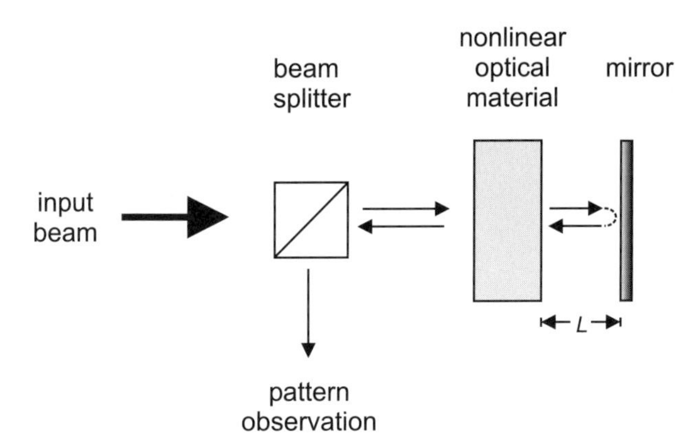

Fig. 8.3. Schematic illustration of a single-mirror feedback system incorporating the basic ingredients of nonlinearity, diffraction and feedback

the counterpropagating laser beams become modulationally unstable against the growth of certain transverse wave numbers. The filaments of the laser beam usually arrange themselves into a geometrical structure with a hexagonal or other regular symmetry. This direct image of the laser beam leaving the nonlinear optical medium will be referred to as the *near field.*

A simple representation of a modulated two-dimensional pattern is given by the spatial Fourier transform of the intensity pattern, which is easily accomplished in optics by a free-space propagation (referred to as the *far field*) or by the use of a single lens. A typical output of a pattern-forming system is shown in Fig. 8.2 for the case of a feedback system with a photorefractive nonlinearity.

The simplicity of the feedback scheme shown in Fig. 8.3 has inspired extensive research on pattern formation in single-mirror feedback systems with a variety of different nonlinear optical materials, namely atomic vapors [16, 17, 18], liquid crystals [19, 20, 21], liquid crystal light valves [22, 23], organic materials [24] and, last but not least, photorefractive crystals [25, 26, 27, 28, 29, 30, 31, 32, 33, 34, 35, 36]. Despite the obvious differences in the nonlinearities, all these systems show similar cooperative behavior when excited beyond the instability threshold. While square patterns and squeezed hexagonal patterns have been reported in a photorefractive single-mirror feedback system [32], the hexagonal pattern is the predominant one. When an atomic vapor is used, polarization effects come into play, and transitions between stable squares, rolls and hexagons have been predicted [37] and experimentally observed [38] when the polarization of the incident light or the magnetic field is varied [39, 40]. If a field transformation is introduced into the feedback loop by rotation, translation or magnification, the number and complexity of different pattern states rise considerably, owing to the nonlocal interaction between the nonlinearity and the light beam. It should be added that these so-called *kaleidoscope* patterns [41], including quasi-crystalline patterns, have to be classified as patterns induced by the field transformation, rather than as a manifestation of the self-organized system itself. We shal focus in our subsequent considerations on the case of a free-running system with purely diffractive feedback and without any field transformations, in order to explore the nature and spontaneous patterns and the conditions under which they are formend in the photorefractive single-mirror feedback system. Figure 8.2a shows the near field, which is a direct image of the laser beam leaving the medium for such a system; in this typical case it is rearranged into a hexagonal spot array. The spatial Fourier transform in Fig. 8.2b achieved by simple propagation of the beam shown in Fig. 8.2a, displays six equally distributed spots placed on a circle corresponding to the first order, six spots corresponding to the second order, and six to the third order. It shows the inherent ability of a spatial Fourier transform to yield a simple, but complete representation of a complex spatial structure. Extensive use has been made of this property for the manipulation and selection of optical patterns a Fourier control method, which will be discussed in Chap. 10.

Much of the interest in pattern formation due to counterpropagation in nonlinear optical media has been stimulated by a number of key experimental observations in liquid-crystalline systems. These systems consist of a thin nonlinear medium and can display Kerr-like nonlinearities for a certain voltage range, and are therefore ideal candidates for mathematical investigations. In the case of these Kerr materials, the instability threshold depends on the laser intensity, thus limiting the size of the plane wave to a certain size in order to achieve the laser power necessary to exceed the threshold. Therefore, in most systems, the aspect ratio Γ, (the ratio of the total beam size to the smallest typical scale in the pattern) has been limited to values of $\Gamma \approx 3-5$. In systems using a liquid crystal light valve as the nonlinear element, this disadvantage can be reduced, because in this case the coupling between the forward- and backward- propagating beams is achieved in an electro-optical way by writing and reading the liquid crystal light valve hybrid module [22], and aspect ratios up to several hundred can be realized. Even in this experiment, the threshold depends on the intensity itself, requiring high laser intensities to obtain patterns above threshold. The situation is different in photorefractive nonlinear materials, where the nonlinear coupling depends on the intensity ratio and not on the intensity itself.

8.2.1 The Talbot Effect Approach to Pattern Formation

The occurrence of periodic structures in the phase (and intensity) of the reflected beam after traversing the nonlinear medium can be understood using the well-known Talbot effect of Fourier optics [42, 43]. Talbot found that diffraction from a periodic grating results in a diffraction pattern that repeats itself after a distance l_{T}. The optical field in vacuum is given in the paraxial approximation by

$$\frac{\partial A}{\partial z} = -\frac{\mathrm{i}}{2k}\nabla_{\perp} A \, , \tag{8.1}$$

where A is the weakly spatially dependent complex amplitude of the optical field, which has benn separated from the carrier wave, which is described by $\exp\left[-\mathrm{i}(kx-\omega t)\right]$. To solve this differential equation, we Fourier transform $A(x,y,z)$ with respect to the coordinates $r=(x,y)$, resulting in

$$\frac{\partial \tilde{A}}{\partial z} = \frac{\mathrm{i}}{2k}|\boldsymbol{k}_{\perp}|^2 \tilde{A} \, , \tag{8.2}$$

where $\tilde{A}$ is the Fourier transform of the optical field, and $\boldsymbol{k}_{\perp}$ is the transverse wave vector in Fourier space. We obtain

$$\tilde{A}(\boldsymbol{k}_{\perp}, z=2L) = \exp\left(\frac{\mathrm{i}2L}{2k}|\boldsymbol{k}_{\perp}|^2\right)\tilde{A}(\boldsymbol{k}_{\perp}, z=0) \, , \tag{8.3}$$

indicating that propagation over a length $2L$ results in a multiplication by $\exp(\mathrm{i}d|\boldsymbol{k}_\perp|^2)$ in Fourier space. This phase term, is in the case of $l_\mathrm{T} = 2L$, equivalent to

$$2L = l_\mathrm{T} = \frac{4\pi k}{|\boldsymbol{k}_\perp|^2} = \frac{k}{\pi}\Lambda^2 \,. \tag{8.4}$$

When the field has Fourier components at wave vectors of equal length, the same field distribution will be reproduced after a distance l_T. However, for $2L = l_\mathrm{T}/4$ or $2L = 3l_\mathrm{T}/4$, the phase factor equals i or $-$i, resulting in a field that experiences a phase shift of $\pm\pi/2$ relative to the carrier wave.

Let us now assume a nonlinear medium that is periodically modulated by the interference of two beams, e.g. a forward- and a backward-travelling wave. If $n = n_0 + \Delta n \cos(\boldsymbol{k}_\perp \cdot \boldsymbol{r}_\perp)$, the field experiences a phase modulation

$$A_0 \exp(-\mathrm{i}k_0 L)[n_0 + \Delta n \cos(\boldsymbol{k}_\perp \cdot \boldsymbol{r}_\perp)] \tag{8.5}$$

$$\approx A_0 \exp(-\mathrm{i}k_0 n_0 L)[1 - \mathrm{i}k_0 L \Delta n \cos(\boldsymbol{k}_\perp \cdot \boldsymbol{r}_\perp)] \,. \tag{8.6}$$

When the plane wave interacts in the medium with the backward-travelling wave, their interference pattern results in a periodic transverse phase modulation (PM). The plane wave travelling in the forward direction is subject to that modulation. During free propagation in vacuum, Talbot fractional imaging causes the periodic phase modulation to be changed into amplitude modulation in a way that is periodic in the propagation direction z. More precisely, if the initial wave is weakly phase-modulated with a phase grating having a spatial period Λ, the free propagation for a distance

$$l_\mathrm{T} = \frac{2\Lambda^2}{L} \tag{8.7}$$

is necessary to restore the initial field at $z = 0$ by self-reproduction at planes $z_m = m l_\mathrm{T}$, m being a positive integer. The length l_T given by (8.7) is called the Talbot self-imaging distance. This remarkable result is a straightforward property of the Fresnel propagator. A closer inspection of the Fresnel integral shows that, after free propagation over a distance $l_\mathrm{T}/4$, the initially phase-modulated wave is changed into an amplitude-modulated wave. The amplitude modulation of the optical field at $z = l_\mathrm{T}/4$ is in phase with the phase modulation of the field at $z = 0$. After further propagation over $l_\mathrm{T}/4$, the wave again becomes phase-modulated, but out of phase with the initial field. After another distance $l_\mathrm{T}/4$, the wave is still amplitude-modulated, but with opposite phase. Eventually, after a total distance l_T the initial phase modulation is restored.

These passages from phase to amplitude modulation are schematically sketched in Fig. 8.4. Assume now that a mirror is placed at the plane $z = l_\mathrm{T}/8$ or at $z = 3l_\mathrm{T}/8$. From Fig. 8.4, it is obvious that the field reflected onto the sample is amplitude-modulated in phase with the phase modulation of the incident wave:

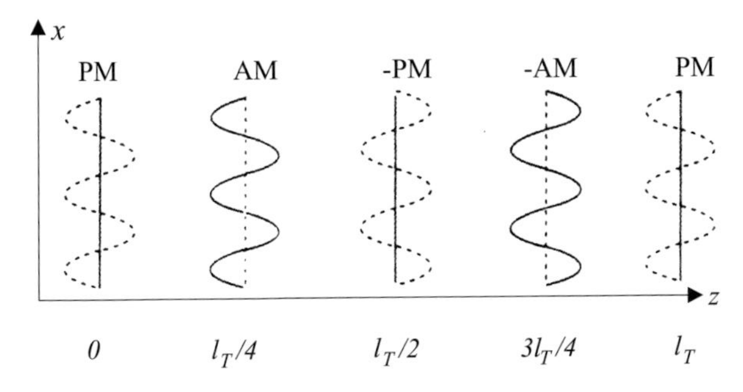

Fig. 8.4. Fractional Talbot effect for a plane wave propagating along z, sinusoidally modulated with the wave vector K. AM = amplitude modulation plane, PM = phase modulation plane. $l_\mathrm{T} = 4\pi k/(|\boldsymbol{k}_\perp|^2) = (k\Lambda^2)/\pi$

$$E_0 \exp(-\mathrm{i}k_0 n_0 L)[1 \pm k_0 L\, \Delta n \cos(\boldsymbol{k}_\perp \cdot \boldsymbol{r}_\perp)] \,. \tag{8.8}$$

The field that is fed back in this case is for $d = l_\mathrm{T}/8$ in phase with the refractive index contribution in the medium, and for $d = 3l_\mathrm{T}/8$, out of phase.

When the mirror distance is fixed, in-phase feedback is possible for a pattern with a wavelength Λ_in, and out-of-phase feedback for a wavelength Λ_out. The corresponding mirror positions

$$L = \frac{l_\mathrm{T}}{8} = \frac{\pi k}{2|\boldsymbol{k}_\perp|^2} = \frac{k\Lambda_\mathrm{in}^2}{8\pi} \quad \text{and} \quad L = \frac{3l_\mathrm{T}}{8} = \frac{3\pi k}{2|\boldsymbol{k}_\perp|^2} = \frac{3k\Lambda_\mathrm{out}^2}{8\pi} \,. \tag{8.9}$$

Owing to the nonlinearity of the medium, the amplitude modulation of the reflected waves induces a corresponding phase modulation in the incoming wave tending to enhance the spatial modulation already present. A positive feedback is achieved only for a feedback in phase, for a self-focusing material. Therefore, a perfectly adjusted mirror allows only patterns of the wavelength Λ_in.

According to the Talbot model, the mirror-to-sample distance L determines the period Λ_T of the spatial phase grating that can suffer positive feedback in the medium, because the following condition must be fulfilled:

$$\Lambda_\mathrm{T} = \sqrt{(\lambda l_\mathrm{T}/2)} = 2\sqrt{(\lambda L)} \,. \tag{8.10}$$

When the incident laser intensity I_0 is above the threshold for self-oscillation, a two-dimensional index grating with a lattice period given by (8.10) is generated in the nonlinear medium. If perfect cylindrical symmetry is assumed around the z axis, only three regular lattices can be realized in the (x, y) plane given a single lattice constant $K = 2\pi/\Lambda$. These are a sinusoidal or ripple lattice (roll or stripe), a square lattice and a hexagonal lattice; the latter is preferred in the case of a photorefractive nonlinear medium.

The Talbot approach is an elegant and efficient model that shows the direct dependence of the transverse modulation on the mirror-to-sample distance. However, it does not explain the onset of a modulation instability starting from noise, and is not capable of giving an adequate explanation of the appearance of different spatial patterns in two transverse dimensions.

8.2.2 Linear Stability Analysis

A linear stability analysis of the equations of motion provides a basic understanding of the system behavior and yields information about the parameter regions of transverse instability. In this analysis, the stationary homogeneous solution is perturbed, with a single or periodic perturbation. Depending on the parameters (one or more *control parameters*), these perturbations either are exponentially damped (stable solution, negative growth rate) or grow exponentially (unstable solution, positive growth rate). The instability *threshold* is determined by the value of the control parameter for which the growth rate goes through zero. In the case considered here, a homogeneous intensity profile is perturbed by a modulation with a certain spatial period. The modulation period for which the growth rate first goes through zero will grow exponentially. This is referred to as a *modulation instability*, the instability of a nonlinear system to periodic perturbations. Thus, the parameter space can be divided into stable and unstable regions, with a threshold curve that is the marginal-stability curve, where the growth rate changes sign. The simplest bifurcations of fixed-point solutions depending on one control parameter are saddle node, transcritical and pitchfork bifurcations. In a Hopf bifurcation, a complex pair of eigenvalues crosses from the stable to the unstable regime. Thus, a stable fixed point loses stability to a stable limit cycle. The details of the theoretical background and applications of linear stability analysis are the subject of a large number of textbooks on nonlinear dynamics and pattern formation (see e.g. [44]).

In order to understand the main features of a linear stability analysis, let us start with the following common expression for a dynamical system:

$$\frac{\mathrm{d}U}{\mathrm{d}t} = f(r, U, t)i \ , \tag{8.11}$$

where U is the state variable of the system, r is the external control parameter (or a set of control parameters), and f is a nonlinear function of the variables and of the control parameter. We add a small perturbation to the stationary solution U_0 so that,

$$U = U_0 + \varepsilon U' \ . \tag{8.12}$$

This new solution is substituted into the original system equation. If we expand the function f into a power series of ε and neglect all nonlinear terms in ε, the linearized problem reads

$$\varepsilon \frac{\mathrm{d}U'}{\mathrm{d}t} = \varepsilon \frac{\partial f}{\partial U} U' = \varepsilon \Lambda U' \,. \tag{8.13}$$

The solution of this linear differential equation for the disturbance U' is generally given by

$$U' \propto \exp\left[\Lambda(r)t\right]. \tag{8.14}$$

In general, $\Lambda(r)$ represents a complex variable, i.e. $\Lambda(r) = \exp(\lambda t)\exp(\mathrm{i}\delta t)$. Equation (8.14) represents a linear equation in U' and implies that the perturbation U' grows exponentially if $\lambda(r) > 0$ and decays if $\lambda(r) < 0$. We have explicitly noted the direct dependence of Λ on the control parameter r, which implies that variation of the control parameter r offers the possibility to manipulate the stability (or instability) of the whole system. This is the reason why r is called the control parameter: one or a set of external control parameters decides whether a complex physical system is stable or driven to instability.

In the framework of the linear stability analysis, $\lambda(r)$ is referred to as the growth rate and decides whether a stationary solution is stable or not. If

$$\lambda(r) < 0 \,, \qquad \text{the stationary solution is stable,} \tag{8.15}$$

$$\lambda(r) = 0 \,, \qquad \text{the instability is neutral or marginal,} \tag{8.16}$$

$$\lambda(r) > 0 \,, \qquad \text{the stationary solution is unstable .} \tag{8.17}$$

The quantity δ represents the oscillation frequency and thus determines the temporal dynamics of the solution. If

$$\delta \neq 0 \,, \qquad \text{we have oscillatory instability;} \tag{8.18}$$

$$\delta = 0 \,, \qquad \text{we have static instability .} \tag{8.19}$$

This short description represents the simplest case of a linear stability analysis. In the case of more than one state variable, Λ is the Jacobian matrix at the fixed point. It is a linear operator and the multivariable analogue of the derivative $\partial f/\partial U$. In the case of a Jacobian operator, it is necessary to obtain the spectrum of eigenvalues λ_i in order to determine the stability of the system (e.g. the nonlinear system is absolutely stable when all eigenvalues $\lambda_i < 0$). Moreover, if there is a supplementary spatial dependence of the function, for example $f = f(r, U, \partial U/\partial x, t)$, the translational invariance of the infinitely extended plane waves requires an expansion into spatial Fourier series. In this case, the spatial derivative $\partial/\partial x$ is transformed to $\mathrm{i}k_x$ in Fourier space, thus reducing the problem from a partial differential equation to a simple differential equation in time. The marginal-stability condition $\lambda(r, k_x) = 0$ then allows one to derive the threshold curve separating the area of stable ($\lambda < 0$) and unstable ($\lambda > 0$) solutions. The aim of the linear stability analysis in this case is to derive the most unstable wave vector k_c, for which the curve $r = r(k_x)$ has a global minimum (see Fig. 8.5). The point $(k_\mathrm{c}, r_\mathrm{c})$ in parameter space defines the onset of spatial instability with a typical wave vector k_c of the structure at the critical control parameter r_c.

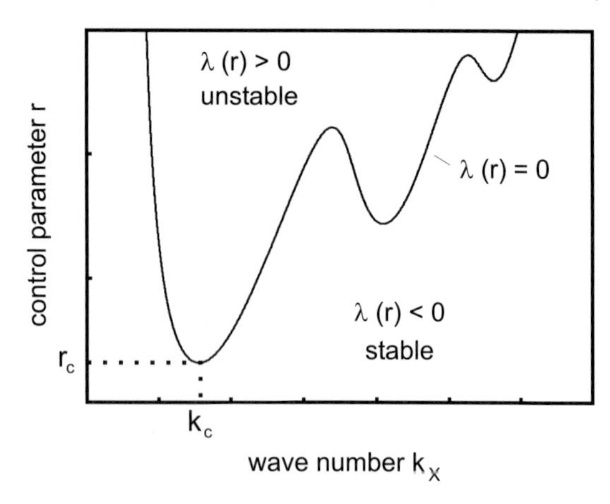

Fig. 8.5. Example of a threshold curve determining the onset of a modulation instability. The absolute minimum of the curve gives the critical wave number k_c, which loses stability at a critical value of the control parameter r_c. Also shown are the regions of stability and instability, which are separated by the marginal-stability curve $\lambda(r, k_x) = 0$

8.2.3 Nonlinear Stability Analysis

The linear stability analysis described in the last section predicts the instability of all wave vectors on a circle in the Fourier plane. These wave vectors possess the same magnitude, but are degenerate with respect to their orientation. In the nonlinear stage of the pattern generation process, a finite set of wave vectors is selected by the system, leading to a specific pattern, e.g. a hexagonal one. Prediction of the pattern type is completely in the realm of nonlinear analysis. We expect that the spectrum of unstable wave vectors derived from the linear stability analysis will interact in a nonlinear way and select some wave vectors to form a symmetric pattern by resonant superposition. As a consequence, the number of wave vectors essential for pattern formation is reduced significantly, allowing one to limit the order of the perturbation amplitudes of the wave vectors.

The basic system behavior is characterized by the interaction of a large number of degrees of freedom. However, in the vicinity of the instability threshold, around $U = U_0 + \varepsilon U'$, where ε is the stress parameter describing the deviation from the threshold, only a limited number of degrees of freedom may be operating, allowing one to describe the system by a limited number of *order parameters* (see [45, 44]). A weakly nonlinear analysis therefore leads to an order parameter equation. At threshold, the perturbation $\Delta U = U - U_0$ can be expanded into a finite sum of N Fourier modes, yielding

$$\Delta U(\boldsymbol{r}, t) = U(\boldsymbol{r}, t) - U_0 = \sum_{i=1}^{N} \left[A_i(\boldsymbol{r}, t) \exp\left(\mathrm{i}\boldsymbol{k}_i \cdot \boldsymbol{r}\right) + \text{c.c.} \right], \qquad (8.20)$$

where the wave vectors $\boldsymbol{k}_i$ all lie on a circle and their orientation is given by the scalar product $\boldsymbol{k}_i \cdot \boldsymbol{r}$. The complex amplitudes A_i are proportional to the Fourier amplitudes of the expansion. In this context, they are called order parameters of the nonlinear system [45].

The principle of the nonlinear stability analysis using this approach is similar to that of the linear stability analysis discussed earlier. The ansatz (8.20) is inserted into the system equation, with the deciding (and mathematically challenging) aspect that nonlinear terms are also taken into account. In general, one has to include both temporal and spatial dependences of the order parameter(s). This can be accomplished using new, slowly varying time and space variables (*multiple-scale analysis*, see [46]). Applying the so-called Fredholm alternative (see [47]), leads to the order parameter equations. However, even in the case of the simplest optical nonlinearity, the Kerr-nonlinearity, this approach leads to a complicated set of partial differential equations. Experiments with nonlinear materials under optical feedback suggest that a hexagonal pattern represents a generic solution for these pattern forming systems, analogous to what is found in pattern formation experiments in hydrodynamics [44]. This also holds for a photorefractive nonlinearity in the case of a positive diffraction length [31]. Therefore, instead of taking an infinite sum of Fourier modes in the expansion of the perturbation, we restrict ourselves to a triad of Fourier modes to describe the nonlinear stage of the pattern-forming process:

$$\Delta U(\boldsymbol{r}, t) = \sum_{i=1}^{3} A_i \exp(\mathrm{i}\boldsymbol{k_i} \cdot \boldsymbol{r}) + \text{c.c.} \tag{8.21}$$

This superposition of three rolls is visualized for an angle of $2\pi/3$ between the three modes in Fig. 8.6. Owing to the three-wave interaction $\boldsymbol{k_1}+\boldsymbol{k_2}+\boldsymbol{k_3} = 0$, a hexagon is formed by resonant generation of the remaining wave numbers. Using this triad mode ansatz (8.21), the temporal evolution of the order parameters A_i can be described by a set of three Ginzburg–Landau-type equations (also called the Ginzburg–Landau hexagon equations)

$$\frac{\partial A_1}{\partial t} = \nu_0 A_1 + \mu A_2^\star A_3^\star - \left[g_0|A_1|^2 + 2g_{\pi/3}(|A_2|^2 + |A_3|^2)\right] A_1 \,, \tag{8.22}$$

$$\frac{\partial A_2}{\partial t} = \nu_0 A_2 + \mu A_1^\star A_3^\star - \left[g_0|A_2|^2 + 2g_{\pi/3}(|A_1|^2 + |A_3|^2)\right] A_2 \,, \tag{8.23}$$

$$\frac{\partial A_3}{\partial t} = \nu_0 A_3 + \mu A_1^\star A_2^\star - \left[g_0|A_3|^2 + 2g_{\pi/3}(|A_1|^2 + |A_2|^2)\right] A_3 \,, \tag{8.24}$$

The original system of equations reduces to a set of three amplitude equations. Similar equations have been obtained in a number of systems in hydrodynamics [44]. Hexagon formation is universally described by these coupled equations of Ginzburg–Landau type; the system properties enter solely via the coefficients ν_0, μ, g_0 and $g_{\pi/3}$. The universality of the observations of

Fig. 8.6. Hexagon as a resonant superposition of three modes with an angle of $2\pi/3$ between them. Resonant three-wave interactions via $\boldsymbol{k_1} + \boldsymbol{k_2} + \boldsymbol{k_3} = 0$ lead to the remaining spots of the hexagon. Higher-order spots $2k_i$ (indicated by a *dashed arrow*) and $\sqrt{3}k_i$ (indicated by *dotted arrows*) are generated by four-wave interactions between the basic wave vectors

hexagons is reflected in the simplicity of the amplitude equations (8.22)–(8.24). This is not surprising since these equations describe the simplest form of three-wave interactions $\boldsymbol{k_1} + \boldsymbol{k_2} + \boldsymbol{k_3} = 0$.

The first term, $\nu_0 A_1$, describes the exponential growth or decay of the wave and is proportional to the distance to the linear instability threshold $r - r_c$. The second term with the quadratic coefficient μ, initiates the hexagon formation with $\boldsymbol{k} = \boldsymbol{k_1} + \boldsymbol{k_2}$ and can be regarded as a phase-sensitive hexagon source term. It represents a special feature of two-dimensional systems, since it requires all three waves to be present. The remaining self-cubic and cross-cubic terms cause saturation of the explosive growth generated by the first two terms and lead to the generation of higher harmonics in the Fourier plane. This is illustrated in Fig. 8.6. The self-cubic interaction $g_0|A_i|^3$ leads to the second order spots $2k_i$, while four-wave interactions such as $\boldsymbol{k} = \boldsymbol{k_1} + \boldsymbol{k_2} - \boldsymbol{k_3}$ contribute to the $\sqrt{3}$ terms and are described by the factor $g_{\pi/3}$ in (8.22)–(8.24). These higher-order spots are clearly visible in the experimental picture Fig. 8.6, and validate the triad mode ansatz (8.21) and the subsequent Ginzburg–Landau hexagon (GLH) equations (8.22)–(8.24).

The coupled equations (8.22)–(8.24) determine the behavior of the three sets of rolls describing the hexagonal structure. Since a single roll pattern ($A_2 = A_3 = 0$) is a subset of the hexagonal solution, (8.22)–(8.24) are useful in particular for describing hexagon–roll competition [48]. They possess stationary solutions of four pure kinds: the *homogeneous state* (O), with all amplitudes vanishing, $A_i = 0$, *rolls* (R), with $A_2 = A_3 = 0$; and *positive*

(H^+) and *negative* (H^-) *hexagons*, depending on the relative phases of the three interacting waves. For the steady state, where $\partial A_i/\partial t = 0$, the stability, instability and bifurcation behavior of these solutions are determined by the coefficients μ, g_0 and $g_{\pi/3}$ [48]. Unfortunately, the computation of these factors is a big challenge, even for "simple" nonlinearities such as Kerr slices [17].

The Ginzburg-Landau hexagon (GLH) equations (8.22)–(8.24) are a special case of a more general form of amplitude equation

$$\frac{\partial A_j}{\partial t} = \varepsilon A_j + \sigma_{jlm} A_l^* A^* m - \sum_{l=1}^{N} \gamma_{jl} A_l A_l^* A_j \ , \tag{8.25}$$

where $j, l, m = 1, \cdots, N$. Here, ε is the linear, σ_{jlm} is the quadratic and γ_{jl} is the cubic interaction coefficient. These equations are the *Ginzburg–Landau amplitude equations* and determine the behavior of the complex amplitudes A_j of the set of rolls describing the hexagonal structure. The quadratic coefficient σ_{jlm} can only be nonzero when $\boldsymbol{k}_j + \boldsymbol{k}_l + \boldsymbol{k}_m = 0$ ($N = 3$). In rotationally degenerate problems (where the critical modes lie on a circle), as in our case, this condition is only satisfied when each pair of wave vectors forms an angle of $2\pi/3$, thus forming a hexagonal pattern. If σ_{jlm} is nonzero, the quadratic term will initially dominate close to threshold, leading to a hexagonal pattern. Therefore, the formation of hexagons can be directly concluded from the presence of the quadratic nonlinear term in the order parameter equation and this simple phase-matching criterion on the critical circle. Since there is no saturation at this order, we cannot draw conclusions about the long-term behavior of the system. A symmetry such as an inversion symmetry ($A_i \rightarrow -A_j$) may force $\sigma_{jlm} \equiv 0$. As a consequence, the cubic terms arising from the three-wave interaction $\boldsymbol{k}_j + \boldsymbol{k}_l - \boldsymbol{k}_l$, will play the dominating role in the pattern formation process in this case.

The central importance of the above considerations lies in the existence of a Ginzburg–Landau equation, which constitutes a generic and generalized equation for pattern-forming systems. A complex system is reduced to a simple set of equations, where the specific physical properties enter only into the parameters of the equations. Equations (8.22)–(8.24) are also known as order parameter equations. As a consequence, the Fourier amplitudes A represent order parameters in the sense of the traditional theory of self-organization [45].

8.3 The Photorefractive Single-Feedback System

The first observation of hexagonal patterns due to counterpropagating beams in a photorefractive $KNbO_3$ crystal was reported in 1993 by Honda [25]. In this pioneering experiment, the counterpropagating beams were created by reflection at the crystal surface or, alternatively, by using a phase-conjugate replica of the incoming beam. Honda observed the formation of hexagonal patterns above a certain threshold for the photorefractive coupling strength. These results initiated a whole new field in photorefractive optics – transverse pattern formation due to counterpropagating beams. We shal highlight throughout the following chapters the most important of the contributions to this field that lead to a general understanding of the origins of pattern formation in photorefractive materials, but will also present new and fascinating effects that have so far not been explained.

Pattern formation in photorefractive feedback systems occurs through modulation instabilities that arise owing to the formation of reflection gratings. Above a certain threshold for the photorefractive coupling strength γl, satellite beams are generated with a particular angle θ relative to the central beam. Photorefractives are well suited for experimental pattern observation, since their intrinsically slow dynamics simplify time-resolved measurements. These systems offer a number of important advantages and meet the requirements for a robust and versatile pattern-forming device. The direct access to the control parameters of the pattern-forming process and to the Fourier spectrum of the pattern offers a variety of possibilities for steering the pattern-forming process. We shall learn more about the possibilities to manipulate the pattern states in Chapter 10.

The principle of the interaction geometry of a photorefractive feedback system is depicted in Fig. 8.7a. As in the previously described general approach to optical pattern formation, single-mirror feedback is easily created by placing a high-reflectivity mirror behind the photorefractive medium. A plane wave $\mathcal{F}$ of complex amplitude F is incident on a thick photorefractive medium of length l, where it interacts with the counterpropagating feedback beam $\mathcal{B}$ (amplitude B), which is provided by reflection of the forward wave $\mathcal{F}$ from a feedback mirror. These two beams couple via a reflection grating of wave vector $2k_0n_0$ inside the photorefractive medium, where k_0 is the vacuum wave number of beams $\mathcal{F}$ and $\mathcal{B}$, and n_0 is the linear refractive index of the photorefractive medium.

8.3.1 The Virtual-Mirror Concept

Despite the simplicity of the system described above, two different imaging systems that are advantageous for the creation of a whole variety of patterns, as well as for the realization of novel techniques for manipulation and control of pattern formation, were used in our experiments. These optical imaging systems are shown in Figs. 8.7b,c and depict ways to realize a feedback system with one (Fig. 8.7b) or two (Fig. 8.7c) lenses in the feedback path. The three systems shown in Fig. 8.7 are referred to as (a), (b) and (c) in what follows. The advantage of these systems in contrast to the simple standard setup (a) is the possibility to create a *virtual mirror* inside the crystal, which provides negative diffraction lengths [49]. From an experimental point of view, these mirror positions are of central importance to accessing spontaneous patterns

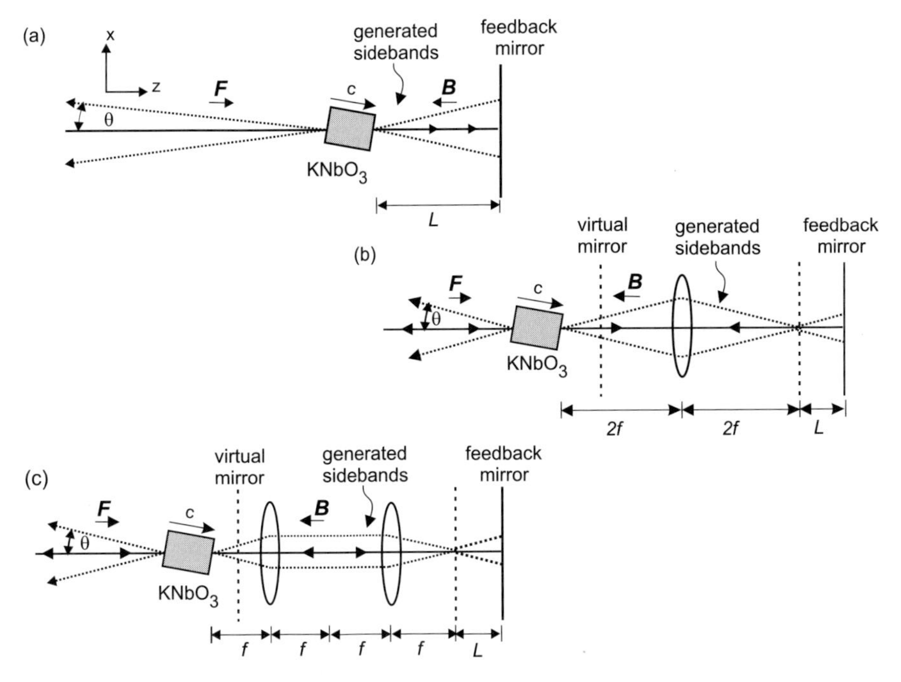

Fig. 8.7. Different possibilities for a single-mirror feedback setup with an additional diffraction length $2L$. (**a**) Without any lenses, pure free-space diffraction length $2L$. (**b**) With one lens creating a $2f$–$2f$–$2L$–$2f$–$2f$ feedback system. (**c**) 1:1 imaging f–f–f–f–$2L$–f–f–f–f feedback system with two lenses in the optical feedback path. A Fourier plane is created between the two lenses

other than the hexagonal one, which is the only solution obtainable when the setup (a) is used without imaging lenses. Another reason for using one of the two setups (b) and (c) is of a purely experimental nature and is simply related to the convenience of placing a mirror behind the crystal without touching the latter.

During a round trip in the feedback path, the sideband beams experience a phase lag of $\varphi = k_0\theta^2 L$ relative to the carrier beam [31, 50], where θ is the angle between the carrier and the sidebands. A negative diffraction length thus corresponds to a negative phase lag. This in turn allows one to investigate pattern formation symmetrically around the zero-diffraction-length plane.

The usefulness of these configurations can be shown by comparing the modes of operation of the three different configurations by use of the standard $ABCD$ matrix formalism [51, 52]. This formalism is based on the fact that an optical system can be represented by a ray transfer matrix, which describes the propagation through an optical system and links the transverse coordinate x (calculated relative to the optical axis) and the angle θ relative to the optical axis at the input to the corresponding values at the output of the optical system according to

$$\begin{pmatrix} x_{\mathrm{out}} \\ \theta_{\mathrm{out}} \end{pmatrix} = \begin{pmatrix} A & B \\ C & D \end{pmatrix} \begin{pmatrix} x_{\mathrm{in}} \\ \theta_{\mathrm{in}} \end{pmatrix} . \tag{8.26}$$

Ray transfer matrices of this kind can be calculated for every combination of optical components [51] and represent idealized constructions corresponding to the mathematical limit $\lambda \to 0$. More complex systems are described by simple matrix products of the individual matrices.

The ray transfer matrix $\mathbf{M}$ for the simple feedback system (a) can be shown to be completely equivalent to the system (c) with two lenses in the optical feedback path:

$$\mathbf{M}_{(\mathrm{c})} = \mathbf{M}_{(\mathrm{a})} = \begin{pmatrix} 1 & 2L \\ 0 & 1 \end{pmatrix} . \tag{8.27}$$

The two rays are related to each other as if a real mirror was placed behind the crystal. The system (c) has the supplementary advantage that the Fourier spectrum of the pattern is automatically created between the two lenses of the feedback arm, which was used for pattern control using spectral techniques as described in Chap. 10.

This equivalence does not hold in a mathematical sense for a comparison of the single-mirror feedback scheme with the one-lens setup (b). In this case, the multiplication of the matrices for the subsequent free-space propagations and lens passes yields a different ray transfer matrix $\mathbf{M}$ [50]:

$$\mathbf{M}_{(\mathrm{b})} = \begin{pmatrix} 1 + 2L/f & 2L \\ 2(1 + L/f)f^{-1} & 1 + 2L/f \end{pmatrix} . \tag{8.28}$$

This results in nonconcentric imaging and a change of position and angle compared with the systems (a) and (c). However, the deviations between (8.28) and (8.27) are of the order L/f. Since in our experiments the condition $L \ll f$ is generally fulfilled, the correction term can be neglected and justifies the use of feedback system (c) in some of the experiments described later.

8.4 Linear Stability Analysis of the Equations of Motion

Our analysis of the photorefractive feedback system to obtain the threshold condition for pattern formation is based on the formation of a transverse modulation instability of the two counterpropagating waves, which couple through a single reflection grating in a thick photorefractive medium. A simple representation of our system is depicted in Fig. 8.8 and introduces the notation used in the subsequent calculations. Two plane waves with complex amplitudes F_0 (forward) and B_0 (backward) enter the photorefractive crystal from

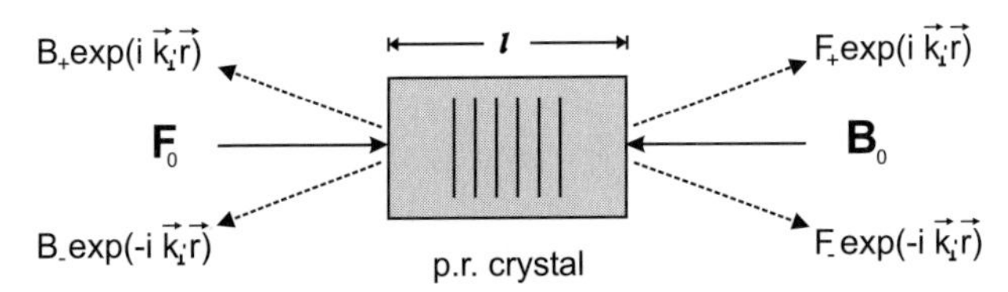

Fig. 8.8. Basic interaction geometry, introducing the variables for the linear stability analysis. F_0 and B_0 denote the amplitudes of the plane pump waves. The *dashed arrows* depict the directions of the spatial sidebands, which have corresponding transverse wave numbers $\boldsymbol{k}_\perp$

opposite sides. In order to analyze the widely used single-feedback-mirror configuration, the second beam B_0 is is assumed to be created by reflection at a mirror (not shown here) placed at a distance L from the medium. The homogeneous solutions F_0 and B_0 are assumed to be perturbed with periodic modulations with wave numbers $\boldsymbol{k}_\perp$, leading to spatial sidebands in the optical far field with transverse wave numbers $\boldsymbol{k}_\perp$ and $-\boldsymbol{k}_\perp$. The small amplitudes of the perturbations are denoted by F_+ and F_- for the incoming beam and by B_+ and B_- for the back-reflected beam. Reflection at the mirror connects the perturbations $F_+(l)$ and $F_-(l)$ with $B_-(l)$ and $B_+(l)$ in a defined way, as we shall see later. The photorefractive two-wave mixing equations in such a reflection geometry are given by (3.76) and (3.77) and can be written, for the case of two waves with amplitudes $F = A_1$ and $B = A_2$, as

$$\frac{\partial F}{\partial z} - \frac{\mathrm{i}}{2k_0 n_0}\nabla_\perp^2 F = \gamma \frac{|B|^2}{|F|^2 + |B|^2} F \, , \tag{8.29}$$

$$\frac{\partial B}{\partial z} + \frac{\mathrm{i}}{2k_0 n_0}\nabla_\perp^2 B = \gamma^* \frac{|F|^2}{|F|^2 + |B|^2} B \, , \tag{8.30}$$

where k_0 is the wave number in vacuum, n_0 is the linear refractive index of the photorefractive medium, and γ is the complex photorefractive coupling coefficient as defined in (3.75). $\nabla_\perp^2 = \partial^2/\partial x^2 + \partial^2/\partial y^2$ is the transverse Laplace operator, representing the effects of transverse coupling. The different signs of the terms containing the Laplace operator originate from the different incidence directions of the forward and the backward beam.

The linear stability analysis for the case of two counterpropagating beams and mirror boundary conditions follows the calculations of Honda and Banerjee [31]. This analysis was performed for the completely frequency degenerate case, i.e. not only the pump beams, also the spatial sidebands have the same optical frequency ω. The assumption of frequency degeneracy was motivated by the absence of experimental observations of the dynamics in the exact counterpropagating case. Possible dynamic instabilities were taken into account in [50] and will be shown to be a crucial element in the case of frequency-detuned pump beams.

The right-hand sides of (3.76) and (3.77) are based on the material response originally derived for two plane waves. We assume that these equations are valid if the wave number of the transverse pattern is much smaller than that of the reflection grating. In most of the experiments reported here, the wave number of the transverse patterns was 1/400 to 1/200 of that of the reflection grating, which justifies this assumption.

Let us now assume that two weak sidebands appear owing to the coupling of the forward and the backward beam, and that these sidebands are arranged symmetrically around the central carrier beam:

$$F(\boldsymbol{r}) = F_0(z)[1 + F_{+1}(z)\exp(\mathrm{i}\boldsymbol{k}_\perp \cdot \boldsymbol{r}_\perp) + F_{-1}(z)\exp(-\mathrm{i}\boldsymbol{k}_\perp \cdot \boldsymbol{r}_\perp)] \,, \tag{8.31}$$

$$B(\boldsymbol{r}) = B_0(z)[1 + B_{+1}(z)\exp(\mathrm{i}\boldsymbol{k}_\perp \cdot \boldsymbol{r}_\perp) + B_{-1}(z)\exp(-\mathrm{i}\boldsymbol{k}_\perp \cdot \boldsymbol{r}_\perp)] \tag{8.32}$$

Here, $\boldsymbol{r}_\perp$ is the position in the transverse plane, z is the position along the beam axis, $\perp$ denotes the transverse direction, and $F_{\pm 1}$ and $B_{\pm 1}$ are the relative amplitudes of the spatial sidebands. $F_0(z)$ and $B_0(z)$ represent the plane-wave solutions, i.e. the solutions of (3.76) and (3.77). For simplicity, we assume that the reflectivity of the feedback mirror is 1 and that absorption in the medium is negligible.

8.4.1 Theoretical Approach to the Linear Stability Analysis

For the following linear stability analysis, several assumptions need to be made. The transverse wave vector should fulfill $|\boldsymbol{k}_\perp| \ll 2k_0 n_0$, which implies that the sideband angles are small. The amplitudes of the sidebands, which are considered as the disturbances in the linear stability analysis, are also assumed to be small compared with the carrier beam, giving $F_{\pm 1}$ and $B_{\pm 1} \ll 1$. These assumptions are justified by our experimental observations. Finally, the forward and backward beams are assumed to be exactly counterpropagating. Noncollinearity between the sidebands because of an angular deviation is equivalent to a frequency shift between the forward and the backward beam [29, 53], and may lead to moving gratings, which in turn may lead to supplementary drift instabilities in the feedback system. This case will be treated later on in Sect. 10.2.

As in the general procedure for the linear stability analysis, (8.31) and (8.32) are inserted into the beam-coupling equations. When the assumptions stated above are made, we obtain achieve

$$\left(\frac{\partial}{\partial z} + \mathrm{i}k_\mathrm{d}\right) F_{+1} = \gamma A(-F_{+1} - F_{-1}^* + B_{+1} + B_{-1}^*) \,, \tag{8.33}$$

$$\left(\frac{\partial}{\partial z} - \mathrm{i}k_\mathrm{d}\right) F_{-1}^* = \gamma^* A(-F_{+1} - F_{-1}^* + B_{+1} + B_{-1}^*) \,, \tag{8.34}$$

$$\left(\frac{\partial}{\partial z} - \mathrm{i}k_\mathrm{d}\right) B_{+1} = \gamma^* A(F_{+1} + F_{-1}^* - B_{+1} - B_{-1}^*) \,, \tag{8.35}$$

$$\left(\frac{\partial}{\partial z} + \mathrm{i}k_\mathrm{d}\right) B_{-1}^* = \gamma A(F_{+1} + F_{-1}^* - B_{+1} - B_{-1}^*) \,, \tag{8.36}$$

where

$$k_{\mathrm{d}} = \frac{k_{\perp}^2}{2k_0 n_0} = \frac{k_0 \theta^2}{2n_0} \tag{8.37}$$

i.e. k_{d} includes the sideband angle θ.

The right-hand sides of (8.34)–(8.36) are proportional to the parameter $A(z)$, which represents the strength of transverse coupling, and is given mainly by the modulation depth m:

$$A(z) = \frac{|F_0(z)|^2 |B_0(z)|^2}{(|F_0(z)|^2 + |B_0(z)|^2)^2} = \frac{1}{4} m(z)^2 \,. \tag{8.38}$$

Because A is an explicit function of z, (8.34)–(8.36) cannot be solved analytically in a general way. As long as r is close to 1, the transverse coupling $A(z)$ depends only weakly on the intensity ratio $r = |B_0(l)|^2/|F_0(l)|^2$. This corresponds to the case of a high reflectivity of the feedback mirror. For $1 < r < 0.66$, a value that corresponds to the maximum reflectivity in the experiments, including losses due to Fresnel reflections at the surfaces of the crystal, the dependence of A is almost a constant and allows us to set $A = 1/4$ for all propagation distances z inside the crystal. In a more generalized analysis, it is necessary to take account also of intermediate modulation depths, which can be done in a first approximation by the renormalization $f(r) = r/\rho$, where ρ is a constant factor of about 3 for standard photorefractive feedback systems [54]. In a complete, exact derivation, $f(r)$ is a complex function of the intensity and the coupling constant (see Sect. 3.5.3). For the present idealized case, however, the system (8.34)–(8.36) can be solved analytically.

Defining the amplitudes of the sidebands in a vector representation

$$\boldsymbol{u} = \begin{pmatrix} F_{+1} \\ F_{-1}^* \\ B_{+1} \\ B_{-1}^* \end{pmatrix} , \tag{8.39}$$

we can write (8.34)–(8.36) in a matrix formalism as

$$\frac{\partial}{\partial z} \boldsymbol{r} = D \cdot \boldsymbol{u} \,, \tag{8.40}$$

where

$$D = \begin{pmatrix} -\gamma/4 - \mathrm{i}k_{\mathrm{d}} & -\gamma/4 & \gamma/4 & \gamma/4 \\ -\gamma^*/4 & -\gamma^*/4 + \mathrm{i}k_{\mathrm{d}} & \gamma^*/4 & \gamma^*/4 \\ \gamma^*/4 & \gamma^*/4 & -\gamma^*/4 + \mathrm{i}k_{\mathrm{d}} & -\gamma^*/4 \\ \gamma/4 & \gamma/4 & -\gamma/4 & -\gamma/4 - \mathrm{i}k_{\mathrm{d}} \end{pmatrix} . \tag{8.41}$$

In general, the solution of an equation of the type (8.40) is given by

$$\boldsymbol{u} = \exp(Dz) \cdot \boldsymbol{u}_0 \,, \tag{8.42}$$

where $\boldsymbol{u}_0$ represents the boundary condition at $z = 0$. This equation can be solved either using a Laplace transformation method [55], as stated in [31], or with a base transformation method [56], applied here in order to find the threshold condition for modulation instability. The phase retardation of beam B due to the diffraction in the feedback path is given by [31, 33]

$$\varphi = 2k_{\mathrm{d}} n_0 L. \tag{8.43}$$

This phase factor emerges from simple free-space propagation and can be easily derived using the Fresnel–Kirchhoff diffraction theory [50]. Therefore, the boundary conditions for the case of a mirror placed behind the medium at a distance L can be written as

$$F_+(0) = F^\star_-(0) = 0 \ , \tag{8.44}$$

$$B_+(l) = B_+(l) \exp(-2\mathrm{i}k_{\mathrm{d}} n_0 L) \ , \tag{8.45}$$

$$B^\star_-(l) = F^\star_-(l) \exp(2\mathrm{i}k_{\mathrm{d}} n_0 L) \ . \tag{8.46}$$

Using these boundary conditions, an equation of the form

$$\mathcal{B} \cdot \begin{pmatrix} b_+(0) \\ b^*_-(0) \end{pmatrix} = 0 \tag{8.47}$$

can be obtained, where the threshold condition for modulation instability is determined by $\det \mathcal{B} = 0$. The threshold condition for modulation instability in the case of two counterpropagating beams subject to mirror boundary conditions can be written as

$$\begin{aligned} &\cos(\chi l)\cos(k_{\mathrm{d}} l) + \frac{k_{\mathrm{d}}}{\chi}\sin(\chi l)\sin(k_{\mathrm{d}} l) \\ &+\frac{\mathrm{Re}(\gamma)}{2\chi}\left\{\sin(\chi l)\cos\left[k_{\mathrm{d}}(l + 2n_0 L)\right]\right\} \\ &+\frac{\mathrm{Im}(\gamma)}{2\chi}\left\{\sin(\chi l)\sin(k_{\mathrm{d}} l) - \sin(\chi l)\sin\left[k_{\mathrm{d}}(l + 2n_0 L)\right]\right\} = 0 \ . \end{aligned} \tag{8.48}$$

This equation defines the conditions for a certain vector $\boldsymbol{k}$ to become unstable, and depends on the coupling constant γ, on the propagation length L and on k_{d}, which includes the transverse wave vector and thus the sideband angle θ. Therefore, the threshold equation is of the form $f(\gamma, L, \theta) = 0$.

In general, γ is a complex-valued measure. In the analysis of our experimental realization, however, we shall focus on photorefractive $KNbO_3$, a crystal with a high coupling constant dominated by diffusion-induced charge transport effects. Therefore, the coupling constant in this case is a purely real variable, and only energy coupling arises, i.e. $\mathrm{Im}(\gamma) = 0$. Setting $\gamma \rightarrow \mathrm{Re}(\gamma)$, the threshold condition (8.48) simplifies to

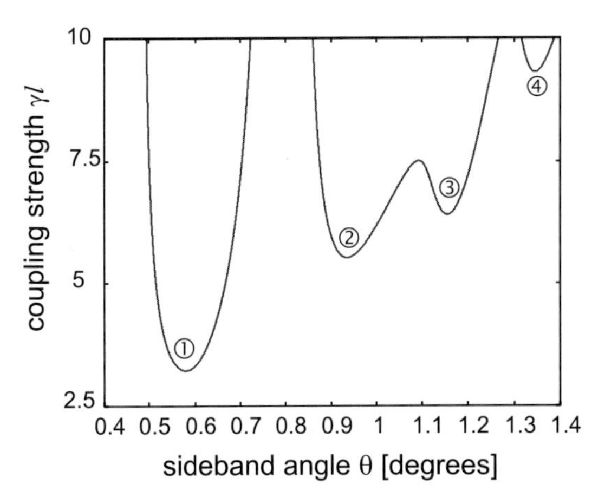

Fig. 8.9. Example of a threshold curve for $n_0 L/l = 0.6$. The absolute minimum of this curve gives information about the wave number k_d at threshold and determines the sideband angle θ in the optical far field. The relative minima are numbered consecutively from 1 to 4

$$\cos(\chi l)\cos(k_\mathrm{d} l) + \frac{k_\mathrm{d}}{\chi}\sin(\chi l)\sin(k_\mathrm{d} l)$$
$$+\frac{\gamma}{2\chi}\{\sin(\chi l)\cos[k_\mathrm{d}(l + 2n_0 L)]\} = 0\;, \tag{8.49}$$

where $\chi = \sqrt{k_\mathrm{d}^2 - \gamma^2/4}$ and $k_\mathrm{d} = k_0\theta^2/(2n_0)$.

8.4.2 Results of the Linear Stability Analysis

In general, $f(\gamma, L, \theta) = 0$ is a real-valued equation of three variables, which need to be represented in three-dimensional space. However, in order to gain a deeper insight into the threshold conditions, it is useful to consider a particular mirror position L, normalized to the crystal length l, and investigate the two-dimensional solution for this specific case. Given a mirror position L, a threshold curve $f(k_\mathrm{d}, \gamma) = 0$ can be derived, where the absolute minimum provides information about the unstable transverse wave number and the threshold coupling strength for that mirror position. One typical example of such a threshold curve is given in Fig. 8.9 for a parameter value $n_0 L/l = 0.6$. The transverse wave number has been converted to the sideband angle θ (the angle between the central spot and one of the hexagon spots in the far field) using $\theta = |\boldsymbol{k}_\perp|/|\boldsymbol{k}_0|$ and the definition of k_d in (8.37) for a crystal of length $l = 5$ mm. The choice of the value of $n_0 L/l$ normalizes the diffraction length to the optical crystal length. With this choice, the crystal is located in the range $-1 \leq n_0 L/l \leq 0$.

The areas above the threshold curve denote unstable wave numbers. In contrast to [31], we shall explicitly consider the various relative minima here

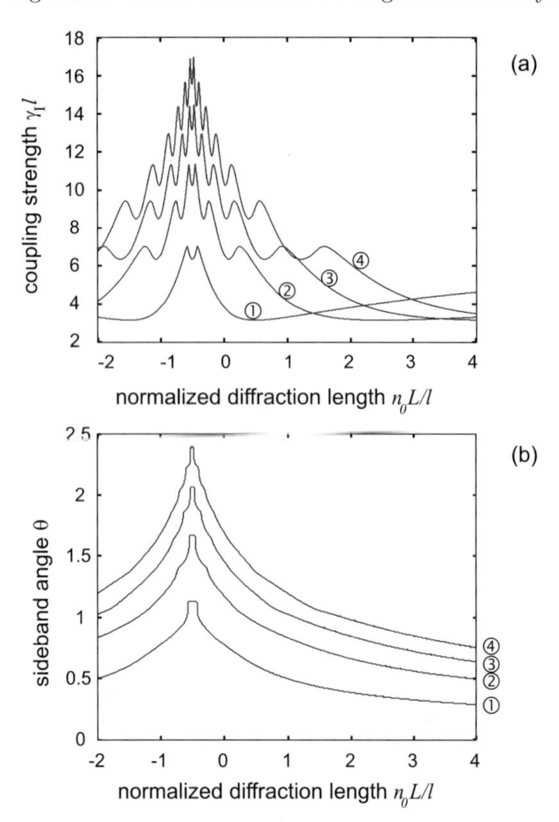

Fig. 8.10. Basic results of the linear stability analysis assuming $\delta = 0$. (**a**) Minimal coupling strength for different normalized mirror positions n_0L/l. (**b**) Theoretical values for the minima of the threshold curves for varying values of n_0L/l. The instability balloons are numbered consecutively and correspond to the numbers in Fig. 8.9

and allow negative diffraction lengths, which were not considered in [33]. The values of the threshold coupling strength $\gamma_I l$ are shown in Fig. 8.10a for the first to the fourth instability "balloon" as a function of the normalized diffraction length $n_0 l/l$. The corresponding sideband angles θ are depicted in Fig. 8.10b.

Note the symmetry around $n_0L/l = -0.5$ of both curves, which is a direct consequence of the symmetry of the threshold condition (8.49) for the special case $\delta = 0$. For a diffraction length around $n_0L/l = 1.4$, the second instability balloon competes with the first and takes over the role of providing the absolute minimum. This is accompanied by a "jump" in the transverse scale [34], which suggests a wave number competition in an experiment. Other important conclusions from the threshold curves in Fig. 8.10 are:

- A coupling strength of approximately $\gamma_I l = 7$ enables us to access patterns for all values of the diffraction length. A lower coupling strength leads to vanishing patterns around $n_0 L/l = -0.5$ [34], as suggested by the shape of the first instability curve in Fig. 8.10a. Both situations have been reproduced in experiments (Sect. 9.3.1).
- Note the nearly vertical leaps of the transverse wave number for $n_0 L/l = -0.43$ and $n_0 L/l = -0.57$. This typical shape is reproduced in the higher-order curves and represents a characteristic property of the photorefractive feedback system.
- The specific shape of the sideband angle curves (Fig. 8.10a) indicates that a whole band of transverse wave vectors becomes unstable, leading to complex competition phenomena or unexpected wave numbers in the region around $n_0 L/l = -0.5$. This coincides with the multiple-pattern region, expressed by a large number of different coexisting patterns with various wave numbers (see Sect. 9).
- For diffraction lengths around $n_0 L/l = 1.4$, the absolute minimum of the threshold curve in Fig. 8.10a is passed from the first to the second branch. This results in a discontinuous jump in the sideband angle from the first to the second branch in Fig. 8.10b. A bistable switching behavior at this critical value of diffraction length can be predicted for the experiment. Even larger diffraction lengths yield consecutive transitions from the second to third, third to fourth, etc., branch as branch that provides the absolute minimum of the threshold curve. The absolute minima of the threshold curves shown in Fig. 8.10a,b are shown in Fig. 8.11.

The linear stability analysis provides information about the unstable wave number and the threshold value for modulation instability. The cooperative existence of and competition between different wave numbers (often referred to as modes) in the transverse plane of the laser beam leads to the creation of spatial structures of different geometries. Thus, a nonlinear analysis is required to explain the pattern formation phenomena encountered.

8.5 Nonlinear Stability Analysis

In order to gain more insight into the reasons for the formation of various patterns in the photorefractive feedback system, let us discuss some aspects of the nonlinear stability analysis of such a system.

A numerical analysis of contradirectional two-beam coupling is – despite the progress that numerical simulations of tranversely resolved photorefractive phenomena have made – still an enormous challenge. The most important problem is that for the case of contradirectional wave mixing in a bulk material, the propagation in the material and the nonlinear interactions cannot be neglected. As a consequence, the boundary conditions are not constant, but change with the result of the coupling and in time, rendering a numerical iterative procedure complicated.

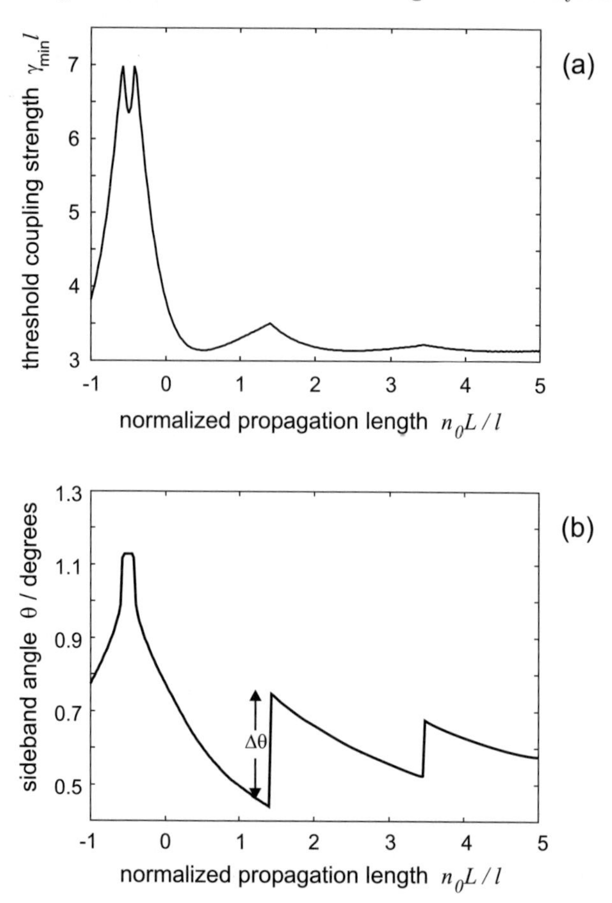

Fig. 8.11. (**a**) Minimum values of the coupling strength for different normalized mirror positions n_0L/l. (**b**) Corresponding values of the sideband angle. Note the consecutive leaps in sideband angle in (**b**) due to the jump from one branch of the threshold curve to the next

In contrast, in other, thin nonlinear materials, the boundary conditions can be handled numerically in a much simpler way, and thus have already led results of to nonlinear stability analysis results and the derivation of amplitude equations, as described in Sect. 8.2.3. Therefore, two-dimensional simulations of pattern formation exist for atomic vapors [16, 17, 18], liquid crystals [21] and liquid crystal light valves [23]. In all these systems, the nonlinear analysis showed, in similar ways, that counterpropagating waves induce a modulation instability above a certain threshold, leading to a breakup of the transverse beam profile into a honeycomb (H^- hexagons) or a hexagonal array of spots (H^+ hexagons). Moreover, the nonlinear stability analysis revealed different parameter regions where various different patterns can be formed. However, despite the fact that two-dimensional simulations exist, the general question of why a hexagonal solution is preferred to a square or roll solution, has still not been clearly answered. Here, we wish to emphasize some plausible explanations in order to provide a better insight into the processes of the selection of different patterns.

For the case of the Bénard instability, Haken could show [45] that the three wave vectors forming a hexagonal structure also represent a minimum of the potential of the system as long as $\boldsymbol{k}_1+\boldsymbol{k}_2+\boldsymbol{k}_3 = 0$ is fulfilled. Therefore, such a hexagonal structure is extremely stable and may appear as the most stable structure in many pattern-forming systems. This can be made plausible in a forces representation. Three force vectors having the same absolute value are in equilibrium, i.e. the potential is at its minimum, if they form a triangle with the same basis lengths.

Inducing a refractive index change in the medium requires the expenditure of work, which must be provided by the incoming light. This work is used to induce the nonlinear response of the material, e.g. the reorientation of the molecules for a liquid crystal or the liberation and migration of charges in a photorefractive material. The index change and, hence, the local work, are very high only near a vertex of the phase lattice created in the medium. Therefore, the total energy that the light must provide to the system is roughly proportional to the number of lattice vertices falling in the illuminated area $\mathcal{A}$. This number is $\mathcal{A}/\zeta^2$ for squares and $\sqrt{3}\mathcal{A}/(2\zeta^2)$ for hexagons, owing to their lower filling factor, where ζ is the length of the basic unit of the hexagonal cell. Because the comparison must be performed with fixed ζ, hexagons require $\sqrt{3/2}$ times less work. Clearly, roll patterns have larger high-intensity areas and are therefore more energy-consuming [57]. Hexagonal patterns require almost isotropic diffusion in the (x, y) plane, i.e. $l_x \approx l_y$. When $l_x \gg l_y$, one-dimensional oscillations prevail, and the resulting periodic pattern assumes the form of rolls orthogonal to the x axis [49].

Although we are not dealing here with forces or classical systems, this interpretation allows us to imagine that a hexagonal pattern, having a basic triangular structure, is a stable solution of a pattern-forming system. For the case of the photorefractive feedback system, Lushnikov [33] derived a Ginzburg–Landau hexagon equation in the form of (8.22)–(8.24). The coefficients μ, g_0 and $g_{\pi/3}$ were calculated in [33]. The corresponding bifurcation diagram is depicted in Fig. 8.12 using the values $\mu = -1$, $g_0 = 3.8$ and $g_{\pi/3} = 2.1$ as given in [33] for a diffraction length $L = 0$.

The stability and instability regions of the solutions were investigated in detail in [48]. Negative hexagons bifurcate subcritically and lose stability far beyond threshold. The positive hexagons bifurcate supercritically, but are always unstable. The roll pattern is unstable at threshold and gains stability at a finite value beyond threshold. In particular, the subcritical bifurcation of hexagons is accompanied by a bistable region of parameters for which a homogeneous and a hexagonal solution coexist. This gives rise to hysteresis phe-

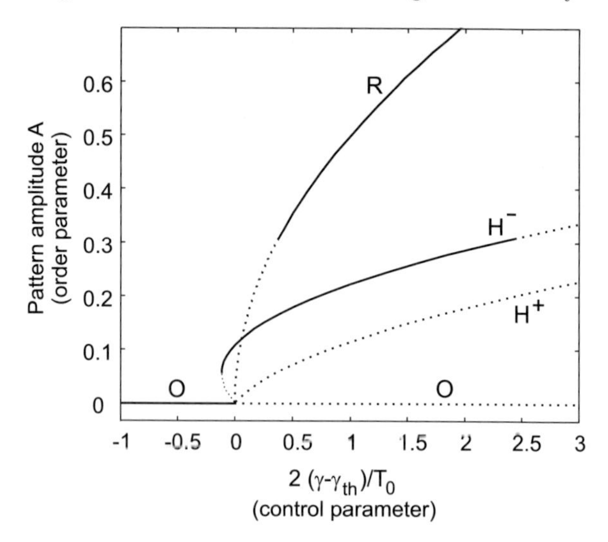

Fig. 8.12. Bifurcation diagram of the GLH equations (8.22–8.24) for the values $\mu = -1$, $g_0 = 3.8$ and $g_{\pi/3} = 2.1$. The *solid lines* depict stable roll (R) or hexagonal (H) solutions. O represents the homogeneous solution, and *dotted lines* indicate unstable solutions. The subcritical bifurcation of negative hexagons is a generic feature of pattern-forming systems that can be described by a set of GLH equations

nomena and interpretations of the bifurcation behavior as first-order phase transitions [58]. The interpretation of the bifurcation diagram in Fig. 8.12 is as follows: when the system reaches the threshold for modulation instability, a discontinuous jump to the H^- branch leads to three resonant wave vectors forming a hexagon, and all other modes are slaved to this triad.

Though this analysis predicts stable hexagons for small positive diffraction lengths, only three possible spatial structures have been taken into account: the homogeneous solution, the roll, and the hexagon solution. Not included in this analysis are, for example squares, since they do not constitute a subset of the hexagon triad wave vectors that were postulated in (8.21).

Sandfuchs et al. [59] discussed the stability and coexistence of substructures of a dodecagonal spot array, which generalizes and improves the above considerations. The dodecagonal ansatz for the modes on the instability circle allows all kinds of mode competition between the substructures, among them hexagons, stripes, squares, dodecagons rhombic patterns. In particular, squares and dodecagons were shown to be stable above threshold for $L = 0$, while the subcritical bifurcation behavior of negative hexagons shown in Fig. 8.12 was reproduced. The subcriticality of hexagons (bringing the system away from the bifurcation point) yields a deviation between experiment and theory of the Landau coefficients (see Eq. (8.22)–(8.24)). This phenomenon is related to a subcriticality dilemma that arises when an asymptotic expansion is performed to obtain the Landau equations for a first-order phase transition. As pointed out in [59], the eigenfunctions of the linear stability problem are determined up to an arbitrary factor. For the hexagon mode a strong renormalization factor is found, which has to be taken into account when making a direct comparison with experimental or numerical resutls. The eigenfunctions were not explicitly calculated in [33], and the normalization was chosen arbitrarily, which renders a direct transfer of these results to our experimental results inadequate.

Summary

We have thoroughly discussed the principles of the linear and nonlinear stability that must be performed to predict the behavior of the system from the fully nonlinear equations of motion. Though the linear stability analysis is limited to one transverse dimension, a number of conclusions could be drawn from a discussion of the interaction of different instability balloons. A complete nonlinear stability analysis for any given set of system parameters is mathematically and computationally extremely demanding and is not available to date. The experimental photorefractive feedback system provides all the necessary tests of our theoretical predictions. We shall see that the unusual shape of the threshold curve that we discussed in this chapter coincides with a region of multiple pattern stability. Access to this region is provided by mirror positions inside the crystal. Therefore, we have discussed different methods to create an image of a real mirror to investigate the effects of negative diffraction lengths. With the theoretical investigations and foundations discussed in this chapter, we are now ready to get to know the experimental properties of the free-running pattern formation system. A photorefractive diffusion-dominated sample of potassium niobate was the crystal of choice for the experiments described in the next chapter.

References

1. C.O. Weiss, *Spatio-temporal structures part II*, Phys. Rep. **219**, 311–338 (1992).
2. F.T. Arecchi, *Space–time complexity in nonlinear optics*, Physica D **51**, 450–464 (1991).
3. L.A. Lugiato, *Spatio-temporal structures part I*, Phys. Rep. **219**, 293–310 (1992).
4. L.A. Lugiato, *Transverse nonlinear optics: introduction and review*, Chaos Solitons Fractals **4**, 1251–1258 (1994).
5. L.A. Lugiato (ed.), *Nonlinear optical structures, patterns, chaos*, Chaos, Solitons and Fractals **4** (1994).
6. G.J. de Valcárel, E. Roldán and R. Vilaseca (guest eds.), *Special issue on Patterns in Nonlinear Optical Systems (part 1)*, J. Opt. B: Quantum Semiclass. Opt. **10** (1998).

7. G.J. de Valcárel, E. Roldán and R. Vilaseca (guest eds.), *Special issue on Patterns in Nonlinear Optical Systems (part 2)*, J. Opt. B: Quantum Semiclass. Opt. **1** (1999).
8. R. Neubecker and T. Tschudi (guest eds.), *Pattern formation in nonlinear optical systems*, Chaos Solitons Fractals **10** (1999).
9. S. Boccaletti F.T. Arecchi and P.L. Ramazza, *Pattern formation and competition in nonlinear optics*, Phys. Rep. **318**, 1–83 (1999).
10. W. Lange and T. Ackemann (guest eds.), *Special issue on Complex Behaviour in Optical Systems and Applications*, J. Opt. B: Quantum Semiclass. Opt. **2** (2000).
11. M. Belić (guest ed.), *Special issue on Nonlinear Dynamics in Optics*, Asian J. Phys. **7** (1998).
12. A.M. Turing, *The physical basis of morphogenesis*, Phil. Trans. R. Soc. B **237**, 37 (1952).
13. G. Grynberg, E. Le Bihan, P. Verkerk, P. Simoneau, J.R.R. Leite, D. Bloch, S. Le Boiteaux and M. Ducloy, *Observation of instabilities due to mirrorless four-wave mixing oscillation in sodium*, Opt. Commun. **67**, 363–366 (1988).
14. J. Pender and L. Hesselink, *Degenerate conical emission in atomic sodium vapor*, J. Opt. Soc. Am. B **7**, 1361–1373 (1990).
15. A. Petrossian, M. Pinard, A. Maître, J.Y. Courtois and G. Grynberg, *Transverse pattern formation for counterpropagating beams in rubidium vapor*, Europhys. Lett. **18**, 689–695 (1992).
16. G. D'Allesandro and W.J. Firth, *Spontaneous hexagon formation in a nonlinear optical medium with feedback mirror*, Phys. Rev. Lett. **66**, 2597–2600 (1991).
17. G. D'Allesandro and W.J. Firth, *Hexagonal spatial pattern for a Kerr slice with a feedback mirror*, Phys. Rev. A **46**, 537–547 (1992).
18. W. Lange, Yu.A. Logvin and T. Ackemann, *Spontaneous optical pattern in an atomic vapor: observation and simulation*, Physica D **96**, 230–241 (1996).
19. R. Macdonald and H.J. Eichler, *Spontaneous optical pattern formation in a nematic liquid crystal with feedback mirror*, Opt. Commun. **89**, 289–295 (1993).
20. M. Tamburrini, M. Bonavita, S. Wabnitz and E. Santamato, *Hexagonally patterned beam filamentation in a thin liquid-crystal film with single feedback mirror*, Opt. Lett. **18**, 855–857 (1993).
21. M. Tamburrini and E. Ciaramella, *Hexagonal beam filamentation in a liquid crystal film with single feedback mirror*, Chaos Solitons Fractals **4**, 1355–1367 (1994).
22. B. Thüring, R. Neubecker and T. Tschudi, *Transverse pattern formation in an LCLV feedback system*, Opt. Commun. **102**, 111–115 (1993).
23. R. Neubecker, B. Thüring and T. Tschudi, *Formation and characterization of hexagonal patterns in a single feedback experiment*, Chaos, Solitons and Fractals **4**, 1307–1322 (1994).
24. J. Glückstad and M. Saffman, *Spontaneous pattern formation in a thin film of bacteriorhodopsin with mixed absorptive dispersive nonlinearity*, Opt. Lett. **20**, 551–553 (1995).
25. T. Honda, *Hexagonal pattern formation due to counterpropagation in $KNbO_3$*, Opt. Lett. **18**, 598–600 (1993).
26. P.P. Banerjee, H.L. Yu, D.A. Gregory, N.V. Kukhtarev and H.J. Caulfield, *Self-organization of scattering in photorefractive $KNbO_3$ into a reconfigurable hexagonal spot array*, Opt. Lett. **20**, 10–13 (1995).

27. T. Honda and H. Matsumoto, *Buildup of spontaneous hexagonal patterns in photorefractive $BaTiO_3$ with a feedback mirror*, Opt. Lett. **20**, 1755–1757 (1995).
28. T. Honda, *Flow and controlled rotation of spontaneous optical hexagon in $KNbO_3$*, Opt. Lett. **20**, 851–853 (1995).
29. M. Saffman, A.A. Zozulya and D.Z. Anderson, *Transverse instability of energy-exchanging counterpropagating waves in photorefractive media*, J. Opt. Soc. Am. B **11**, 1409–1417 (1994).
30. A.I. Chernykh, B.I. Sturman, M. Aguilar and F. Agulló-López, *Threshold for pattern formation in a medium with local photorefractive response*, J. Opt. Soc. Am. B **14**, 1754–1760 (1997).
31. T. Honda and P. Banerjee, *Threshold for spontaneous pattern formation in reflection-grating-dominated photorefractive media with mirror feedback*, Opt. Lett. **21**, 779–781 (1996).
32. T. Honda, H. Matsumoto, M. Sedlatschek, C. Denz and T. Tschudi, *Spontaneous formation of hexagons, squares and squeezed hexagons in a photorefractive phase conjugator with virtually internal feedback mirror*, Opt. Commun. **133**, 293–299 (1997).
33. P.M. Lushnikov, *Hexagonal optical structures in photorefractive crystals with a feedback mirror*, JETP **86**, 614–627 (1998).
34. C. Denz, M. Schwab, M. Sedlatschek, T. Tschudi and T. Honda, *Pattern dynamics and competition in a photorefractive feedback system*, J. Opt. Soc. Am. B **15**, 2057–2064 (1998).
35. M. Schwab, M. Sedlatschek, B. Thüring, C. Denz and T. Tschudi, *Origin and control of dynamics of hexagonal patterns in a photorefractive feedback system*, Chaos Solitons Fractals **10**, 701–707 (1999).
36. M. Schwab, C. Denz and M. Saffman, *Multiple-pattern stability in a photorefractive feedback system*, Appl. Phys. B **69**, 429–433 (1999).
37. A.J. Scroggie and W.J. Firth, *Pattern formation in an alkali-metal vapor with a feedback mirror*, Phys. Rev. A **53**, 2752–2764 (1996).
38. A. Aumann, E. Büthe, Yu.A. Logvin, T. Ackemann and W. Lange, *Polarized patterns in sodium vapor with single mirror feedback*, Phys. Rev. A **56**, R1709–R1712 (1997).
39. A. Aumann, E. Grosse Westhoff, T. Ackemann and W. Lange, *Magnetic field control over microscopic symmetry properties of an optical pattern forming system: experiment*, J. Opt. B: Quantum Semiclass. Opt. **2**, 421–425 (2000).
40. Yu.A. Logvin, A. Aumann, M. Tegeler, T. Ackemann and W. Lange, *Magnetic field control over microscopic symmetry properties of an optical pattern forming system: theory*, J. Opt. B: Quantum Semiclass. Opt. **2**, 426–431 (2000).
41. M.A. Vorontsov and W.B. Miller. *Self-Organization in Optical Systems and Applications in Information Technology.* Springer, Berlin, Heidelberg, 1995.
42. H.F. Talbot, *Facts relating to optical science IV*, Philos. Mag. **9**, 401 (1936).
43. K. Patorski, *The self-imaging phenomenon and its applications*, Prog. Opt. **27**, 3 (1989).
44. M.C. Cross and P.C. Hohenberg, *Pattern formation outside of equilibrium*, Rev. Mod. Phys. **65**, 851–1112 (1993).
45. H. Haken. *Synergetics. An Introduction. Nonequilibrium Phase Transitions & Self-Organization in Physics, Chemistry and Biology.* Springer, Berlin, Heidelberg, 1978.

46. A.C. Newell, T. Passot and J. Lega, *Order parameter equations for patterns*, Ann. Rev. Fluid Mech. **25**, 399 (1993).
47. P. Manneville. *Dissipative Structures and Weak Turbulence.* Academic Press, Boston, Mass., 1990.
48. S. Ciliberto, P. Coullet, J. Lega, E. Pampaloni and C. Perez-Garcia, *Defects in roll–hexagon competition*, Phys. Rev. Lett. **65**, 2370–2373 (1990).
49. E. Ciaramella, M. Tamburrini and E. Santamato, *Observation of non-hexagonal laser-beam patterning in a thin liquid crystal cell in front of a single mirror*, Appl. Phys. Lett. **63**, 1604 (1993).
50. M. Schwab. *Manipulation and Control of Self-Organized Transverse Optical Patterns.* Shaker, Aachen, 2001.
51. A. Siegman. *Lasers.* University Science Books, Mill Valley, Ca, 1986.
52. N. Hodgson and H. Weber. *Optical Resonators - Fundamentals, Advanced Concepts and Applications.* Springer, London, 1997.
53. M. Schwab, C. Denz, A.V. Mamaev and M. Saffman, *Manipulation of optical patterns by a frequency detuning of the pump beams*, J. Opt. B: Quantum Semiclass. Opt. **3**, 318–327 (2001).
54. O. Sandfuchs, F. Kaiser and M.R. Belić, *Spatiotemporal pattern formation in counterpropagating two-wave mixing with an externally applied field*, J. Opt. Soc. Am. B **15**, 2070–2078 (1998).
55. M.R. Spiegel. *Schaum's Outline of Theory and Problems of Laplace Transforms.* McGraw-Hill, New York, 1983.
56. J.B. Geddes, R.A. Indik, J.V. Moloney and W.J. Firth, *Hexagons and squares in a passive nonlinear optical system*, Phys. Rev. A **50**, 3471–3485 (1994).
57. J.B. Geddes, J. Lega, J.V. Moloney, R.A. Indik, E.M. Wright and W.J. Firth, *Pattern selection in active and passive nonlinear systems*, Chaos Solitons Fractals **4**, 1261 (1994).
58. S.G. Odoulov, M.Yu. Goulkov and O.A. Shinkarenko, *Threshold behavior in formation of optical hexagons and first order optical phase transition*, Phys. Rev. Lett. **83**, 3637–3640 (1999).
59. O. Sandfuchs, F. Kaiser and M.R. Belić, *Self-organization and fourier selection of optical patterns in a nonlinear photorefractive feedback system*, Phys. Rev. A **64**, 3809 (2001).

9 Multiple Patterns and Complex Pattern Competition

From the discussion of the linear and nonlinear stability analyses, it is obvious that spontaneous pattern formation in photorefractive nonlinear materials will lead to hexagonal patterns, as well as more complex structures. Therefore, experimental investigations are an excellent means to gain deeper insight into the different patterns available for a given set of external control parameters. On the one hand, our experimental investigations were focused on the question of the creation and stabilization of different patterns. On the other hand, the existence of pattern dynamics is a direct consequence of a coexistence of two quasi-stable solutions and was therefore a major motivation for our experimental approach.

In our experiments on spontaneous pattern formation, we used photorefractive $KNbO_3$ as the material of choice, owing to its large electro-optic coefficient, which allows strong beam coupling in the counterpropagating beam configuration, and its low noise. We used two different crystal samples, which will be referred to as crystal samples 1 and 2. The major difference between these two samples was the different values of the photorefractive coupling strength: $(\gamma l)_1 \approx 5$ for sample 1 and $(\gamma l)_2 \approx 7.75$ for sample 2.

The setup for experimental observation of spontaneous patterns using a photorefractive single-mirror feedback system is depicted in Fig. 9.1. The light obtained from a linearly polarized frequency-doubled cw Nd:YAG laser operating at a wavelength of 532 nm passed through an optical diode, which isolated the laser from unwanted back-reflections, and a variable attenuator consisting of a half-wave plate and a polarizing beam splitter cube. The light was then focused by lens L1, with a long focal length, to provide a Gaussian focus spot about 320 μm in diameter, which gives a good compromise between sufficient intensity on the crystal and a reasonably large aspect ratio of the pattern (defined as the transverse size of the beam divided by the wavelength of the modulation). Another advantage of the long focal length of lens L1 is the long beam waist, which can be calculated to be several millimeters long, providing a constant intensity within the crystal. The crystal was an iron-doped photorefractive sample of $KNbO_3$. In order to reduce the influence of back-reflections from the crystal and to prevent Fabry–Perot oscillations, the crystal was inclined by $4° - 10°$ degrees. The c axis of the crystal was oriented in such a way that the incoming beam was attenuated while the

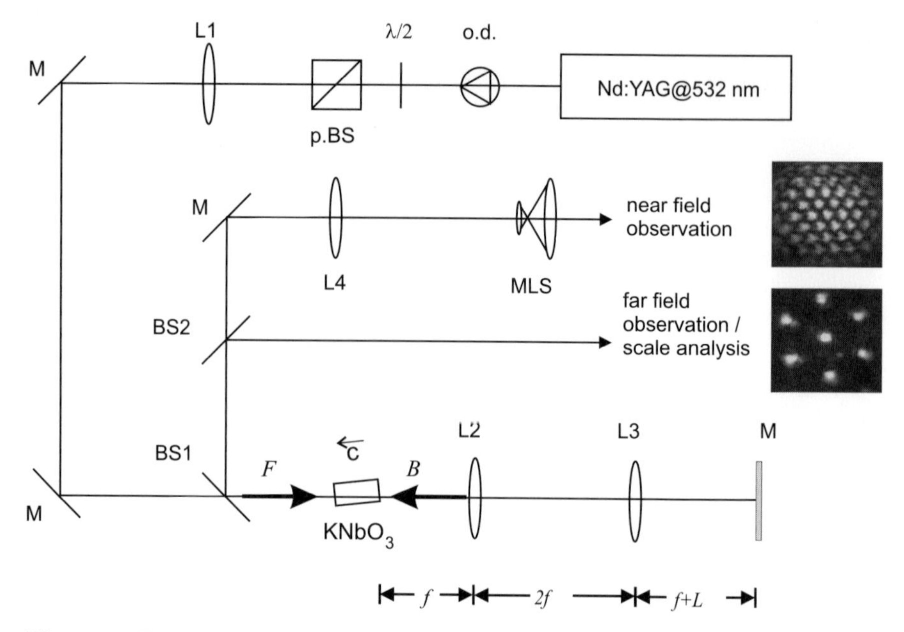

Fig. 9.1. Typical experimental setup for the observation of spontaneous pattern formation in a counterpropagating geometry with mirror boundary conditions: o.d. = optical diode, $\lambda/2$ = half-wave plate, p.BS = polarizing beam splitter, L = lens, M = mirror, BS = (nonpolarizing) beam splitter, MLS = microscope lens system

reflected beam was amplified by the photorefractive two-wave-mixing process. The a axis of the crystal was parallel to the polarization of the laser beam in order to exploit the large r_{13}-component of the electro-optic tensor (see (3.8)), which leads to $r_{\text{eff}} = r_{13}$ for $\varphi = 0$. In fact, the pattern formation process occurs only for a polarization along the crystal a axis, and not for the orthogonal polarization, for which the effective electro-optic tensor and, as a consequence, the photorefractive coupling constant are not sufficient to drive the system to instability.

The power incident on the crystal was in the range of several tens of milliwatts, allowing us to vary the photorefractive time constant τ. The buildup time of the pattern coved more than two decades, from one second down to some tens of milliseconds for large intensities. Higher/lower laser power lowers/raises the time constant τ of the crystal. The minimum laser power for pattern formation was 0.5 mW for crystal sample 2. For crystalsample 1, the corresponding minimum laser power was determined to be 11 mW for a comparable beam size, which is a direct consequence of the different values of the photorefractive coupling strengths of the two crystal samples. The feedback system consisted of two identical lenses with focal lengths f of typically 100 mm and a highly reflective metal mirror positioned at a location $f + L$ from the lens L3. This system produces a virtual mirror at a distance L from the crystal and allows one to achieve negative diffraction lengths by choosing $L < 0$, resulting in a virtual mirror position inside the crystal. The net reflectivity of this feedback system was $r \approx 0.5$, where the Fresnel losses due to two crystal–air and air–crystal transmissions, four lens passes and the mirror reflectivity have been taken into account. The counterpropagating beams F and B interacted inside the photorefractive medium. If the photorefractive coupling strength exceeded the threshold value for modulation instability, spontaneous pattern formation occured and the laser beam was modulated, under its Gaussian envelope. In general, the laser beam filaments arranged themselves in a geometric manner, where hexagons were the predominant solutions for mirror positions $L > 0$. By means of a beam splitter between lens L1 and the crystal, this modulated laser beam was coupled out and projected onto a screen. Free-space propagation gave the optical Fourier transform of the modulated beam, and an additional lens–microscope-lens system were used to monitor the optical near field, which as an actual image of the beam exiting the crystal at $z = 0$. The near and far fields were usually projected onto a screen and monitored by a CCD camera, which was connected to standard video equipment and frame-grabbing devices for long-time recording or snapshots of the spatial distribution of the laser beam.

9.1 Spontaneous Hexagon Formation

The typical output of this standard pattern-forming setup was depicted earlier in Fig. 8.2 in the optical near and far fields. The near-field pattern consists of spots located on a hexagonal grid (this is also known as an H^+ hexagonal pattern). Since we have to deal with a modulated beam and a thick medium, the Talbot effect comes into play [1]; this describes the self-imaging effect of a modulated structure after a Talbot length $l_T = 2\Lambda^2/\lambda$ (see (8.4)), where λ is the wavelength of the carrier wave and Λ is the spatial modulation. For a beam diameter $d = 320$ µm and eight modulations, which is a typical as 6 mm, which corresponds approximately to the length of the crystal. After a distance $l_T/2$, the modulation depicted in Fig. 8.2a is inverted to the structure given in Fig. 9.2a, which is also referred to as an H^- hexagonal or honeycomb pattern. The Fourier transforms of the H^+ and H^- patterns look identical and only the imaging plane of the lens–microscope-lens system in the experimental setup determines which pattern is observed. Owing to the thickness of the nonlinear medium, a distinction between the two types of hexagonal patterns is not appropriate for the system considered here.

Fig. 9.2a shows a typical near-field image for a diffraction length around $L = 0$. An intensity profile in the direction indicated by the dotted line is given on the right and shows the high-contrast modulations of such a honeycomb structure. The intensity is concentrated in the honeycomb "bridges", while the intensity in the holes goes down nearly to the background level. Figure 9.2b shows the results of a numerical simulation performed by Sandfuchs et al. [2] for a photorefractive single-mirror feedback system and a super-Gaussian beam. The intensity profile shows the large modulation that occurrs above the threshold for modulation instability. The correspondence between the experimental picture (Fig. 9.2a) and numerical simulations (Fig. 9.2b) validates the starting equations (8.29), (8.30).

9.2 Multiple-Pattern Region and Complex Pattern States

For negative diffraction lengths $n_0L/l \approx -0.3$, the hexagonal pattern collapses and gives way to a rich variety of different pattern states and transverse wave numbers. In a small parameter region of diffraction lengths $-0.7 \leq n_0L/l \leq -0.3$, square, squeezed hexagonal, rectangular, parallelogram-shaped and even dodecagonal quasi-crystalline pattern states can be found as states of the system besides hexagonal patterns with large aspect ratios (see Fig. 9.3). The patterns either are stable for a long period of time or lose stability to a different pattern state on a timescale much larger than

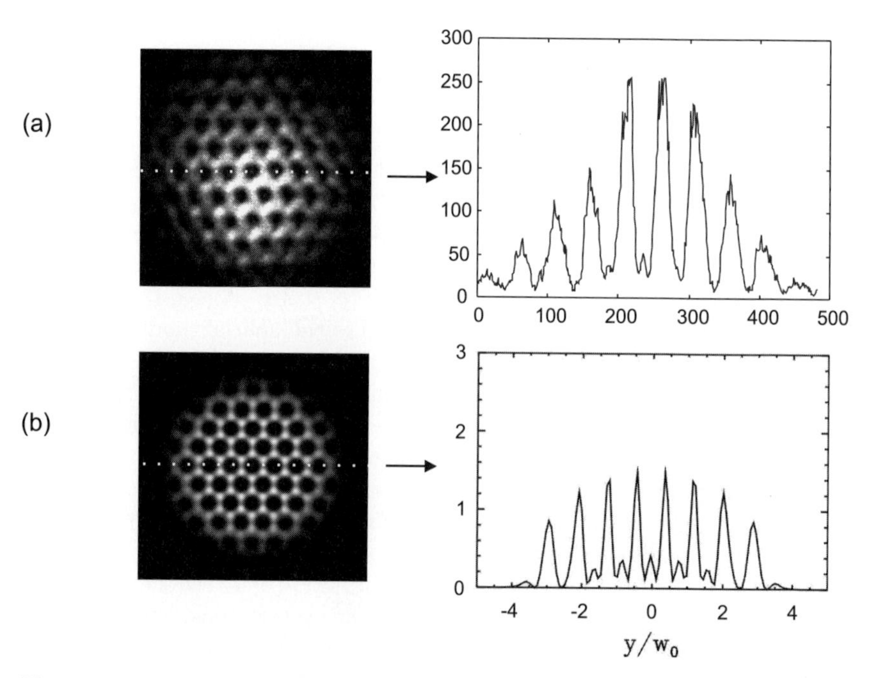

Fig. 9.2. Typical hexagonal honeycomb structures for a mirror position of $L = 0$. (**a**) Experimental result, with an intensity profile along the *white dotted line*. (**b**) Numerically obtained honeycomb structure, obtained for the same parameters as in the experiment. Contributed by O. Sandfuchs, Institute of Applied Physics, TU Darmstadt

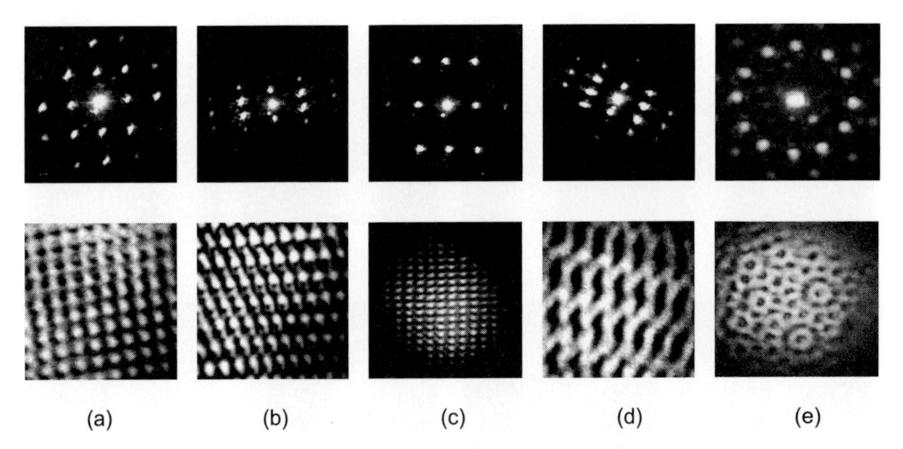

Fig. 9.3. Examples of the variety of patterns observed in the multiple-pattern region: (**a**) square, (**b**) squeezed hexagonal, (**c**) rectangular, (**d**) parallelogram, (**e**) dodecagonal pattern. *Upper row*, far field, *lower row*, near field

the time constant of the system ($t > 200\tau$). For the same value of the parameter n_0L/l and with other external parameters kept constant, more than one stable pattern can be observed, which is a clear indication of a region of *multiple stability*, such regions are also known from pattern observations in other physical systems far from thermal equilibrium [3], but have not been observed before in an optical pattern-formation experiment. Besides these pure pattern states, mixed states can also be observed, where different patterns coexist on the same or different length scales. Another scenario is a temporal alternation of two or more patterns, which appear on long but not reproducible timescales. Figure 9.4 gives an impression of the complexity of possible pattern states in this parameter region.

The name *multiple-pattern region* for this region represents the complexity of pattern observations in this area. It is important to note that the different patterns do not necessarily appear with the same transverse wave number. The observed sideband angles vary from $\theta = 0.9°$ to $\theta = 2.7°$ for the rectangular pattern. The latter is far beyond the predictions of the linear stability analysis, as can be deduced from Fig. 8.10b. The plateau of the hat-shaped curve corresponds to a value of the sideband angle of $\theta = 1.1°$, which is far below the observed length scales. The threshold values for modulation instability for the other balloons are far beyond the threshold value for the first balloon. Thus, interactions with higher-order balloons have to be excluded as a possible reason for the occurrence of a variety of different wave numbers. The wave number with the lowest value of the threshold coupling strength slaves all other wave numbers, making predictions for other wave numbers worthless. Obviously, the predictions of the linear stability analysis concerning the most unstable wave number fail for this system in the multiple-pattern region. The various nonlinear interactions leading to the complex pure and

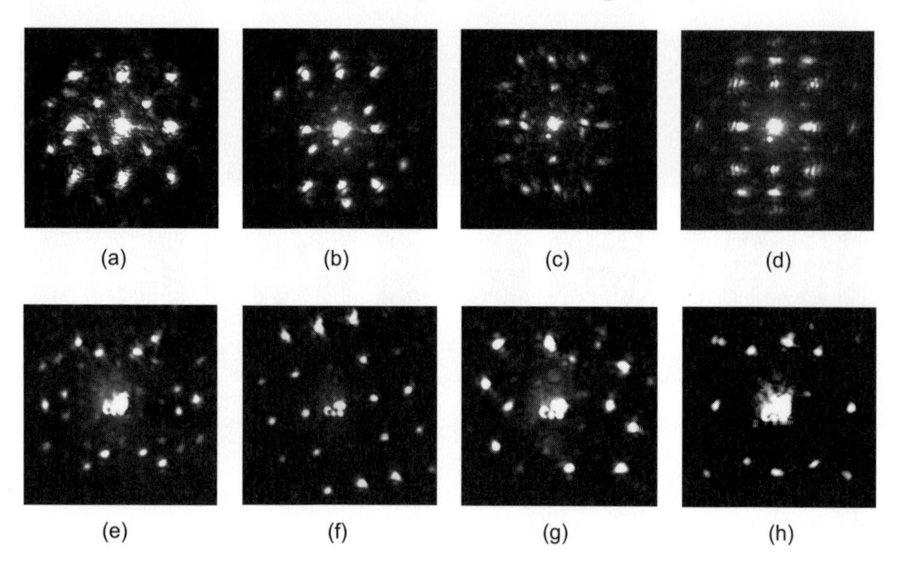

Fig. 9.4. Examples of the occurrence of mixed pattern states in the multiple-pattern region: (**a**) two square patterns with different length scales and orientations, (**b**) two rectangular patterns on the same scale, but different orientations, (**c**) two rectangular patterns and a square pattern, (**d**) square pattern and rectangular pattern, (**e**) three coexisting hexagonal patterns, (**f**) square pattern with two missing spots and pronounced second orders, (**g**) hexagon and rhomboid patterns, (**h**) coexisting square and hexagon patterns

mixed pattern states indicate that nonlinear terms have to be taken into account for an appropriate description of the system behavior. Figures 9.3 and 9.4 visualize the complexity of the observations in the parameter region $-0.7 \leq n_0L/l \leq -0.3$ when the virtual mirror is located around the middle of the crystal. For one and the same set of parameters, two or more stable solutions can be identified, hence the name multiple-pattern region.

Multiple stability is a well-known phenomenon and is observed in a variety of different fields ranging from optics [4, 5] to chemistry and even neuroscience [6]. On the one hand, a detailed analysis of the stability regions and transverse scales of various patterns as a function of the system parameters of diffraction length and coupling strength is vital for a closed understanding of the pattern-forming system. On the other hand, the motivation for a closer analysis of the multiple-pattern region goes beyond basic investigations of bistability and multiple-pattern stability in photorefractives and directly involves the control methods presented in the next two chapters. Bistabilities provide potential control tools for switching between two system states. A nonlinear system operated in a bistable regime is extremely sensitive to small perturbations and represents an excellent model for a detailed investigation of the impact of noise on the output of the pattern-forming system. Consequently, a detailed understanding of the multiple-pattern region is an important precondition for successful implementation of active pattern control schemes.

Multiple stability implies that minor perturbations induced by sources of noise in the system (thermal noise or phase fluctuations in the feedback path) may cause the system to perform random transitions from one state to another. Transitions appear in the multiple-pattern region on a timescale much larger than the timescale of the crystal, in the second, minute or even hour range. Owing to the complexity of the system, statistical methods have to be applied in order to obtain improved insight into the stability regions of any given pattern.

9.3 Quantitative Measurements of Sideband Angles

The linear stability analysis presented in Sect. 8.4 provides information about the unstable wave number or, alternatively, the transverse extension of the pattern in the far field. Measurements of the sideband angle therefore constitute important tools for checking the validity of the linear stability analysis. The measurements of the sideband angle described here are split into two parts. The first part deals with the complete two-dimensional pattern-forming system and incorporates the measurement of the hexagon sideband angles. The second part of this section is devoted to a system forced into one transverse dimension and showing roll patterns as the output. Though we prevent pattern complexity, the loss of some spatial degrees of freedom enables us to perform a quantitative comparison between the measurements and the results of the linear stability analysis even inside the multiple-pattern region.

9.3.1 Two Transverse Dimensions: Hexagons

Figure 9.5 shows the results of a measurement of the sideband angle together with the theoretical curves for the first stability balloon (solid line) and parts of the second (dotted line). The multiple-pattern region is excluded here for clarity (the number of different sideband angles is too numerous to be displayed here). The measured values of the hexagon sideband angles agree well with the predictions of the linear stability analysis, indicated by the solid line. For larger positive ($n_0L/l \geq 0.65$) or negative ($n_0L/l \leq -1.7$) diffraction lengths, a competition between two hexagons on different scales appears; an intermediate picture is shown in the inset of Fig. 9.5. Either the larger hexagon (denoted by squares) or the smaller hexagon (circles) survives this complex pattern competition. It is noteworthy that the larger-scale hexagon coincides with the corresponding curve for the second instability balloon, which is shown in Fig. 9.5 as a dashed curve. For this parameter region, the values of the minima of the first and second instability balloons are comparable (see Fig. 8.10b) making a complex competition between different modes

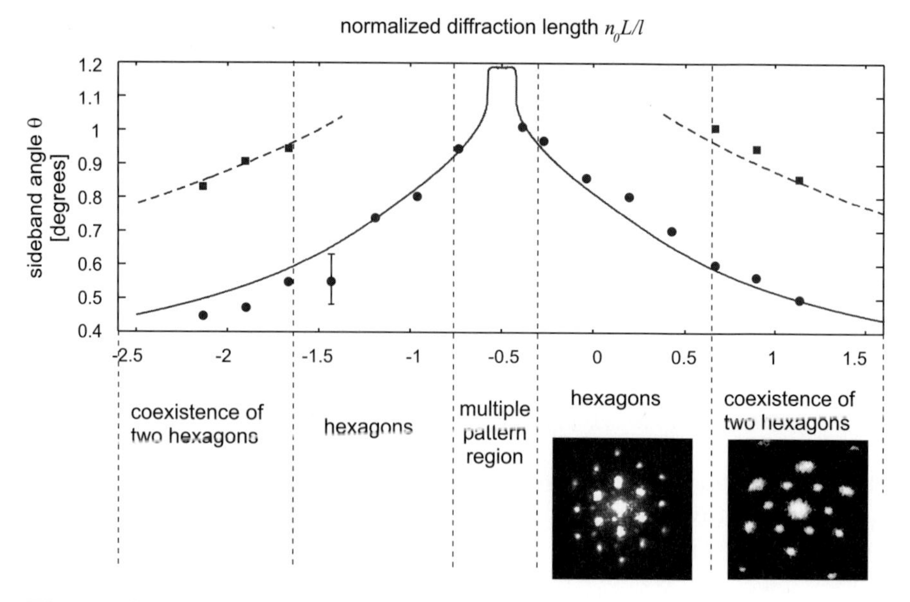

Fig. 9.5. Sideband angle θ as a function of the normalized diffraction length n_0L/l. The theoretical curve is displayed together with experimental values for the transverse scale (*circles* and *squares*)

plausible. This is especially supported by the fact that the ratio of the transverse wave numbers is around $\sqrt{3}$, which is exactly the value for a three-wave interaction leading to a higher-order hexagon spot, as discussed in Sect. 8.2.3. The two transverse wave numbers show resonance effects, leading to complex pattern competition scenarios. The hexagons are tilted by 30° relative to each other, which is necessary to support the $\sqrt{3}$ mode interaction. This resonant behavior was predicted in [7] and observed experimentally for the first time in the photorefractive single-mirror feedback system. It is clearly apparent that for small positive and negative diffraction lengths, the only stable solution the system selects is a hexagonal one, as pointed out in [8].

It should be noted that the observation of a multiple-pattern region was restricted to crystal sample 2. Crystal sample 1 did not show any modulation instability in the parameter region considered above [7]. When the virtual mirror was moved into the crystal, the far-field hexagon vanished for $-0.75 \leq n_0L/l \leq -0.3$, with a maximum achievable sideband angle of $\theta_{\max} = 0.9$ (see [7] for details). The lack of patterns inside the multiple-pattern region for this crystal can be fully explained by the lower coupling constant γl of this crystal. As can be deduced from the shape of the threshold coupling strength depicted in Fig. 8.10b (the curve for balloon 1), a minimum coupling strength of $\gamma l \approx 7$ has to be present in order to sustain modulation instability for all virtual-mirror positions inside the crystal. This is illustrated in Fig. 9.6, where the positions of the threshold curves

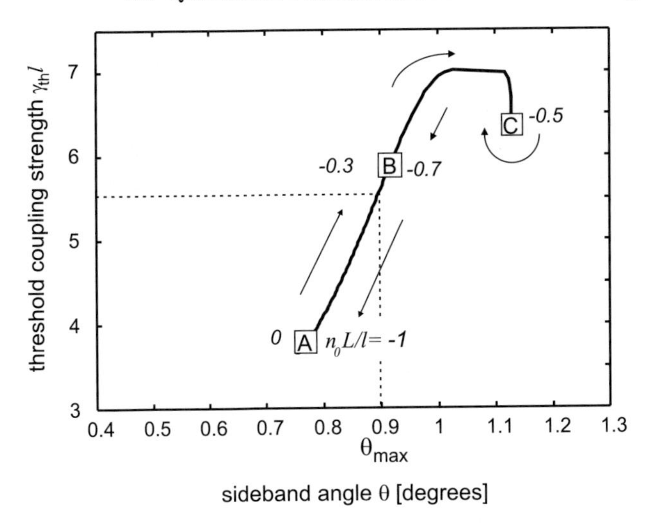

Fig. 9.6. Minima of threshold curves as a function of the virtual-mirror position for the parameter region $-1 \leq n_0L/l \leq 0$. Some selected parameter values are $n_0L/l = -1, 0$ (point A), $n_0L/l = -0.3, -0.7$ (point B) and $n_0L/l = -0.5$ (point C). The largest observable sideband angle $\theta_{\max}$ is indicated, corresponding to a coupling strength of $\gamma l \approx 5.5$

are depicted as a parameter curve in the $(\gamma l, \theta)$ space. This parameter curve starts at $n_0L/L = 0$ (point A) and increases via point B ($n_0L/l = -0.3$). Point C, at $n_0L/L = -0.5$ represents the turning point. The direction is then reversed; the curve passes through point B at $n_0L/l = -0.7$ and ends at point A at $n_0L/L = -1$. One can directly conclude from Fig. 9.6 that the maximum sideband angle of $\theta_{\max} = 0.9$ in the experiment [7] is connected to a value of the photorefractive coupling strength $\gamma l = 5.5$ for this crystal sample 1. The branches above this value require a larger coupling strength and are therefore not excited in this case. Consequently, the lack of patterns for virtual-mirror positions around the middle of the crystal is explained by an insufficient coupling strength, and represents an alternative method to determine the strength of the nonlinearity of a photorefractive crystal. We performed two-beam coupling experiments with both crystals and estimated a value of $(\gamma l)_1 = 5$ for this crystal, which is in reasonable agreement with the value $\gamma l = 5.5$ determined here. For crystal sample 2, a lack of pattern observation related to a maximum sideband angle was not found, pointing to a coupling strength $(\gamma l)_2 \geq 7$.

9.3.2 One Transverse Dimension: Rolls

A feedback system exploiting both transverse dimensions allows all kinds of resonant wave interactions in the transverse plane. In order to compare the results of the linear stability analysis with the experiment, it is instructive to

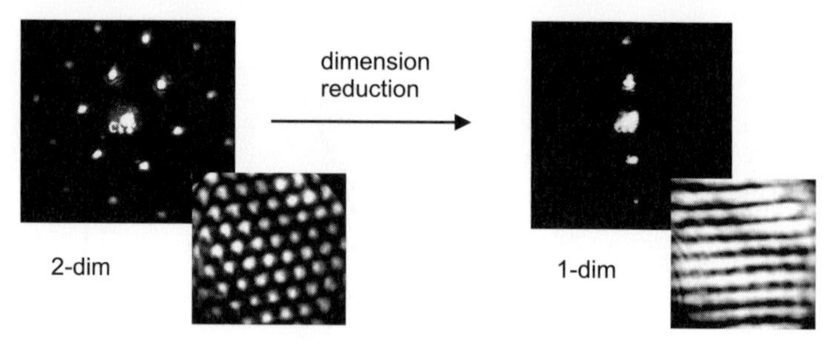

Fig. 9.7. The roll pattern is the stable system output in the case of a system forced into one transverse dimension. *Left*: hexagonal pattern in the far field (*main picture*) and in the near field (*inset*), as the solution of the free-running two-dimensional system. *Right*: stripe pattern in the far and near fields for the system constrained into one transverse dimension

reduce the number of spatial degrees of freedom and exclude resonant wave interactions in two transverse dimensions. To do this, slit filter with a width of around 0.25 cm was inserted in the Fourier plane of the experimental setup depicted in Fig. 9.1 and restricted the feedback system to one transverse dimension. The spatial filtering method used here is the topic of Chap. 10, where it is utilized for selection of a defined pattern. The only pattern appearing in this artificially constrained system was a spatial roll pattern, as shown in Fig. 9.7 (right). The name "roll" originates from circular motions in pattern formation experiments in hydrodynamics [3] and is used to refer to the spatial stripe pattern throughout the remainder of the book.

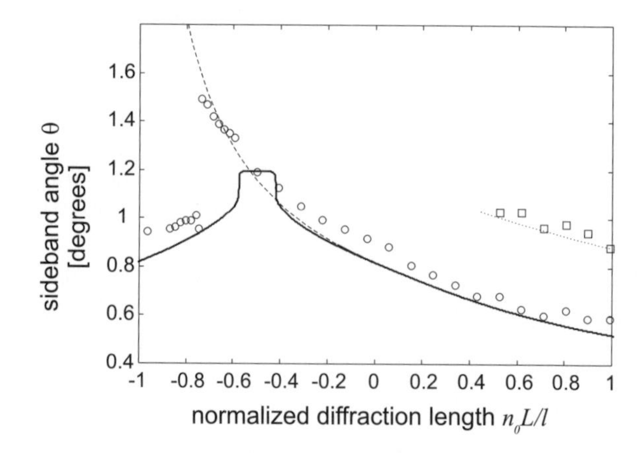

Fig. 9.8. Experimentally found sideband angle θ for the one-dimensional system. Also shown are the theoretical threshold curves derived from the linear stability analysis (*solid curve* for instability balloon 1 and *dotted curve* for balloon 2) and the theoretical curve assuming a real mirror inside the crystal (*dashed line*)

The experimental results (performed with crystal sample 2) are depicted in Fig. 9.8. For positive diffraction lengths, the qualitative and quantitative results are identical to those obtained for the case of hexagonal patterns in two transverse dimensions. For larger diffraction lengths, a second length scale coinciding with the wave number determined by the second instability balloon appears, and competition between the two wave numbers can be observed. The lack of the second transverse dimension prohibits a multiple-pattern region. Nevertheless, the hat-shaped profile suggested by the theoretical curve is not reproduced, even in the one-dimensional case. For a diffraction length around $n_0L/l = -0.7$ (which also marks the end of the multiple-pattern region in the two-dimensional case), a jump from a large- to the small-scale pattern can be observed, with an intermediate competition of these two scales. The system falls back to the branch determined by the linear stability analysis with decreasing wave numbers, for larger negative diffraction lengths.

A physical interpretation can be given by considering the contributions of nonlinear coupling from different regions of the medium. This was first suggested in [9] and was applied to the photorefractive feedback system to account for the deviations between the experimental results and the theoretical predictions. The dashed line in Fig. 9.8 shows the dependence of θ on the diffraction length if only the coupling to the left of the virtual mirror is taken into account, i.e. if the virtual mirror is taken as a real mirror at $L = 0$, neglecting the nonlinear coupling from the right-hand region of the crystal behind the mirror. To calculate this curve, the value of the wave number at the instability threshold for $L = 0$ was taken, while the length l of the crystal was replaced by $l(1 + n_0L/l)$. The pattern formation process is obviously dominated by the contributions from the left side of the crystal in front of the virtual mirror. This can be understood by taking into account the intensity dependence across the crystal as illustrated in Fig. 3.5. Owing to the beam-coupling process, the intensity decreases exponentially from $z = 0$ to the exit face of the crystal at $z = l$.

As a consequence, the time constants increase across the crystal, while the photorefractive coupling strength decreases with the decreasing pump intensity. This dependence suggests a combined effect of the time constant and the photorefractive coupling strength. If the coupling strength at the right-hand end of the crystal is below the threshold for modulation instability, that region does not take part in the pattern formation process. The right-hand region contributes only to the beam coupling, which is taken into account by the unchanged coupling strength γl. A control measurement showed that the curve depicted in Fig. 9.8 is valid for both directions of the diffraction length. A pronounced hysteresis effect around $n_0L/l = -0.7$, where the jump in the transverse scale appears, was not observed in the experiment.

9.4 Complex Pattern Coexistence, Competition and Dynamics

9.4.1 Pattern Competition for Collinear Pump Beams

A feature of central importance of the single-mirror photorefractive feedback system is the complexity of pattern states in particular parameter regions. We have already discussed the appearance of two hexagons on different scales and possible mixed pattern states when two patterns compete and coexist on the same or a different transverse scale. Especially in parameter regions containing a number of different stable pattern states, spontaneous switching or alternation between different pattern states can be observed without the external parameters being changed. The transitions from one pattern to another are obviously initiated by experimental noise due to minimal air movements or temperature fluctuations in the feedback system. They are not induced by any means and have to be classified as inherent in the system. In particular, non-collinearity was excluded as a possible source of the dynamics by carefully checking the system alignment.

Visualization Method

Complex pattern dynamics, as described here, requires an appropriate method for visualization in order to extract the important information from a series of images describing the pattern transition. The most appropriate method for our purposes is depicted in Fig. 9.9 and was proposed by Thüring et al. in [10]. The method is based on the unfolding of the intensity information on the unstable circle in Fourier space into a linear dimension. The distribution of far-field spots on a circle of radius k_c (the critical wave number) is projected onto a linear axis, with the polar coordinate ϕ as the horizontal axis and time as the vertical axis. As a result, every movement of the spots on the circle leads to a corresponding pattern change in Cartesian coordinates. An example is given in Fig. 9.9 of a static hexagon, represented by six parallel stripes extending in the positive y direction (time axis). A hexagon rotating at constant angular velocity would lead to six tilted stripes, with the slope incorporating the information about the angular velocity. The major advantage of this method is that a single image provides us with all the information about the dynamics, instead of a sequence of many individual pattern images, leading to a reduction of the information by considering the unstable circle in Fourier space only.

Examples of Pattern Transitions

Pattern transitions occur either on the same transverse scale or on completely different scales, indicating highly nonlinear coupling between the spatial wave

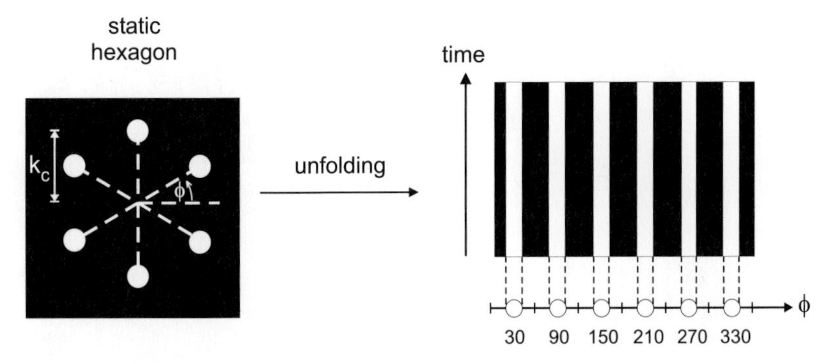

Fig. 9.9. Visualization method for illustrating pattern dynamics based on the unfolding of the angular spectrum of the far-field pattern. The example given here shows the reduced information given by a static hexagon

numbers. We restrict ourselves to the analysis of pattern transitions appearing on the same transverse scale in order to use the visualization method described above. This also restricts the number of possible patterns. Rectangles and squeezed hexagons, which each possess two different wave numbers, are therefore unsiutable for this analysis (unless the two wave numbers are treated independently or the projection of one wave number on another is considered). Two examples of pattern transitions in the uncontrolled, free-running system are shown in Fig. 9.10, for the case of a dodecagon–hexagon competition (Fig. 9.10a) and a complex square–dodecagon hexagon transition (Fig. 9.10b). Even small perturbations due to diffraction length variations caused by temperature changes or due to air movement are sufficient to move the system from one marginally stable state to another quasi-stable state. The dodecagonal pattern shows a clear tendency to relax into a hexagonal pattern; transitions from dodecagons to squares were not observed in the experiment. It is also clearly apparent from both transition sequences that certain wave numbers survive the competition, i.e. in Fig. 9.10a the hexagonal pattern was already a part of the dodecagonal pattern, and in Fig. 9.10b, the four spots of the initial square pattern survive the transition to the dodecagonal pattern. Nevertheless, only two of those four spots appear in the final hexagonal pattern. The underlying roll pattern in this spatial direction is obviously the most stable one, since it is not affected by the pattern competition. This has to be attributed to anisotropies in the photorefractive medium that favor one particular spatial direction. Pattern transition sequences like those depicted in Fig. 9.10 can be found throughout the whole multiple-pattern region. In principle, any two or more patterns that exist as stable solutions for one set of parameters are potential candidates for pattern transition phenomena. This also includes spontaneous oscillations between two patterns as reported in [7]. The vertical lines in Fig. 9.10 are not tilted, which confirms that an angular misalignment was not present and

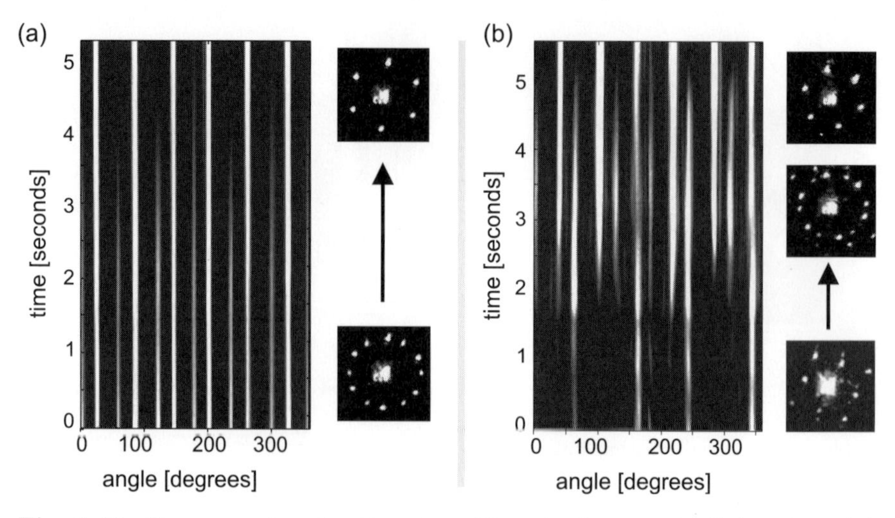

Fig. 9.10. Two examples of pattern transitions in the uncontrolled system. (**a**) Dodecagonal quasi-pattern relaxing to a simple hexagon. (**b**) Pattern sequence proceeding from a square via a dodecagonal to a hexagonal structure. All sequences were recorded for $n_0 L/l \approx -0.5$, inside the multiple-pattern region

was therefore not responsible for the pattern transitions in our experiment. However, the case of noncollinear pump beams is also of general interest since it also provides interesting, complex patterns – in this case, the complexity lies more in the dynamic processes.

9.4.2 Pattern Dynamics for Noncollinear Pump Beams

So far, we have only considered the case of collinear pump beams and have shown that even in this case, complex pattern dynamics and competition occur. Nevertheless, those dynamics were always restricted to competition effects between different patterns; the instabilities were always static. The situation changes dramatically when an asymmetry is introduced into the system, which can be achieved most easily by tilting the feedback mirror. Such a tilt induces a flow of the pattern in the near field and causes complex movement scenarios in the optical far field [7]. One chosen scenario is illustrated in Fig. 9.11a and serves as an example of the complexity of possible movements in the case of noncollinear pump beams.

This drift instability was predicted in [11] and demonstrated experimentally [12] in a single-mirror feedback system where rubidium vapor provided the optical nonlinearity. For a photorefractive single-mirror feedback system, the flow in the near field as a result of a drift instability was first reported in [13], where an additional erase beam was used to control the speed of rotation of the far-field hexagon. The intensity gradient of the additional beam could be used as a control parameter in this case. Obviously, an intensity gradient

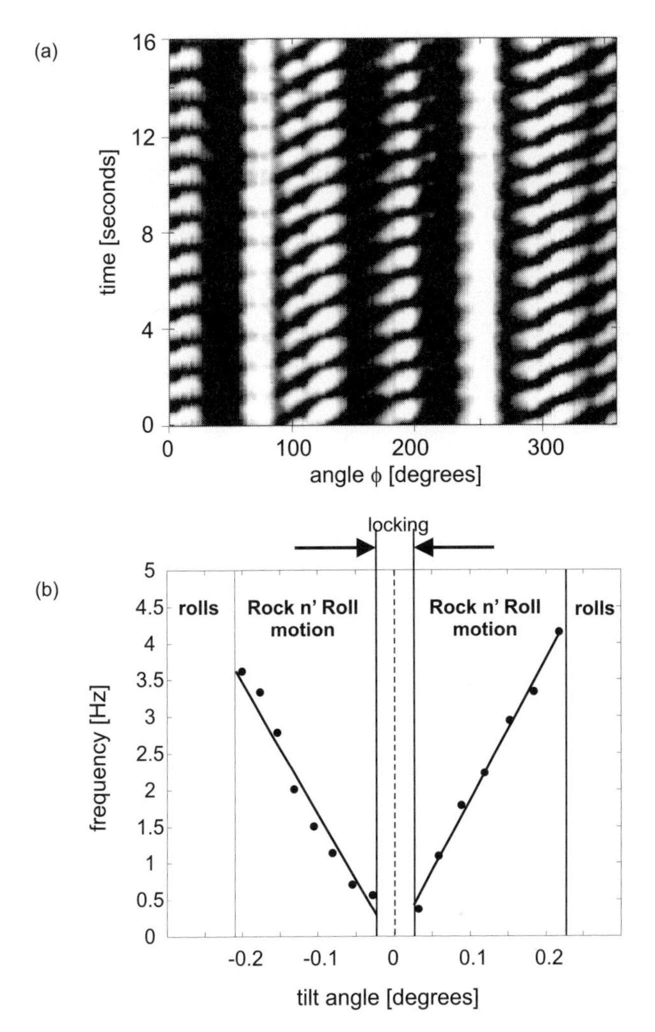

Fig. 9.11. (**a**) Complex motion scenario obtained with noncollinear pump beams. Two spots (rolls) are stationary in space, but not in time. The four remaining spots rotate and then jump back to the initial position. (**b**) Motion frequency of the spots in the case of a "Rock'n'Roll" motion. Also indicated is the locking region and the transition to rolls

serves as a trigger for a drift instability. This is exactly the case for feedback with a tilted mirror. The exact reproduction of the pattern onto itself is no longer achieved, but the pattern seeks a reproduction of itself and consequently flows in the direction of large intensity; this large intensity may be in the form of bright hexagonal spots. The flow of the pattern can therefore be interpreted as a result of the self-adjustment of the system to compensate for the phase change introduced by the misalignment. The experimental picture is the following: a small mirror tilt leads to a flow of the near-field pattern in the plane of the mirror tilt, where the direction of the flow can be reversed by reversing the sign of the mirror tilt angle. This serves as an excellent adjustment criterion for a pattern-forming system. The collinearity of the counterpropagating beams can be easily achieved by steering the flow through tilting the mirror. The direction and velocity of the flow give important information about the nature and size of the angular misalignment.

Figure 9.11b shows the dependence of the frequency of motion of the far-field spots for the case of a motion of the kind depicted in Fig. 9.11a. The frequency of motion depends linearly on the tilt angle, and the slow response of the photorefractive medium leads to stationary roll patterns for larger mirror tilts – in this region, the medium is too slow to follow the fast flow of the optical near field. The linear dependence of the drift frequency on the mirror tilt has already been proposed theoretically and demonstrated experimentally for the case of atomic vapors [11, 12]. It is also apparent from Fig. 9.11b that a locking phenomenon exists which prevents the patterns from drifting for small tilt angles. Since a residual misalignment is always present in an experimental system, this locking is an important feature of pattern-forming systems. Otherwise, no static patterns could be observed. This locking phenomenon has also been observed in a single-mirror feedback system using sodium vapor to provide the optical nonlinearity [14, 15] and can be interpreted as a general phenomenon in optical pattern formation. In particular, this phenomenon is regarded as a direct effect of the finite size of the Gaussian beam [14, 15]. A larger aspect ratio in the optical far field leads to a reduction of the range of angular tilts where the locking phenomenon exists. This was observed in experiments using different diffraction lengths to vary of the aspect ratio.

Summary

A detailed understanding of the free-running pattern-forming system is a basic condition for the successful implementation of sophisticated pattern control schemes. This chapter has dealt with the pattern formation properties of the single-mirror feedback setup without any external means to control or manipulate the output of the pattern-forming system. To the best of our knowledge, the multiple-pattern region provided by the photorefractive single-mirror feedback system is unique and constitutes an excellent field for exploring and testing the effectiveness of various approaches to the control and manipulation of the pattern. In certain parameter spaces of the multiple-pattern region, it is impossible to address a wanted pattern in a defined way by choosing system parameters, owing to the coexistence of several quasi-stable solutions. At these points in the parameter space, only a sufficiently powerful control method can enable us to select and address a particular pattern. As it turns out, a frequency detuning applied to the beam reflected from the mirror represents an excellent tool for selecting one of two patterns at a bistability. This detuning is easily applied by attaching the feedback mirror to a piezoelectric translation stage so as to obtain a mirror movement with constant velocity. The influence of a frequency detuning on the patterna-forming process, the transverse wave number and the threshold for modulation instability is one of the topics of the next chapter. This investigation represents an initial step towards active control of patterns by choosing system parameters. The additional system parameter available directly affects and manipulates the feedback signal.

The question arises of whether it is possible to obtain patterns of a certain geometry in parameter regimes where only one pattern is the predominant one, and where even a frequency detuning is unable to switch the system to a different pattern. A spatial filtering process working in the Fourier domain fulfills these requirements in an impressive way, as we shall also show in the next chapter.

References

1. H.F. Talbot, *Facts relating to optical science IV*, Philos. Mag. **9**, 401 (1936).
2. O. Sandfuchs, M.R. Belić and F. Kaiser, *Wave mixing in a bulk photorefractive medium: spatiotemporal structures and amplitude equations*, Int. J. Bif. Chaos **11**, 2823 (2001).
3. M.C. Cross and P.C. Hohenberg, *Pattern formation outside of equilibrium*, Rev. Mod. Phys. **65**, 851–1112 (1993).
4. G. Puccioni F.T. Arecchi, R. Meucci and J. Tredicce, *Experimental evidence of subharmonic bifurcations, multistability, and turbulence in a Q-switched gas laser*, Phys. Rev. Lett. **49**, 1217–1220 (1982).
5. M. Brambilla, F. Battipede, L.A. Lugiato, V. Penna, F. Prati, C. Tamm and C.O. Weiss, *Transverse laser patterns II: variational principle for pattern selection, spatial multistability, and laser hydrodynamics*, Phys. Rev. A **43**, 5114–5120 (1991).
6. J. Foss, F. Moss and J. Milton, *Noise, multistability and delayed recurrent loops*, Phys. Rev. E **55**, 4536–4543 (1997).
7. C. Denz, M. Schwab, M. Sedlatschek, T. Tschudi and T. Honda, *Pattern dynamics and competition in a photorefractive feedback system*, J. Opt. Soc. Am. B **15**, 2057–2064 (1998).
8. P.M. Lushnikov, *Hexagonal optical structures in photorefractive crystals with a feedback mirror*, JETP **86**, 614–627 (1998).
9. T. Honda, H. Matsumoto, M. Sedlatschek, C. Denz and T. Tschudi, *Spontaneous formation of hexagons, squares and squeezed hexagons in a photorefractive phase conjugator with virtually internal feedback mirror*, Opt. Commun. **133**, 293–299 (1997).
10. B. Thüring, A. Schreiber, M. Kreuzer and T. Tschudi, *Spatio-temporal dynamics due to competing spatial instabilities in a coupled LCLV feedback system*, Physica D **96**, 282–290 (1996).
11. G. Grynberg, *Drift instability and light-induced spin waves in an alkali vapor with feedback mirror*, Opt. Commun. **109**, 483–486 (1994).

12. A. Petrossian, L. Dambly and G. Grynberg, *Drift instability for a laser beam transmitted through a rubidium cell with feedback mirror*, Europhys. Lett. **29**, 209–214 (1995).
13. T. Honda, *Flow and controlled rotation of spontaneous optical hexagon in* $KNbO_3$, Opt. Lett. **20**, 851–853 (1995).
14. T. Ackemann, B. Schäpers, J.P. Seipenbusch, Yu.A. Logvin and W. Lange, *Drifting and locking behavior of optical patterns – an investigation using Fourier filtering*, Chaos Solitons Fractals **10**, 665–673 (1999).
15. W. Lange and T. Ackemann, *Alkaline vapors with single-mirror feedback – a model system for pattern formation*, Asian J. Phys. **7**, 439–452 (1998).

10 Manipulation and Control of Self-Organized Patterns by Spatio-Temporal Techniques

The formation of spatially extended optical patterns is a highly attractive research field owing to the prospects for technologically relevant applications. From the point of view of application, it is of interest to access the whole range of possible solutions and to control the structures inherent in the system. Manipulating or controlling such a system by suppressing the predominantly excited solutions or by encouraging underlying solutions to become stable offers an opportunity to stabilize, select and manipulate these patterns in a defined way.

10.1 Spatio-Temporal Control Techniques

There are several reasons for wishing to control pattern formation processes in optical systems that exhibit spontaneous formation of patterns above a certain threshold. First, suppressing pattern formation or chaos can offer the possibility to operate systems in the highly nonlinear regime, while retaining attractive and desirable features such as temporal and/or spatial coherence without undesired nonlinear dynamical effects. Complex, modern optical-transmission schemes require devices working in the nonlinear regime. A large energy flux is definitely related to an increasing importance of nonlinearity and, along with it, nonlinear dynamics. Detailed understanding helps us to control or prevent these pattern formation processes in order to construct high-performance optical-transmission. Second, stabilizing a desired state out of the manifold of possible chaotic or spontaneously formed states is attractive for extending the area of application of optics to information processing far beyond the linear regime. The potential information capacity of nonlinear optical systems is determined by the number of different pattern states accessible to the system.

In all these cases, it is desirable that the influence of the control on the system is as small as possible, to guarantee that at the end of the control process the old system is regained – the only difference being a desired and actively selected pattern state.

10.1.1 Control Techniques for Temporally Chaotic Systems

The aim of controlling a nonlinear dynamical system is as old as the studies of effects connected to nonlinear dynamics and chaos. Research devoted to the "control of chaos" is very lively, and numerous different control schemes have been suggested and investigated. The pioneering work by Ott, Grebogi and Yorke [1, 2] dealt with the stabilization of periodic orbits in a chaotic attractor by carefully chosen small, discrete perturbations in an accessible system parameter. Unstable periodic orbits in the chaotic attractor can be stabilized by determining the stable and unstable directions in the Poincaré section. As long as the attractor is known, this technique can be proven to be valid for any nonlinear dynamical system. A number of publications [3, 4, 5] followed, aimed at gaining improved insight into the control of low-dimensional chaos. While the method described in [1] (often referred to as the OGY method) is difficult to realize experimentally, two other techniques have turned out to be ideal for experimental implementation.

- The delayed-feedback control scheme of Pyragas [5, 6, 7] is based on the principle of a self-controlling feedback perturbation using a simple time delay in the output signal. This is known as the *Pyragas* or *feedback method.* The feedback signal can either be derived from the system signal itself, changed by a time-delay operator, or be proportional to the difference between the output of the system and the target pattern. In both cases, the magnitude of the control signal decreases gradually as the system approaches the desired target state, therby providing a noninvasive method to push the system into a desired state. However, naive application of the method of delayed optical feedback [8, 9] leads to the observation of spatio-temporal complexity rather than the suppression or control of chaotic behavior. Such instabilities caused by time-delayed feedback have been known for a long time [10]. The technological relevance of semiconductor lasers in optical information transmission is one of the major factors that motivates ongoing research. In practical applications, instabilities caused by delayed optical feedback from an optical disk, an external mirror or an optical fiber are known to decrease the performance of semiconductor lasers considerably. Consequently, the stabilization or complete removal of instabilities is a topic of great interest and technological relevance [11, 12, 13, 14]. Semiconductor lasers are just one example of devices, whose performance is limited by the appearance of complex spatio-temporal behavior.
- Another method is to modulate a system parameter with a small periodic perturbation or to add a weak periodic forcing to the system [15, 16, 17] – the *nonfeedback control method.* This method is based on the breaking of the symmetry of the current state, thus enhancing the stability of the desired pattern. In this technique, the intrusion into the system may be large, causing the system to change its state, but allowing a desired state to be stabilized. Although the methods of nonfeedback control are generally

not as effective as the feedback methods, they have the advantage that they do not require a priori knowledge of the system behavior. Therefore, they are particularly appealing for systems whose state is difficult to measure in real time and where feedback control is hard to realize.

10.1.2 Control Methods for Spatially Extended Systems

The use and application of pattern-forming nonlinear systems require the regularization of the spatio-temporal behavior and, for the operator, the possibility to choose from different spatial patterns. The challenge of pattern control schemes is the defined selection of one spatial pattern, while other underlying, potentially stable solutions are suppressed. As a consequence, the number of different pattern states accessible in one parameter region is increased, because control is not necessarily restricted to stable states. The inclusion of unstable pattern states increases the number of available states; at the same time, the stabilization of otherwise unstable patterns has a decisive fundamental relevance, because it gives access to unstable branches of theoretical bifurcation diagrams, thereby improving the understanding of the pattern-forming system. A versatile control tool is required, which is capable of providing pattern selection but at the same time not altering the system and its solutions. The challenge is to access the maximum number of system solutions – various different stable or even unstable pattern states – even in the presence of a predominant system solution, for a wide range of parameters. In contrast to those types of control applied to temporal dynamical systems which require chaotic dynamics to provide a large number of unstable states, the control of spatio-temporal patterns focuses on the selection of states by exploiting the breaking of the symmetry due to the pattern-forming process.

Owing to the complexity of spatio-temporal patterns and the vast number of possible states, in early efforts towards a control of these patterns, the methods focused mainly on the suppression of spatio-temporal disorder in chaotic regimes [18, 19, 20, 21, 22, 23, 24], some of them by using an extension of the OGY algorithm [20, 25]. A related aim, which is comparable to that of the purely temporal methods mentioned above, to stabilize and address unstable spatial states [25, 23, 24] rather than just to suppress unwanted chaotic states.

Nevertheless, methods for the control and manipulation of spatio-temporal pattern formation have only been suggested recently, although optical systems display, on fast timescales, phenomena common to many spatially extended systems. Lu et al. extended the Pyragas method to an optical system [26], and Martin et al. developed a Fourier space technique to stabilize and track unstable patterns in a two-level system in an optical cavity [27, 28]. The main objective of these researchers was to control the nonlinear optical system in such a way that the desired target state was an unstable state of the original pattern-forming system. The system, and thereby its spectrum of solutions, is not changed by the intervention. Though these theoretical suggestions offered promising prospects for the control of spatio-temporal patterns in nonlinear optics, experimental implementations of these techniques are few at present, although they have some attractive and promising features.

In this chapter, we shall present several different experimental approaches to the idea of controlling and manipulating spatially extended patterns. A first, very simple technique for manipulation of patterns is to change the frequencies of the interacting counterpropagating beams, thereby altering the threshold and pattern formation conditions considerably. We shall discuss this approach in the first part of this chapter and demonstrate the effectiveness of frequency detuning for the active control of the selection process of spontaneous optical patterns. A second technique, which is straightforward to implement, is that of Fourier space manipulation. We shall show that different degrees of invasion of the system are possible, and that it is possible to realize spatial control techniques that do not influence the system state in real experimental situations.

10.2 Manipulation of Optical Patterns by Frequency Detuning

The majority of publications concerning pattern formation in photorefractive media cover the case of mirror boundary conditions, where the two interacting beams acquire certain amplitude and phase relationship owing to the coupling of the two beams via the reflection mirror. The main results of these investigations have been the observation mainly of hexagonal patterns, but also of many other symmetric patterns in the region of multistability. This led to the conclusion that two independent, counterpropagating beams are not able to induce formation of a symmetric pattern, even if they differ in other parameters, such as their frequency. Consequently, theoretical investigations have to date been restricted to static instabilities, explicitly excluding dynamic instabilities [29, 30].

However, in our experimental realization soon after our first pattern formation observation [31], we could show that there was evidence for a modulation instability in photorefractive counterpropagating two-wave mixing without feedback. The key element for spontaneous-pattern observation in this case is the frequency detuning of the pump beams. This detuning enables us to drive the system to instability.

Therefore, we extended the existing linear stability analysis to the case of frequency-detuned beams. As soon as frequency differences between the two counterpropagating beams are allowed, dynamic instabilities play a key role in pattern formation with open boundary conditions [31]. A dynamic grating component (manifesting itself as a sideband frequency detuning in the spatial far field) is always necessary for obtaining a solution of the threshold condition for modulation instability. Even in the case of frequency-degenerate beams, dynamic instabilities have to be considered, since they might possess a lower threshold for modulation instability compared with the static instabilities [31].

As we shall demonstrate in this section, not only does detuning the pump frequency provide only a new mechanism for pattern formation that complements mirror boundary situations, but it also represents an excellent means of actively controlling the output of the pattern-forming system. Detailed analysis of the impact of a frequency detuning on the spatial scale of the pattern and on the threshold values is therefore vital and is an essential precondition for understanding the mechanism of pattern selection under the influence of a frequency detuning. Since our main focus in this chapter is the active manipulation and control of patterns, we skip the details of the mathematically demanding calculation of the threshold of the linear instability for mirror boundary conditions and frequency-detuned pump beams. This calculation is left for the interested reader and can be found in [32, 33].

In this analysis the coupling constant is chosen to be complex, thus covering modulation instability for any type of charge carrier transport mechanism. Owing to the sluggish response of the photorefractive effect, a frequency detuning of the interacting optical beams changes both the amplitude and the phase of the complex coupling constant (see Sect. 3.6.2). A detailed study incorporating a complex coupling constant and a frequency detuning in the sidebands was performed in [34] for a set of equations neglecting the diffraction term for computational simplicity. For the case of mirror boundary conditions and a diffusion-dominated photorefractive material, an infinitely large threshold for modulation instability was found, contradicting experimental results. However, this result clearly indicates the relevance to the occurence of modulation instability of the diffraction term, which describes the transverse dependence of the laser beam profile as a spatial coupling process. Subsequent theoretical investigations took into account the transverse variation of the beam profile, but focused on a photorefractive nonlinearity of the drift type [35], or the analysis was restricted to the special case of static instabilities [29, 36, 30, 37].

The lack of experimental investigation of dynamic instabilities in the case of frequency-degenerate beams in the single-mirror feedback setup can be partially explained by the intensity dependence of the grating relaxation parameter τ, which is kept constant in the theoretical considerations for simplicity. This intensity dependence was recently identified as preventing dynamic instabilities in certain parameter regions [38]. A detailed theoretical analysis of the case of two counterpropagating beams, with an additional externally applied electric field and the grating dynamics taken into account, was performed in [39].

The mode of action of a frequency detuning of the pump beams is similar to that of an electric field applied across the crystal [32]. The advantage of the method of frequency detuning is the easier implementation in an experiment, which is achievable by using a simple piezoelectric translation stage. Application of an electric field in the direction of propagation of a focused laser beam is a challenge and has not been implemented in pattern formation experiments.

The experimental setup that we used was basically the same as the one depicted in Fig. 9.1; the only difference was that the feedback mirror was attached to a piezoelectric translation stage for controlled application of a frequency detuning to the counterpropagating beam.

Figure 10.1 shows examples of the results of the measurements of the transverse wave number together with the corresponding theoretical curves. Figure 10.1a illustrates the dependence of the transverse wave number $k_{\mathrm{d}}l$ on the frequency detuning for $n_0 L/l = 0$ and $\tau_0 = 75$ ms. The photorefractive time constant τ is defined as the grating relaxation time [39], the subscript zero indicating the case of no frequency detuning; the inherent intensity dependence of the relaxation time changes this value when a frequency detuning is applied. The theoretical curves are given for three different drift components γ_{I} of the photorefractive coupling constant. Good agreement between theory and experiment for positive frequency shifts Ω is obtained. The asymmetry of the experimental curve is also reproduced in the theory; the only difference is that the step in the transverse wave number for small negative frequency detunings Ω in the theory does not appear in the experiment. This step is due to the existence of a competing minimum taking over the role of absolute minimum of the threshold coupling strength. The experimentally found transverse wave numbers remain constant for negative values of $\Omega\tau$. Since all values on the threshold curve $\gamma(k_{\mathrm{d}}l)$ are a solution of the threshold equation, the values of γl corresponding to a constant transverse wave number were calculated, as suggested by the experiment. These values differed from the values determined by the absolute minima by only 5% for $\Omega\tau \geq -4$. A constant transverse wave number $k_{\mathrm{d}}l$ is therefore obtainable without a large value of the coupling strength, which is consistent with our experimental results.

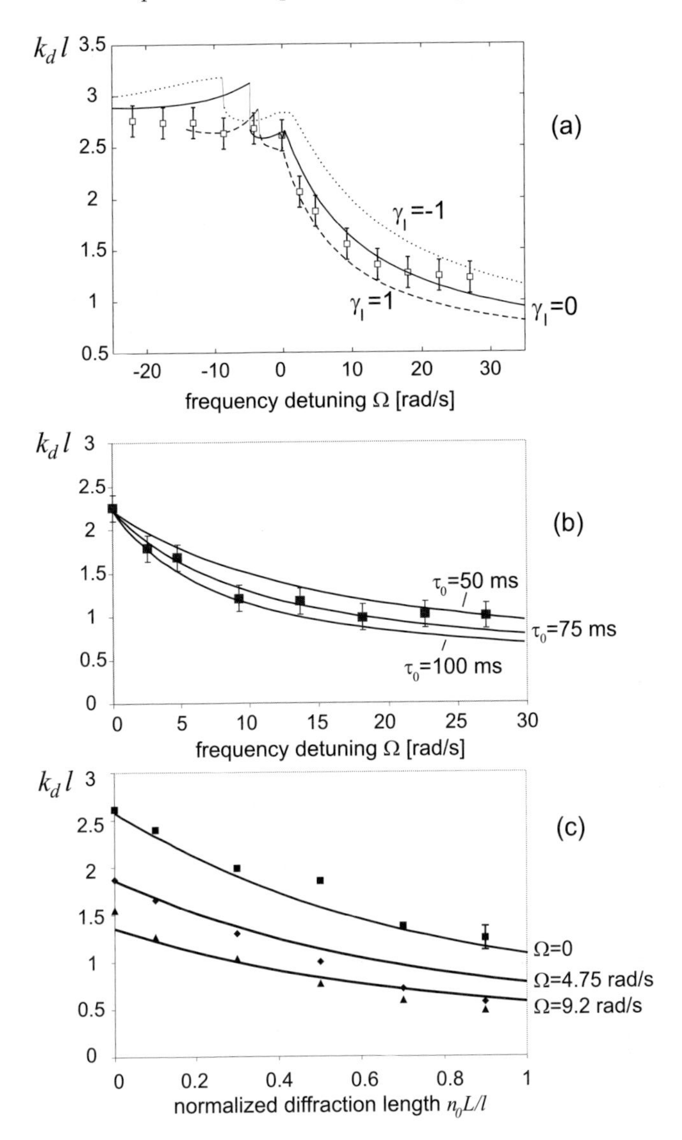

Fig. 10.1. Comparison of theoretical and experimental results for the transverse wave number $k_d l$. (**a**) Results as a function of the frequency detuning for $n_0 L/l = 0$. *Solid curve*, theory for $\gamma_I = 0$; *dashed curve*, theory for $\gamma_I = 1$; *dotted curve*, theory for $\gamma_I = -1$. All curves are given for a value of the time constant $\tau_0 = 75$ ms. *Squares*, experimentally obtained values. (**b**) Results as a function of positive frequency detuning for $n_0 L/l = 0.25$, shown together with the theoretical curves for different values of τ_0. (**c**) Results as a function of the diffraction length $n_0 L/l$ for $\Omega = 0$ (*squares*), $\Omega = 4.75$ rad/s (*diamonds*), and $\Omega = 9.2$ rad/s (*triangles*). Theoretical curves are given for $\tau_0 = 95$ ms

The jump appearing in transverse scale for certain negative frequency detunings $\Omega\tau$ is due to an intraballoon competition, where a new global minimum is created at the expense of the old one. Owing to the special shape of the threshold curves, a whole band of transverse wave vectors can be excited, and nonlinear interactions come into play. In Fig. 10.1b, measurements and theoretical curves for a diffraction length $n_0 L/l = 0.25$ are depicted for different values of τ_0 showing a qualitative agreement of the experimental results with the theoretical analysis. Fig. 10.1c shows the transverse wave number $k_{\mathrm{d}} l$ as a function of the normalized diffraction length $n_0 L/l$ for three values of Ω and $\tau_0 = 95$ ms.

The deviations between theory and experiment for larger diffraction lengths in Fig. 10.1c can be attributed to several effects. First, in our theoretical analysis, absorption is not taken into account, in order to obtain an analytic threshold condition. The presence of absorption raises the experimental transverse wave numbers, as shown in a numerical analysis of counterpropagating two-wave mixing with inclusion of absorption effects [39]. Second, to ensure that an analytic treatment of this problem was possible, an intensity reflection coefficient of $R = 1$ (compared with about $R = 0.5$ in the experiment) was assumed in the linear stability analysis. Another important point is the intensity dependence of the grating relaxation parameter τ. The experiment was performed with a Gaussian beam entering the crystal, which implies a continuous distribution of time constants τ within the Gaussian beam.

In order to investigate the effects of a frequency shift on the pattern state, a diffraction length of about $n_0 L/l = -0.30 \pm 0.10$ was chosen, which is inside the multiple-pattern region. Using a circular spatial filter between the two lenses in the feedback path in order to suppress instabilities with larger transverse wave numbers, a square pattern was achieved as the stable solution of the system, as depicted in Fig. 10.2a.

By applying a small frequency shift $\Omega = 2.78$ rad/s to the feedback beam, it was possible to switch to a hexagonal pattern with a smaller transverse wave number (Fig. 10.2b). When the frequency detuning was turned off, the system relaxed back to the square-type solution. A whole sequence of transitions between hexagons and squares could be obtained by switching the frequency generator on and off. The switching time was some tens of milliseconds, which corresponds to the relaxation time of the photorefractive grating at the intensity of approximately 4 W/cm^2 used in the experiment. For a different diffraction length, $n_0 L/l = -0.60 \pm 0.10$, where a squeezed hexagonal pattern was found to be the stable solution, a switch from the squeezed hexagon to a rhomboid solution could be performed using the same frequency detuning $\Omega = 2.78$ rad/s. It was also possible to continuously switch between a hexagonal and a square pattern by applying a triangular voltage that yielded a frequency shift of $\Omega = \pm\ 4.35$ rad/s for a mirror position of $n_0 L/l = -0.35 \pm 0.10$. Figure 10.3 visualizes the transition from a

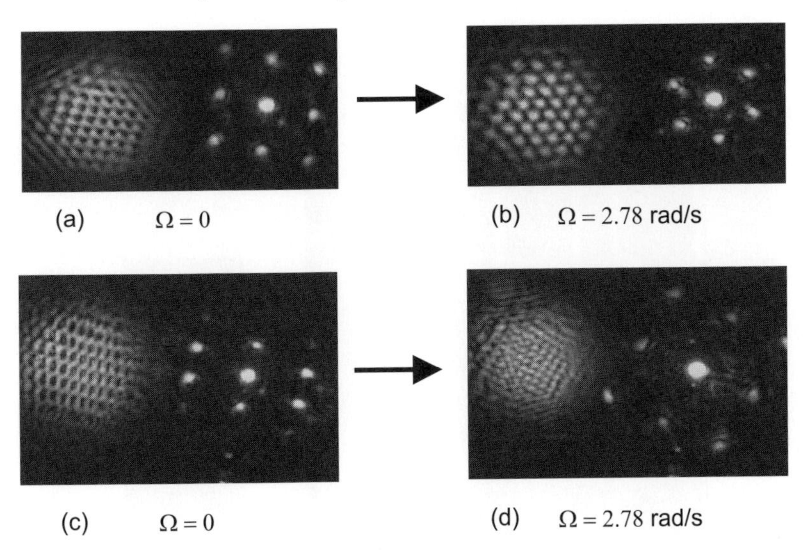

Fig. 10.2. Pattern switch obtained by introducing a frequency detuning into the beam reflected from the feedback mirror. (**a**) Square pattern state without frequency detuning, i.e. $\Omega = 0$, and $n_0 L/l = -0.30$. Measured transverse wave number $k_\mathrm{d} l = 3.5$. (**b**) Controlled hexagonal pattern state with frequency-detuned pump beams, $\Omega = 2.78$ rad/s, yielding a smaller transverse wave number of $k_\mathrm{d} l = 2.7$. (**c**) Squeezed hexagonal pattern state without frequency shift, i.e. $\Omega = 0$, and $n_0 L/l = -0.60$, with two different transverse wave numbers. (**d**) Controlled rectangular pattern state with frequency-detuned pump beams, $\Omega = 2.78$ rad/s

square to a squeezed hexagonal pattern for $n_0 L/l = -0.64$, which was identified as a parameter region of bistability with identical probabilities for these two patterns. In this parameter region, a frequency shift in the sub-Hz region was sufficient to induce a pattern transition from the squeezed hexagonal to the square-type solution. The transition times were around one to two seconds in this case, which indicates a clear dependence of the transition time on the frequency detuning applied. Reversing the sign of the detuning enabled us to select and track the square pattern. In this case, switching off the control signal left the system in the last chosen pattern, which is a clear indication of a hysteresis effect in a bistable system. In another region of parameters, where the hexagonal pattern was known to be the only solution that the system selected (i.e. the parameters were outside the multiple-pattern region), a change to a different pattern could not be obtained. In that case, a frequency shift affected only the transverse wave number of the hexagon.

We have demonstrated in this section the selection of optical patterns by detuning the frequency of one of the two waves taking part in the two-wave-mixing process. With the results of this detailed investigation in mind, we are able to choose from a wealth of different patterns by a suitable choice of system parameters. We have demonstrated that an additional frequency

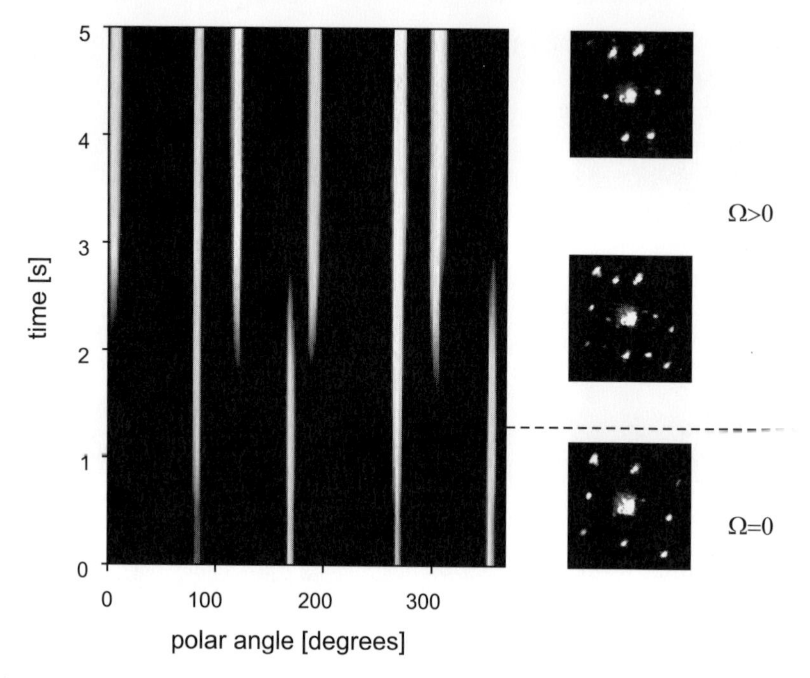

Fig. 10.3. Visualization of the selection of a squeezed hexagonal pattern by controlled application of a frequency shift $\Omega < 1$ Hz. The initial square pattern is suppressed and the system makes a transition to the squeezed hexagonal solution. The inner, unstable circle of the squeezed hexagon is projected onto the outer circle in this visualization

detuning works especially well when we have a bistable state and enables us to choose one pattern or the other by an appropriate choice of the detuning strength. The question arises of whether it is possible to obtain patterns of a certain geometry in a parameter regime where only one pattern is predominant, and where even a frequency detuning is unable to switch the system to a different pattern. A spatial filtering process working in the Fourier domain fulfills these requirements in an impressive way, as we shall show in the next sections.

10.3 Phase-Sensitive Control of Optical Patterns

Optics provides a unique possibility to access the spatial-frequency domain in real time in the back focal plane of a convex lens or by looking at the far-field intensity distribution (compare Figs. 8.2a,b and note the convenient representation in the far field). Therefore, phase-sensitive techniques that are based on the manipulation and control of the Fourier plane can be implemented easily in optical systems.

Extended spatial optical patterns are represented by a limited number of localized spots in Fourier space. Because of the spontaneous breaking of the translational symmetry, pattern forming systems possess a huge number of stable or unstable states for any given parameter value. Since the pattern selection process is determined by nonlinear interactions among a large (possibly infinite) number of wave vectors, direct manipulation of the wave vectors in Fourier space is an extremely attractive tool and constitutes a highly parallel scheme for controlling the system in an all-optical manner. The direct access to the wave vectors enables us to directly influence the pattern selection process, by steering the growth or suppression of certain Fourier modes. Control methods based on spatial filtering are especially attractive for pattern-forming systems which are based on feedback, methods since the two basic conditions for Fourier space control, access to the Fourier spectrum and feedback are inherently available.

The idea of *direct* phase-only filtering in the feedback loop was used in [40, 41] to suppress phase distortions and to change the system in order to generate new patterns. This was definitely a massive intervention in the system, since the phases of the spectral components were directly manipulated. To give another example of an invasive intrusion into the system: going from two transverse dimensions to one by introducing a slit (amplitude) filter into the system changes the system as a whole, and thereby its solutions. If a hexagonal pattern collapses into a roll pattern as a reaction to this invasion, that solution remains the (only) solution of the one-dimensional system: other patterns, among them the hexagonal one, are no longer solutions to the system, since they require the second transverse dimension. This direct amplitude filtering in the feedback arm has an *invasive* character, since it changes the system and the solutions as a whole. Nevertheless, invasive amplitude Fourier filtering has been implemented in a number of different configurations [37, 42, 43, 44, 45] and shows an ability to excite different pattern states by using simple amplitude masks in the Fourier plane of the optical feedback path. A schematic representation of the method of invasive Fourier filtering is given in Fig. 10.4a. The method of direct Fourier filtering is capable of suppressing defects and spatio-temporal disorder when a circular low-pass filter [43]. In addition, information about bifurcation characteristics can be deduced by using Fourier space filtering. Ackemann et al. [45] showed the supercritical bifurcation of an (otherwise unstable) roll pattern and contrasted it with the subcritical bifurcation of the hexagonal pattern. If interest is focused on basic investigations of this kind or on the simple "output" of a "black box" pattern-forming system independent of the question of active alteration of the system, direct Fourier space filtering is an extremely flexible and versatile tool for studying basic mechanisms of pattern formation by controlled introduction of an asymmetry. Pattern manipulation can be performed "online" in an experiment, e.g. by varying the width of a slit in the Fourier plane.

10.3.1 Spatial Control in the Fourier Spectrum

A control method based on spatial filtering in the Fourier spectrum of a spontaneous optical pattern was first proposed theoretically in [27, 28]. This feedback scheme is based on an indirect filtering in an additional feedback (control) loop. A schematic representation of this idea is depicted in Fig. 10.4b. The additional control arm incorporates the two basic ingredients of Fourier filtering with an appropriate mask, described by a filter function $\mathcal{F}$, and reinjection of this spatially filtered signal into the original feedback system. Depending on the phase of the control signal relative to the signal in the main loop, positive or negative control can be realized by in-phase (+) feedback of wanted spectral components or out of phase (-) feedback of unwanted components. Both methods suppress unwanted solutions and enhance the desired pattern geometry. If negative control is used the pump field acquires a spatial modulation which is determined only by the modes found in the instability annulus in Fourier space (see Fig. 10.5). The effect of the extra control arm is described by the impact on the system amplitude B:

$$B'(x,y,t) = [1 \pm sf(x,y)]\, B(x,y,t) \;, \tag{10.1}$$

where

$$f(x,y) = \mathcal{F}^{-1}\mathcal{U}(x,y)\mathcal{F} \tag{10.2}$$

which describes the sequential application of a Fourier transformation $\mathcal{F}$, a Fourier filter transmission function $\mathcal{U}$ and an inverse Fourier transformation. The Fourier-filtering transmission function $\mathcal{U}$ describes the amplitude- and

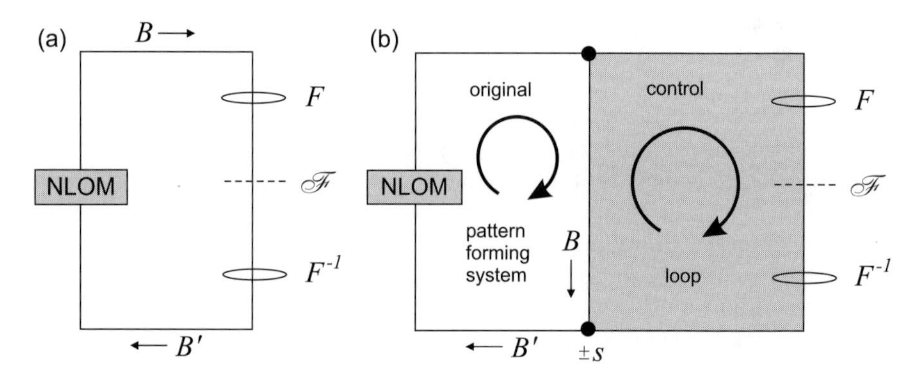

Fig. 10.4. Schematic representation of an invasive (**a**) and a noninvasive (**b**) Fourier filtering method, as proposed by Martin et al. [27] $\mathcal{F}$ represents the spatial filter function, F represents a Fourier transformation, which can be easily accomplished by use of a single lens, and s represents the strength of the control signal (B), (B') are the system amplitudes before and afeter the filtering process. NOLM = nonlinear optical material. The filtering is performed either in the original pattern-forming arm (**a**) or in an extra control loop (**b**)

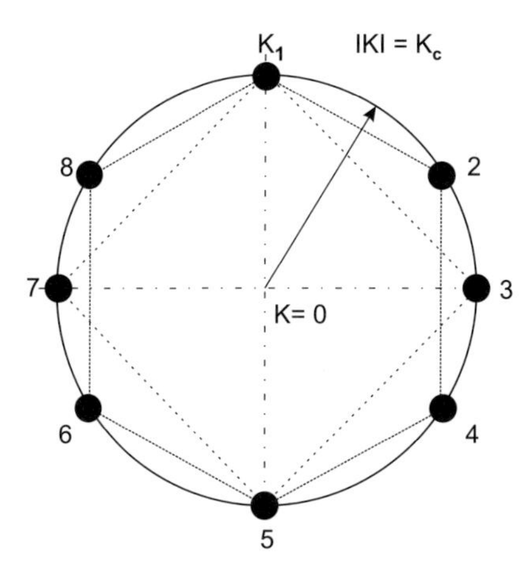

Fig. 10.5. Schematic diagram of the Fourier modes necessary to form the patterns to which control is applied. The modes lie on the critical circle $|\boldsymbol{k}| = k_c$. The *short-dashed lines* indicate the modes which constitute a hexagonal solution, and the *long dashed lines* indicate those necessary for the formation of a square pattern

phase-sensitive Fourier filtering process, and s represents a measure of the strength of the control and is related to the fraction of the beam amplitude which is split into the control loop. As a consequence, s^2 is the ratio of the power in the control loop to that in the main loop. The + sign in (10.1) refers to positive control, whereas the − sign describes negative feedback.

The new variables, manipulated according to (10.1), need to be substituted into the propagation equations of the system. In the case of the stabilization of the homogeneous state, the modulation of B vanishes, since the Fourier transform of the field contains no excited wave vectors in the annulus. This is the most desirable feature of the technique. When the desired state is reached, the control signal tends to zero, thus leaving the system as it was and therefore allowing a completely noninvasive control technique.

In order to selectively control different pattern states, $\mathcal{U}(x, y)$ needs to be modified. To stabilize rolls, for example, we have to remove two diametrically opposite modes $(\boldsymbol{k}_1, \boldsymbol{k}_5)$ on the annulus. With this intervention, the stabilization of a hexagonal pattern is no longer possible. This allows the formation of rolls by suppressing the growth of all modes except the desired ones. Again, the feedback vanishes as the rolls stabilize, ensuring that these rolls are indeed a solution of the system. Because of the rotational degeneracy, we are free to choose the orientation of the stabilized rolls.

The stabilization of a pattern with more than two wave vectors requires an extra degree of control. If we try to stabilize a square pattern (four wave vectors) by removing the wave vectors $\boldsymbol{k}_1, \boldsymbol{k}_3, \boldsymbol{k}_5, \boldsymbol{k}_7$ from the feedback, we end up with the stabilization of rolls in one of the two possible orientations.

The technique needs to be modified in order to ensure the presence of all four wave vectors. Martin et al. [27] suggested a method in which one calculates the control parameters

$$\begin{aligned} u_1 &\propto -A_1 + A_7 \,, & u_5 &\propto -A_5 + A_3, \\ u_3 &\propto -A_3 + A_1 \,, & u_7 &\propto -A_7 + A_5 \,, \end{aligned} \tag{10.3}$$

where A_i is the amplitude of the wave vector $\boldsymbol{k}_i$. This can be achieved experimentally by filtering the Fourier field to obtain the amplitudes A_i and then passing the field through an interferometer with a field-rotating element in one arm to obtain the amplitudes u_i. We then take the inverse Fourier transform to construct $f_2(x, y)$, which is now a positive feedback function. Therefore, the feedback modulation of B is now given by

$$f(x, y) = -s_1 f_1(x, y) + s_2 f_2(x, y) \,, \tag{10.4}$$

where $s_2 > 0$ is the strength of the positive feedback. Equation (10.4) not only suppresses unwanted modes, but also distributes the energy among all four wave vectors necessary for the formation of a square pattern, via a simple rotation in Fourier space. The desired pattern is thus stabilized with a feedback control, which again disappears when stabilization is achieved. This is a particularly interesting result, since square patterns normally do not exist as stable output states of our pattern formation system, and were found, for the first time, only in the region of small negative propagation length in the photorefractive feedback system. It is important to note that, from a theoretical point of view, the width of the filter in Fourier space must be chosen in such a way that the spatial harmonics of the desired pattern are not included in the feedback. These modes form a part of the exact solution to the system equations and therefore must not be suppressed. However, we shall see later on that in experiments, even theoretically insufficient control signals may still be able to stabilize a pattern owing to adaptation processes of the pattern.

A straightforward extension of the model also allows one to control hexagonal patterns. As mentioned in Sect. 8.2.3, there are two classes of hexagonal pattern, distinguished by the sum of the phases of the complex amplitudes, which may be an even (H^+ pattern) or odd (H^- pattern) multiple of π. The control method discussed above stabilizes a hexagonal pattern of either kind, without being able to distinguish between them. In order to add the ability to distinguish between H^+ and H^- patterns to the control mechanism, we can set [27]

$$\begin{aligned} u_1 &\propto -A_1 - A_8 \,, & u_5 &\propto -A_5 - A_4 \,, \\ u_2 &\propto -\mathrm{i}A_2 - A_1 \,, & u_6 &\propto -\mathrm{i}A_6 + A_5 \,, \\ u_4 &\propto -A_4 - \mathrm{i}A_2 \,, & u_8 &\propto -A_8 + \mathrm{i}A_6 \,. \end{aligned} \tag{10.5}$$

In this case, it is possible to stabilize the H^+ solution by choosing $\Phi_1 = \Phi_4 = \pi/2$ and $\Phi_6 = -\pi$. This is one of a two-dimensional manifold of possibilities, but different choices of $\Phi_{1,4,6}$ simply correspond to translating the hexagonal pattern.

To stabilize a H^- hexagon solution, with $\Phi_1 = \Phi_4 = 0$ and $\Phi_6 = \pi$, we need [27]

$$\begin{aligned} u_1 &\propto -A_1 + A_8 , & u_5 &\propto -A_5 + A_4 , \\ u_2 &\propto -A_2 - A_1 , & u_6 &\propto -A_6 - A_5 , \\ u_4 &\propto -A_4 - A_2 , & u_8 &\propto -A_8 - A_6 . \end{aligned} \tag{10.6}$$

By applying this technique, Martin et al. could show by numerical simulations that all stable and unstable solutions of a two-level medium in an optical cavity could be found and tracked throughout the parameter space. Therefore, this method of control not only allows one to adjust desired pattern states or to speed up the process of the natural pattern formation to the stable pattern, but also provides a tool for obtaining an idea of the stable and unstable solution branches (as shown in Fig. 8.12) in areas where even approximate analytical or numerical descriptions fail.

The filters proposed in [27] were designed and feedback was provided in such a way that the control signal vanished when the desired target state was reached, in analogy to what is done in electronic control circuits. Such a system is basically noninvasive, since it leaves the system and the manifold of solutions unchanged. The method theoretically proposed by Martin et al. [27] and later realized experimentally [46, 47] was originally designed for the tracking and stabilization of unstable pattern states.

However, we do not want to restrict ourselves to the stabilization of otherwise unstable states. Our motivation is of a more general nature: the aim is to manipulate the system in such a way that other patterns become stable in parameter space regions where, without any control, the hexagonal pattern is the only experimentally obtainable solution. Nevertheless, the intervention in the system should be as small as possible, motivating us to use the term "minimally invasive". We shall show that an interferometric feedback configuration enables us to distinguish between the original pattern arm and the additional control arm. Thus, the original pattern-forming system remains unchanged. The structured control signal can be fed back either constructively (positive feedback) or destructively (negative feedback). In this sense, "minimally invasive" implies the minimization of the control signal to the lowest possible value in the case of positive control, in order to keep the manipulation of the system as small as possible.

10.4 Invasive Fourier Control of Patterns

The $4f$–$4f$ feedback system used for the majority of the observations of patterns in the last chapter provides us with a Fourier plane between the two lenses. Various amplitude masks were placed into the Fourier plane in order

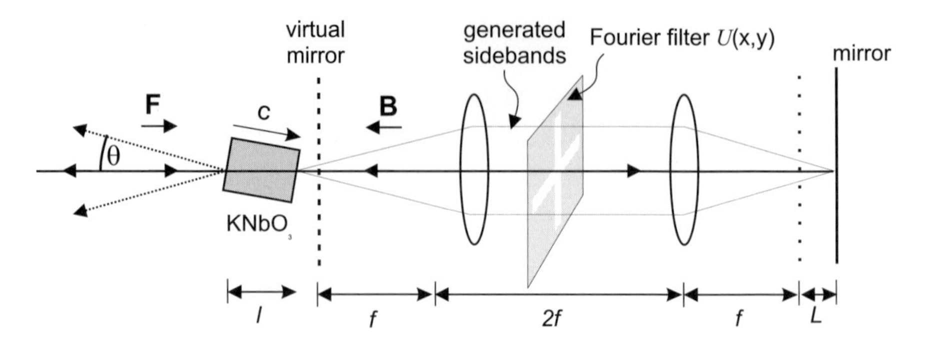

Fig. 10.6. Experimental configuration for invasive manipulation of patterns in the single-feedback mirror arrangement. The Fourier mask, with amplitude transmission function $\mathcal{U}(x, y)$, is positioned at the exact location of the Fourier plane for both lenses

to select the wave numbers participating in the feedback, which determine the spatial symmetry of the self-organized pattern. The basic configuration is depicted in Fig. 10.6. The experiment was performed for $n_0 L/l \approx -0.2$, where the system is in a stable hexagonal solution. Other experimental details have already been given in Sect. 8.2. Selected experimental results for this invasive Fourier filtering are given in Fig. 10.7.

The spatial symmetry of the intended pattern is only suggested by the Fourier filter, not completely determined. There are still sufficient degrees of freedom in the transverse direction. The transverse wave number chosen by the system in the presence of Fourier filtering remains unchanged from the case of a free-running system here. However, this does not hold for the multiple-pattern region, where the observed transverse wave numbers in the one-dimensional system (slit Fourier filter) differ from the available wave numbers in the two-dimensional case under certain circumstances, as explained in Sect. 9.2. It is noteworthy that the available patterns are still restricted to the patterns inherent in the original system. Three-, five- or seven-fold symmetric patterns could not be obtained in the experiments, indicating that the offer of a spatial symmetry does not necessarily lead to an identical pattern with the same symmetry. The chosen patterns have to be members of the family of solutions of the original system.

The influence of a circular Fourier filter for cutting off larger spatial frequencies is known to stabilize temporally unstable patterns or to remove defects in the image plane [43]. Mamaev and Saffman [43] showed that an anisotropic band-pass filtering led to the formation of squeezed hexagons. Using a slit filter with a variable width leads to a sequence from a hexagonal through a squeezed hexagonal to a final roll pattern when the size of the slit is successively reduced (see Fig. 10.8).

The changes in the spectrum can be understood qualitatively by recalling that the dispersion relation for counterpropagation instabilities has a min-

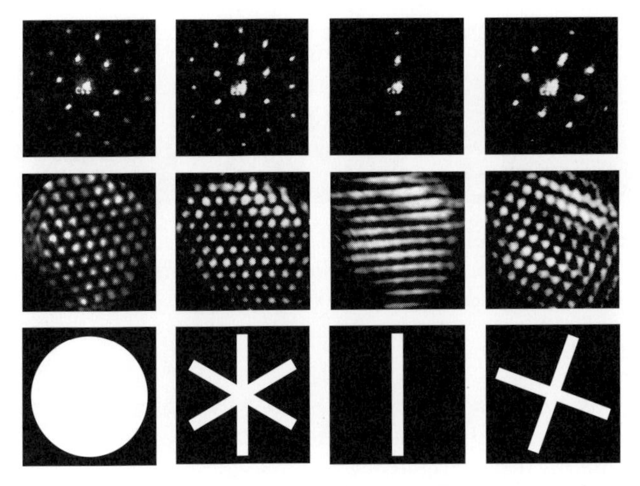

Fig. 10.7. Examples of invasive Fourier filtering. From *top* to *bottom*: far field and near field of the pattern output and the corresponding amplitude mask used for Fourier filtering. *Left column*, hexagonal pattern for the free-running system, *second column*, hexagonal pattern with a different orientation; *third column*, roll pattern; *right column*, square pattern

imum threshold, leading to a well-defined instability angle Θ. Blocking the instability for $\Theta \geq \Theta_0$ leads to instability at angles $\Theta < \Theta_0$, while blocking the instability for $\Theta \leq \Theta_0$ leads to instability at angles $\Theta > \Theta_0$. This effect of anisotropic bandpass filtering is clearly demonstrated in Fig. 10.8.

The direction of the Fourier filter is arbitrary in the sense that the system adjusts to the present situation and uses the spatial orientation proposed. As can be deduced from Fig. 10.7, hexagonal, roll and square patterns are possible in any orientation; it is even possible to continuously rotate a pattern, which proves that the orientation of a pattern is indeed arbitrary and depends only on the initial conditions of the pattern-forming process. This experiment was performed far beyond the instability threshold ($\gamma/\gamma_{\mathrm{th}} \approx 2$). We also measured the relative sideband intensities of the various patterns and found large deviations between the values for the different pattern symmetries. For the uncontrolled system, the hexagonal pattern showed a sideband-to-pump ratio $S/P = 1.05 \pm 0.08$. While this ratio declined to 0.23 ± 0.02 for the spatial roll pattern, a relative sideband intensity of $S/P = 1.37 \pm 0.15$ was found for the square pattern. For a squeezed hexagonal pattern which was obtained by use of an intermediate slit filter size (not shown in Fig. 10.7), the value of S/P was 0.47 ± 0.12. It is noteworthy here that large fluctuations in the sideband intensity were present in all the experiments. In particular, a nonideal pattern in the near field (as shown by the appearance of domains of distinct orientations or by imperfections in the geometrical grid of the pattern) leads to strong variations in the sideband-to-pump ratio and therefore complicates quantitative measurements considerably.

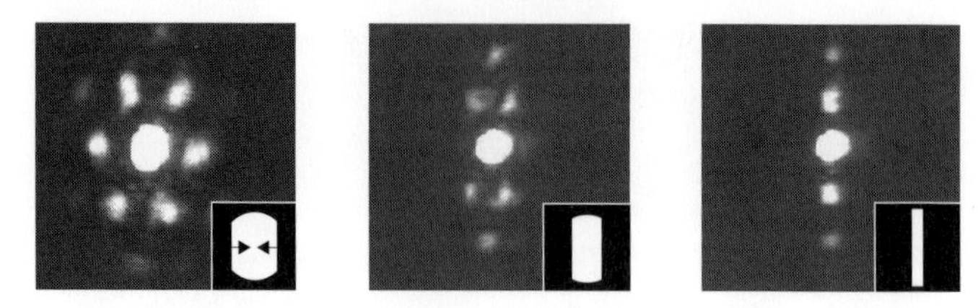

Fig. 10.8. Squeezing of a hexagonal pattern by decreasing the width of a slit filter in the Fourier plane. When the filter reaches a certain threshold width, the squeezed hexagonal pattern jumps directly to the roll solution

A basic feature of the method of Fourier filtering is the minimization of the energy removed by the filter when the targeted state is reached. In the case of the final roll pattern, the whole energy is transferred to the spatial roll pattern, without losses at the filter. For the other patterns, the losses are restricted to the higher-order modes leading to small but nonzero power losses in the system, even in the final equilibrium state. Though the action of the filter on the feedback system is highly invasive, since the feedback system and its solutions are changed, the final pattern state does not require substantial power losses, pointing to the minimally invasive character of Fourier control. In addition, the complete Fourier filter is not necessary for achieving pattern control. For example, in the case of the cross filter,which leads to a square pattern, the symmetry of the feedback system implies that only one-quarter of the filter (one quadrant) is necessary, which minimizes the energy loss caused by the spatial filter and suggests an associative character of Fourier space control.

10.5 Minimally Invasive Pattern Control

The method of invasive Fourier filtering presented above is a versatile and attractive tool for choosing a particular pattern state out of a variety of different solutions. A pattern of a selected geometry, symmetry and orientation can be excited by using an appropriate filter in Fourier space. The functioning of the method is based on a controlled selection of the wave numbers participating in the pattern-forming process. The method constitutes a deep intrusion into the system and changes the feedback system as a whole, since only a limited number of wave numbers are allowed compared with the uncontrolled case. Consequently, the solutions of the system are changed, including the set of different patterns to choose from. The most intrusive pattern manipulation is done by using a slit filter. This reduces the system to one transverse dimension, which diminishes the number of possible pattern states to exactly one – a spatial roll pattern. The aim of the control idea originally put forward in [27] was to achieve stabilization by making the smallest perturbation

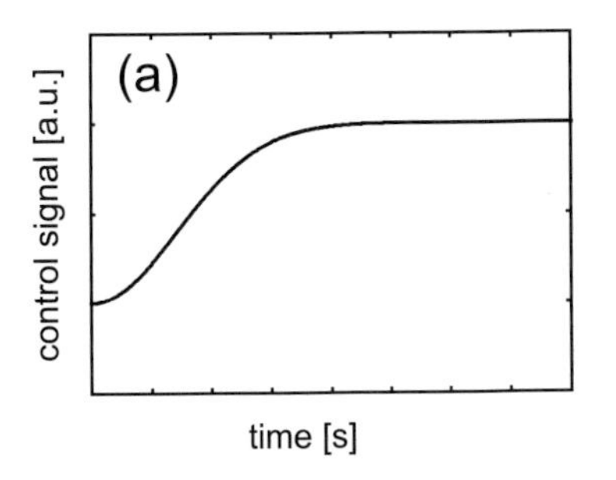

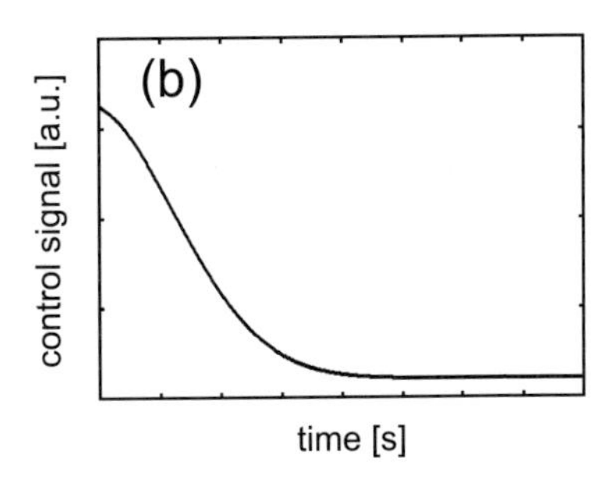

Fig. 10.9. Examples of control signals for positive control (**a**) and negative control (**b**)

possible to the original system. In particular, the perturbation should vanish when the desired target state is reached, which ensures the stabilization of a true system solution rather than the creation of a new solution imposed on the system by an intrusive intervention. The idea is to "persuade" the system instead of forcing it into a new solution.

Our aim is of a slightly different nature and extends the feedback scheme proposed in [27] to the general case of minimally invasive Fourier space control. The particular requirement is the controlled excitation of a pattern in a parameter region where, without control, one specific pattern is the stable output of the system. Since numerical and detailed theoretical studies of the two-dimensional problem of photorefractive counterpropagating two-wave mixing are still few, and the stability and instability regions of the patterns remain undefined, we do not restrict ourselves to the stabilization of otherwise unstable patterns. In order to stay close to the original proposal in [27] and in order to improve the understanding of the basic pattern-forming system, our intention is to leave the basic feedback system unchanged and apply the spatial filtering in an additional control arm. The type of control is then determined by the relative phase of the control signal fed back to the original feedback system: the control signal can be fed back either in phase (positive control) or out of phase (negative control). This artificial separation of control methods clearly distinguishes between the two different ideas of minimally invasive control and noninvasive control, which is limited to the stabilization of otherwise unstable states. In the latter case of negative control, unwanted spectral components are suppressed by destructive feedback of the control signal, which nearly vanishes in the desired state. Figure 10.9 illustrates examples of positive and negative control signals. In the case of positive control, the control signal (the energy fed back to the system) tends to a certain steady-state level, whereas negative control implies that the control signal declines to a very small level (in the ideal case zero) in the desired pattern state. In this context, the word "minimally" suggests that the positive control signal should be as small as possible in the very first stage of control and in the asymptotic steady state. Positive control can also be interpreted as an invasive method with a grey-level amplitude mask. The aim of performing minimally invasive control is equivalent to the minimization of the gray level to a value that is still capable of controlling the output of the pattern-forming system. The system should be slightly pushed to the new solution, rather than forced by a binary invasive mask.

For the negative control method, the minimally invasive property implies the minimization of the initial intervention before the system relaxes to a small control signal level. This can be described adequately by the minimization of the control strength s in (10.1). We applied a minimally invasive manipulation and control scheme to three different experimental configurations: a ring control scheme, a linear scheme and a Michelson-like scheme.

10.5.1 Ring Control Scheme

Our first experimental realizations of a minimally invasive pattern control scheme are shown in Fig. 10.10, applied in the form of a ring resonator and a linear configuration. In order to realize a minimally invasive feedback control unit in the ring control scheme, the small transmission signal of mirror M1 at the end of the feedback path was utilized as the input to a feedback control ring. This control ring was equipped with a lens of focal length $f_3 =$ 300 mm, which provided confocal imaging of the near-field pattern available at mirror M1 onto the crystal input face. Near lens L3, the far field of the pattern was clearly visible, and represented an approximate Fourier plane. Fourier filtering was conducted at a plane near lens L3. The mirror PDM was mounted on a piezoelectric translation stage in order to continuously vary the relative phase of the control signal. To avoid undesired resonance effects in the control arm, the central spot was always blocked in our experiments, which led to a low intensity control signal of about 1–2% of the energy contained in the original feedback arm. Note that the ring geometry of the control loop is bidirectional and that the control signal approaches the feedback system from both sides. Nevertheless, minimally invasive Fourier space control with a Fourier-filtered low-intensity control signal was demonstrated experimentally. Details of the first implementation of a minimally invasive control scheme are given in [48], which shows the principal functions of this control scheme. In a mirror reflectivity region where the system was at the border between hexagonal and roll patterns, the positive-control unit offered the opportunity to excite pure rolls in any desired orientation by using a slit filter as the spatial Fourier filter. An initial realization of negative control enabled us to rotate the spatial orientation of a hexagonal pattern by destructively feeding back two spots of the hexagon signal. As a consequence, the hexagonal pattern orthogonal to the original pattern appeared, minimizing the control signal in the steady state, which is a basic condition for a negative control scheme of the kind proposed in [27]. Pictures from the first experimental demonstration of the principle of negative control are given in Fig. 10.11.

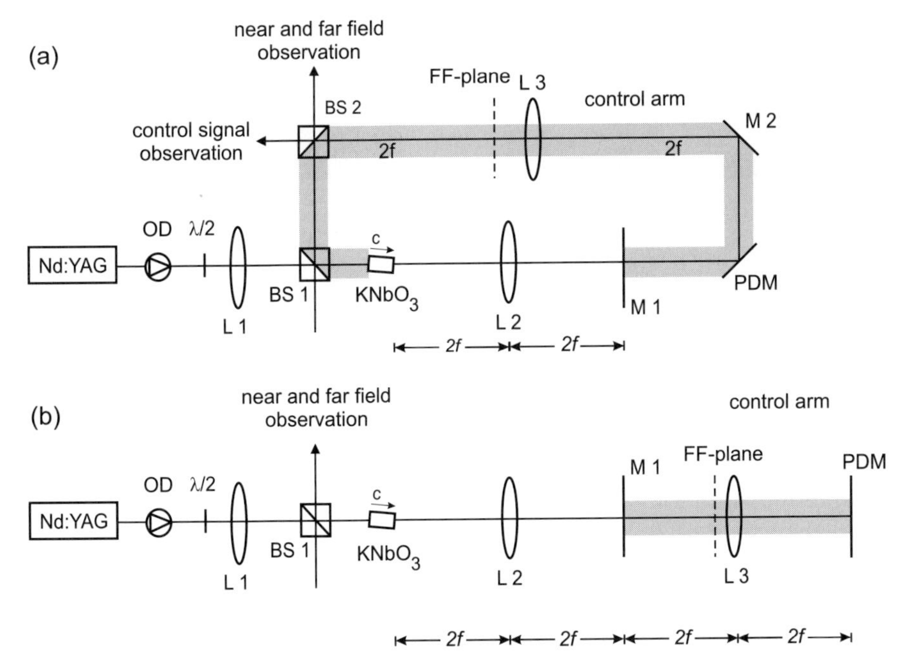

Fig. 10.10. Two setups for the experimental realization of a minimally invasive Fourier control scheme: (**a**) ring control scheme and (**b**) linear control scheme. The crystal sample used for these experiments was crystal sample 1. The original feedback system consisted of a $2f$–$2f$ imaging system; the focal lengths of lenses L1 and L2 were $f_1 = 100$ mm and $f_2 = 120$ mm, respectively. The additional control arms are shaded *gray* for clarity. The lenses in the control arms had focal lengths of $f_3 = 300$ mm in (**a**) and 120 mm in setup (**b**)

In a region of mirror reflectivity chosen high enough to provide a stable hexagonal solution ($r = 40\%$ for crystal sample 1), the control method enables us to choose between different hexagon orientations by in-phase and out-of-phase feedback. When positive feedback is used, the direction of the hexagon adjusts to the position of the filter so that two spots of the pattern pass the filter (Fig. 10.11a). Negative feedback leads to a different hexagon position, to avoid feeding two spots back destructively as a control signal. Switching between positive and negative feedback by changing the relative phase of the control signal via the piezo-driven mirror PDM allows us to choose between the two spatially orthogonal hexagon orientations. By turning the slit filter, a continuous rotation of the hexagonal pattern was obtained. When the control arm was closed, the hexagon remained in its last chosen position, which is a clear indication that the orientation of a spontaneously formed pattern depends only on small variations in the starting conditions. This leads to a breaking of the rotational symmetry of the system in the early phase of pattern development. The control signal used for the manipulation of the

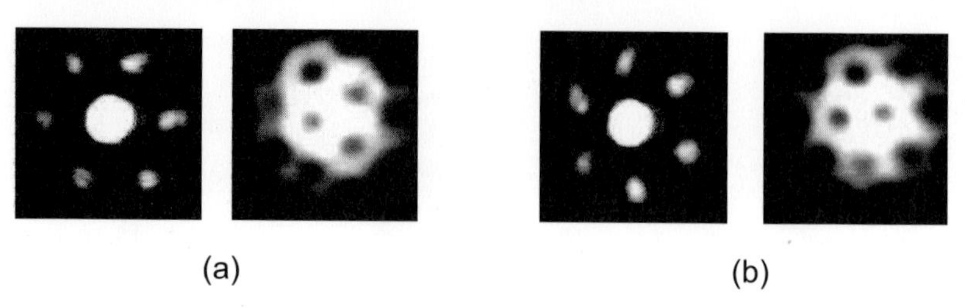

Fig. 10.11. Positive and negative control of the orientation of the hexagonal pattern. (**a**) Positive control, tracking the hexagonal pattern to a desired orientation; (**b**) negative control, suppressing the horizontal direction by destructive interference, thus favoring the spatially orthogonal direction. The Fourier filter was, in both cases, a horizontally oriented slit filter

hexagon orientation was of the order of 1% and declined to an even smaller level ($< 0.1\%$) when the desired orientation was tracked by negative control.

10.5.2 Linear Control Scheme

Another possibility for achieving pattern control with minimal invasion is illustrated in Fig. 10.10b. Here, the small transmission signal of mirror M1 was fed back by an identical $2f$–$2f$ system consisting of lens L3, with a focal length of $f_3 = f_2 = 120$ mm, and a piezo-driven mirror for accurate phase adjustment. The plane of Fourier filtering was again near lens L3, where the far field of the patterns was clearly observable. In order to avoid Fabry–Perot resonances, the central spot was again blocked throughout all measurements. As in the case of the ring control unit described above, the intensity fed back into the system was always in the range of 1–2% of the original intensity in the feedback path. Mirror M1 was a mirror with a variable reflectivity, achieved by wedge-shaped dielectric layers, and provided a means to adjust the feedback reflectivity of the original pattern arm and, therfore, the control signal to the lowest value necessary in the sense of minimally invasive control. The results obtained with the linear control scheme were identical to those obtained with the ring control unit [48]. The absence of bidirectional feedback as in the ring resonator case simplifies the adjustment of the system and improves the quality of the images obtained considerably. However, the results presented apply qualitatively to the ring control scheme also.

We used the Fourier-filtering technique in various parameter regions of the mirror reflectivity r, which represents an important control parameter of the system [37]. Three different reflectivity regions have to be distinguished: a region of pure, stable hexagonal patterns, an intermediate region where a hexagonal and a roll pattern coexist and a reflectivity region where a pure roll pattern is the stable output of the uncontrolled system (see [37]). Figure 10.12a shows the stationary pattern that is obtained in the intermediate

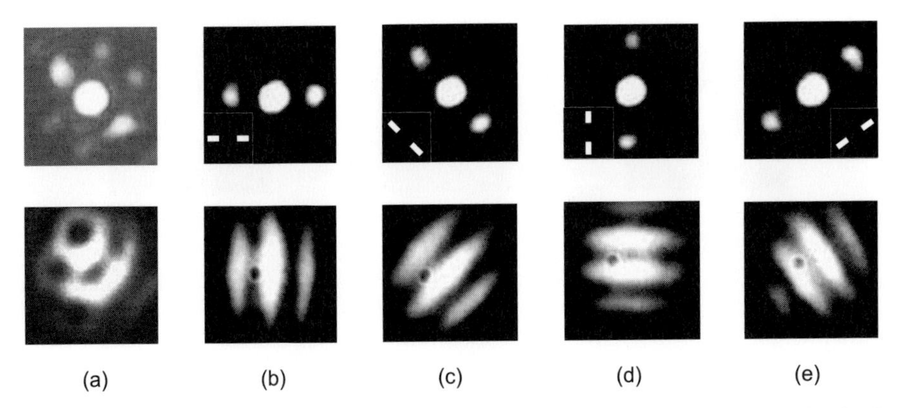

Fig. 10.12. Examples of minimally invasive pattern manipulation. *Top*, far-field pattern; *bottom*: near-field pattern. (**a**) Stationary pattern without control, showing a coexistence of rolls and hexagons. (**b**)–(**e**) Roll pattern in different orientations, obtained with minimally invasive control using the filters shown in the *insets*

reflectivity region (around $r = 25\%$). Using a slit filter in the Fourier plane, it was possible to stabilize the nearest solution available to the system, which is a roll pattern in the filter direction as shown in Figs. 10.12b–e. By changing the orientation of the filter, rolls can be excited in any desired orientation by rotating the slit filter in the Fourier arm. The original feedback system was not perturbed in any of these cases; control was provided only by the additional control arm. If only a single spot was allowed to pass the Fourier filter and was then reinjected into the system, the corresponding roll solution in the appropriate direction was again excited. This proves, as in the case of the invasive control scheme, that even parts of a control signal are sufficient to steer the output of the pattern-forming system, pointing to an associative character of minimally invasive control.

All the results were obtained using positive (in-phase) control. This method suppresses undesired orientations and pattern symmetries and offers the system a preferred direction by feeding back only the correct Fourier components. The control signal provided by the additional arm was not sufficient to excite square patterns with a cross-shaped filter of the type used for invasive control. However, in the intermediate reflectivity region, where rolls and hexagons coexist, this cross filter offers an opportunity to switch between two orthogonal roll directions (Fig. 10.13). The system chooses one of the two directions offered by the Fourier filter, which is sketched in the inset of Fig. 10.13a. A subsequent phase shift of π introduces a switch to the orthogonal roll direction (Fig. 10.13b). For an intermediate phase, a square pattern appeared as a transitional pattern, but was not stabilized (Fig. 10.13c). Feedback reflectivities $r > 40\%$ with a dominant hexagonal solution were also examined regarding the possibility of performing positive and negative control. Though the invasive method suggested the existence of underlying

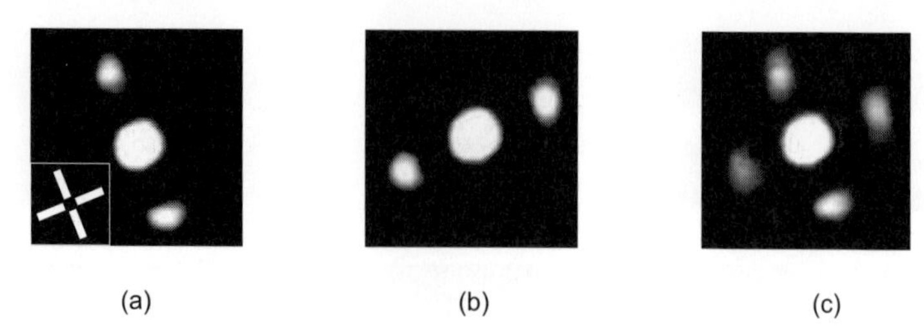

Fig. 10.13. Switching between two orthogonal roll orientations. (a) Roll pattern in the direction of one slit (Fourier filter depicted in the *inset*), (b) roll pattern excited by a phase shift of π with orthogonal orientation, (c) snapshot of an unstable square pattern for an intermediate phase

stable or unstable solutions, no patterns other than the hexagonal one could be excited in this parameter region. Obviously, the control signal was too weak ($\leq 0.5\%$) to induce a pattern transition at the expense of the dominant hexagonal pattern.

10.5.3 Michelson-Like Control Scheme

The realizations of minimally invasive control schemes in ring and linear geometries provide the possibility to switch the control signal on and off in order to provide a desired manipulation of the original pattern-forming system. There is no intervention in the original feedback system, which remains untouched. Nevertheless, the pattern-forming system and the control arm are not completely separated from each other, which causes experimental problems and restrictions. First of all, a change in the reflectivity of the feedback system produced by moving the dielectric mirror M1 affects the relative strength of the control signal at the same time. Another experimental limitation is the problem of diffraction length differences between the pattern-forming and the control arm. In order to have maximum efficiency of the control process, both propagation lengths have to be equal. These problems are circumvented by a Michelson-like control scheme, as shown in Fig. 10.14. Using this setup, the propagation lengths and reflectivities of both arms can be adjusted independently, and, additionally, the relative phase of the two arms can be monitored easily at the appropriate exit of the beam splitter in the feedback path. Crystal sample 2 was used for this experiment with two-arm interferometric feedback. The pattern arm of the interferometer provides the majority of the energy in the feedback loop and serves as the unperturbed pattern-forming system. The other arm, which we call the control arm, filters a small portion of energy for the control and selection of patterns in the pattern arm. The two feedback arms are basically identical $4f$–$4f$ imaging systems containing lenses of focal lengths $f = 100$ mm, and mirrors M1 and M2. A beam splitter right after the first lens in the feedback arm provides

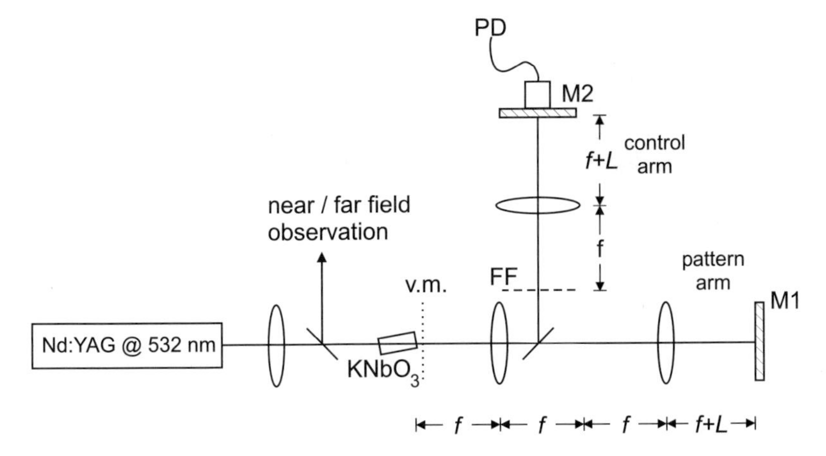

Fig. 10.14. Experimental setup for minimally invasive control with a Michelson-like feedback scheme. M: mirror; v.m.: virtual mirror; PD: piezoelectric driver; FF: Fourier filter; L: diffraction length

access to the control arm, where manipulation in the Fourier space is performed. The lenses were adjusted to give a mirror position of $L = 1$ mm in order to create a hexagonal pattern with a large aspect ratio. Mirror M1 is connected to a piezoelectric translation stage for controlled adjustment of the relative phase of the arms, i.e. for achieving negative (relative phase $\phi = \pi$) or positive (relative phase $\phi = 0$) control. This scheme offers the opportunity (by means of an attenuator in the control arm) to continuously adjust and minimize the strength of the control in the sense of minimally invasive control. The reflectivity ratio of the two feedback arms was $R_1/R_2 = 1.5$, where $R_1 = 15.8\%$ and $R_2 = 10.5\%$; where R_1 (reflectivity of the pattern arm) and R_2 (reflectivity of the control arm) incorporate all losses due to both passes at the beam splitter, lens system and mirrors. The propagation lengths of the arms were adjusted to provide identical transverse scales of the hexagonal patterns created by each arm individually. Control was performed in the Fourier plane of the control arm, where Fourier filters of appropriate shapes were inserted. Binary amplitude masks with different shapes were used: a slit filter for positive control of rolls, a cross filter for positive control of squares and a three-fold slit filter for negative control of hexagons. The central spot of the control signal was always blocked to avoid undesired effects of two-beam interference, which leads to complicated pattern formation effects [49, 50, 51].

Examples of experimental positive control of a roll and a square pattern are illustrated in Fig. 10.15. The system adjusts to the symmetry presented by the Fourier filter (as shown in the insets) and selects the corresponding solution. Clear, rapid switching behavior on timescales comparable to the time constant of the crystal of about 100 ms was observed. When the control path was closed, the predominant hexagonal structure always reappeared,

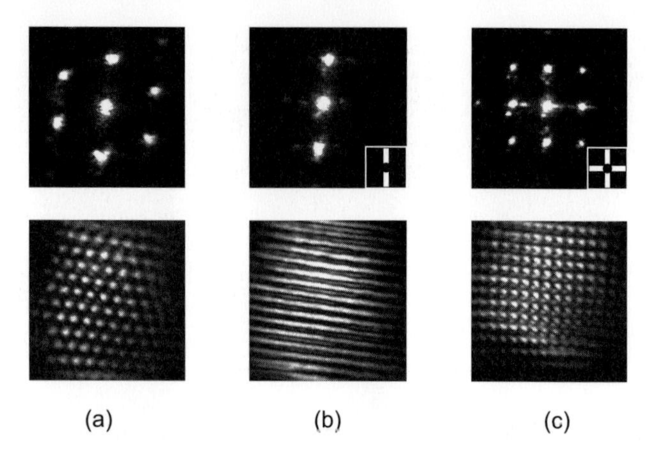

Fig. 10.15. Experimental results for in-phase control. (**a**) Hexagonal pattern appearing without control in the far and near fields, (**b**), (**c**): Results of preparation of a roll pattern and a square pattern with filters indicated in the *insets*

indicating the ability of the Fourier control method to track the system to a certain solution with an appropriate control signal. In the case of positive control, the control signal does not vanish. Therefore, it is not a stabilization of an unstable state, which can only be achieved by negative control and a vanishing control signal. However, the control signal (defined as the control intensity relative to the intensity in the original pattern-forming arm without control) was only 4.7%. This small percentage of energy was able to control the behavior of the whole system, which is exactly the motivation of minimally invasive control. The value of 4.7% was the best value obtainable in an experiment that would still guarantee switching from a hexagonal to a roll or a square pattern. The control strength s as defined in (10.1) is 0.3 for control of roll and square patterns. In both cases, the power absorbed by the filter (apart from the blocked central spot) declines to a small level in the desired state.

As an example of negative control, the orientation of the hexagonal pattern was manipulated by using a three-fold slit mask, where the slits were tilted by 60° relative to each other (see Fig. 10.16). The control signal (the power reinjected into the system) tends to a very small level, as suggested and calculated in [27], with the difference that manipulation of a stable pattern with regard its orientation is performed here, rather than the stabilization of an otherwise unstable pattern. The system remained in the chosen position when the control arm was closed. By rotating the filter, an arbitrary orientation of the hexagon could be obtained.

To compare the experimental results with numerical simulations, an equivalent two-dimensional feedback model, as depicted in Fig. 10.17, was used. A numerical analysis of the photorefractive two-beam coupling equations in the presence of reflection gratings and diffraction is computationally

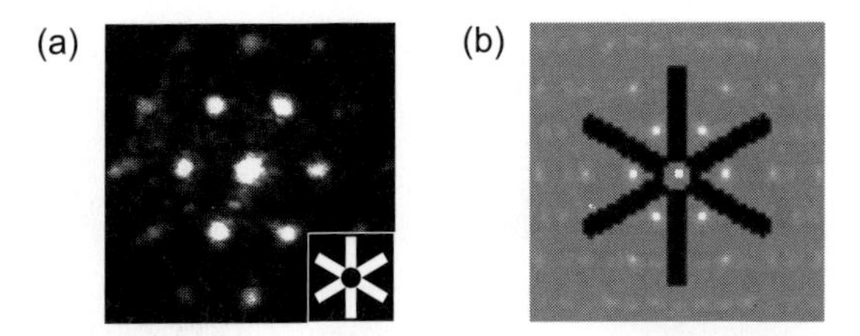

Fig. 10.16. Rotation of the orientation of a haxagonal pattern by out-of-phase control. Experimentally obtained pattern (**a**), with filter in the *inset* and result from a numerical simulation, filter shaded *gray* (**b**)

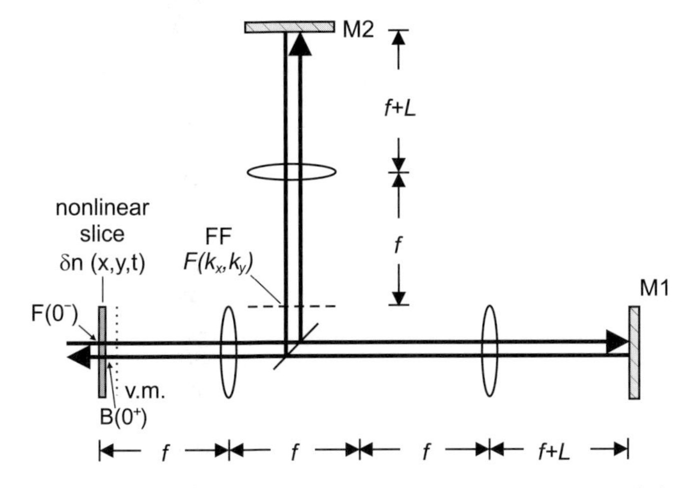

Fig. 10.17. Model system with a nonlinear Kerr slice and forward (**F**) and backward (**B**) beams used for the numerical studies. M = mirror, v.m. = virtual mirror, FF = Fourier filter

expensive. Therefore, a simplified generic model based on a transmission-grating-mediated interaction in a thin slice of a sluggish medium with a cubic nonlinearity [52, 43] was used. With this model, the basic features of pattern formation can be studied without loss of generality. Some experimental results for Fourier filtering in the original feedback arm in the photorefractive system have already been shown to correspond to numerical simulations using the model described above [43]. An extension of an existing code porduced by Saffman was used for the numerical solution of the thin-slice model. This code served as the basis for the investigations in [43]. In the simplified model, the nonlinear refractive index Δn is described by

$$\left(\tau\frac{\partial}{\partial t}+1\right)\Delta n=\mathcal{D}\,\nabla_\perp^2\Delta n+\gamma\frac{|F|^2+|B|^2}{1+|F|^2+|B|^2}\,, \tag{10.7}$$

where τ is the relaxation time of the medium, $\mathcal{D}$ is the diffusion coefficient, and F and B are the amplitudes of the forward- and backward-propagating

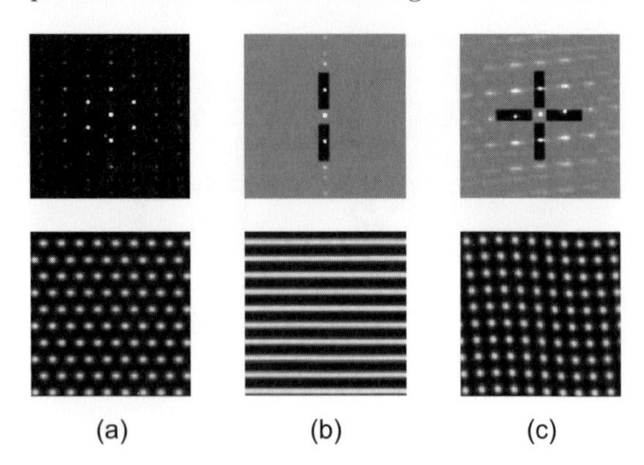

Fig. 10.18. Numerical results from a simulation using a Kerr slice nonlinearity and positive control. (**a**) Generic hexagonal pattern that appears with the control arm closed. (**b**), (**c**) Preparation of a roll and a square pattern by using the amplitude masks shown by *gray shading*. Control strength $s = 0.3$

electric fields (see Fig. 10.17 for notation). The influence of the nonlinearity on the field amplitude is given by $F(x, y, 0^+) = F(x, y, 0^-)\mathrm{e}^{\mathrm{i}\Delta n}$ and $B(x, y, 0^+) = B(x, y, 0^-)\mathrm{e}^{\mathrm{i}\Delta n}$. Free-space diffraction and Fourier filtering are described by the terms

$$\frac{\partial F(x, y, z)}{\partial z} = \frac{\mathrm{i}}{2k} \nabla_\perp^2 F(x, y, z) , \tag{10.8}$$

$$\tilde{B}(k_x, k_y) = \tilde{F}(k_x, k_y) \mathcal{F}(k_x, k_y) , \tag{10.9}$$

$$-\frac{\partial B(x, y, z)}{\partial z} = \frac{\mathrm{i}}{2k} \nabla_\perp^2 B(x, y, z) , \tag{10.10}$$

where the tildes denote Fourier-transformed functions. $\mathcal{F}(k_x, k_y)$ represents the transmission function of the Fourier filter, where blocking is represented by a zero value and complete transmission by a value of unity. The numerical simulation was performed using the same values as used in the experiment ($L = 1$ mm, $\lambda = 532$ nm, and $R_1/R_2 = 1.5$).

The simulation started from an initial hexagonal pattern (as depicted on the left in Fig. 10.18), which is the solution for this system when a Fourier mask is not used. When a spatial filter (gray-shaded region) with a particular geometry was inserted, the system adjusted to the given situation and a roll or square pattern was obtained. Comparison of the experimental results in Fig. 10.15 and the numerical simulations in Fig. 10.18 yields a complete qualitative agreement, despite the difference in the nonlinearities used in the experiment and the numerical simulation. All numerical simulations were performed using a 64×64 grid and checked with a 128×128 grid. In addition, simulations were run for sufficiently long to ensure a steady-state solution.

For negative control of a hexagonal pattern, the same three-fold slit mask as in the experiment (Fig. 10.16) was used. Starting with a hexagonal pattern

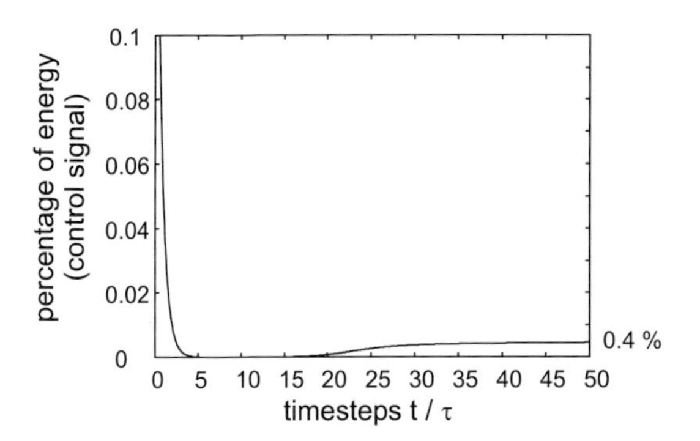

Fig. 10.19. Control signal fed back into the system as a percentage of the total energy in the pattern arm. Results of the numerical simulation leading to Fig. 10.16b

with two spots on the vertical axis, the position of the hexagon was rotated by destructive feedback of the unwanted components. As a result, a rotated hexagonal pattern with two spots on the horizontal axis was obtained. An important property of this control method is that the control signal nearly vanishes when the desired state is selected. In Fig. 10.19, the control signal, i.e. the fraction of energy reinjected into the system relative to the total power in the feedback arm, is shown. The signal declines from the initial value to a very small value of 1.35×10^{-5} at $t/\tau \approx 9$ and then approaches an asymptotic value of about 0.4%, confirming the property of minimal invasion.

Summary

A frequency detuning applied to one of the beams in counterpropagating two-wave mixing enables us to manipulate the output of the pattern-forming system. We have given experimental evidence of switching processes between different pattern states in a bistable parameter region. By explicitly including a further system parameter, we were able to choose from a wealth of different spatial patterns by choosing the set of system parameters appropriately. We have successfully shown that a pattern-forming system can be controlled effectively by manipulating the wave numbers in the Fourier domain. The principle of Fourier filtering was demonstrated by performing invasive Fourier control, where amplitude masks were inserted directly into the feedback path. In order to make the intervention in the system as small as possible, minimally invasive Fourier filtering was performed experimentally in three different configurations. This minimally invasive control scheme enables one to push the system to adopt a square or a roll pattern state in a parameter region where a hexagonal pattern is the only stable output of the pattern-forming system. Only a small control signal was applied to the system, demonstrating the effectiveness and minimally invasive character of the method.

References

1. E. Ott, C. Grebogi and J.A. Yorke, *Controlling chaos*, Phys. Rev. Lett. **64**, 1196–1199 (1990).
2. E. Ott, C. Grebogi and J.A. Yorke, *Using small perturbations to control chaos*, Nature **363**, 411–417 (1993).
3. E.R. Hunt, *Stabilizing high-period orbits in a chaotic system: the diode resonator*, Phys. Rev. Lett. **67**, 1953–1955 (1991).
4. R. Roy, T.W. Murphy, T.D. Maier, Z. Gills and E.R. Hunt, *Dynamical control of a chaotic laser: experimental stabilization of a globally coupled system*, Phys. Rev. Lett. **68**, 1259–1262 (1992).
5. K. Pyragas and A. Tamasevicius, *Experimental control of chaos by delayed self-controlling feedback*, Phys. Lett. A **180**, 99 (1993).
6. K. Pyragas, *Continuous control of chaos by self-controlling feedback*, Phys. Lett. A **170**, 421 (1992).
7. W. Just, T. Bernard, M. Ostheimer, E. Reibold and H. Benner, *Mechanism of time-delayed feedback control*, Phys. Rev. Lett. **78**, 203 (1997).
8. C. Simmendinger and O. Hess, *Controlling delay-induced chaotic behavior of a semiconductor laser with optical feedback*, Physics Lett. A **216**, 97–105 (1996).
9. M. Münkel, F. Kaiser and O. Hess, *Spatio-temporal dynamics of multi-stripe semiconductor lasers with delayed optical feedback*, Physics Lett. A **222**, 67–75 (1996).
10. K. Ikeda, H. Daido and O. Akimoto, *Optical turbulence: chaotic behavior of transmitted light from a ring cavity*, Phys. Rev. Lett. **45**, 709–712 (1980).
11. T. Heil, I. Fischer and W. Elsäßer, *Influence of amplitude–phase coupling on the dynamics of semiconductor lasers subject to optical feedback*, Phys. Rev. A **60**, 634–641 (1999).
12. T. Heil, I. Fischer and W. Elsäßer, *Stabilization of feedback-induced instabilities in semiconductor lasers*, J. Opt. B.: Quantum Semiclass. Opt. **2**, 413–420 (2000).
13. C. Simmendinger, M. Münkel and O. Hess, *Controlling complex temporal and spatio-temporal dynamics in semiconductor lasers*, Chaos Solitons Fractals **10**, 851–864 (1999).
14. G.L. Oppo, R. Martin, A.J. Scroggie, G.K. Harkness, A. Lord and W.J. Firth, *Control of spatio-temporal complexity in nonlinear optics*, Chaos, Solitons and Fractals **10**, 865–874 (1999).
15. R. Lima and M. Pettini, *Suppression of chaos by resonant parametric perturbations*, Phys. Rev. A **41**, 726 (1990).
16. R. Meucci, W. Gadomski, M. Ciofini and F.T. Arrechi, *Experimental control of chaos by means of weak parametric perturbations*, Phys. Rev. E **49**, R2528 (1994).
17. P. Colet and Y. Braiman, *Control of chaos in multimode solid state lasers by the use of small periodic perturbations*, Phys. Rev. E **53**, 200 (1996).
18. H. Gang and H. Kaifen, *Controlling chaos in systems described by partial differential equations*, Phys. Rev. Lett. **71**, 3794 (1993).
19. D. Auerbach, *Controlling extended systems of chaotic elements*, Phys. Rev. Lett. **72**, 1184 (1994).
20. F. Quin, E.E. Wolf and H.-C. Chang, *Controlling spatiotemporal patterns on a catalytic wafer*, Phys. Rev. Lett. **72**, 1459 (1994).

21. I. Aranson, H. Levine and L. Tsimring, *Controlling spatio-temporal chaos*, Phys. Rev. Lett. **72**, 2541 (1994).
22. C. Lourenco, M. Hougardy and A. Babloyantz, *Control of low-dimensional spatiotemporal chaos in Fourier space*, Phys. Rev. E **52**, 1528 (1995).
23. V. Petrov, S. Metens, P. Brickmans, G. Dewel and K. Showalter, *Tracking unstable Turing patterns through mixed-mode spatiotemporal chaos*, Phys. Rev. Lett. **75**, 2895 (1995).
24. A. Hagberg, E. Meron, I. Rubinstein and B. Zalthman, *Controlling domain patterns far from equilibrium*, Phys. Rev. Lett. **76**, 427 (1996).
25. J.A. Sepulchre and A. Babloyantz, *Controlling chaos in a network of oscillators*, Phys. Rev. E **48**, 945 (1993).
26. W. Lu, D. Yu and R.G. Harrison, *Control of patterns in spatiotemporal chaos in optics*, Phys. Rev. Lett. **76**, 3316 (1996).
27. R. Martin, A.J. Scroggie, G.L. Oppo and W.J. Firth, *Stabilization, selection, and tracking of unstable patterns by Fourier space techniques*, Phys. Rev. Lett. **77**, 4007-4010 (1996).
28. G.K. Harkness, G.L. Oppo, R. Martin, A.J. Scroggie and W.J. Firth, *Elimination of spatiotemporal disorder by Fourier space techniques*, Phys. Rev. A **58**, 2577–2586 (1998).
29. T. Honda and P. Banerjee, *Threshold for spontaneous pattern formation in reflection-grating-dominated photorefractive media with mirror feedback*, Opt. Lett. **21**, 779–781 (1996).
30. P.M. Lushnikov, *Hexagonal optical structures in photorefractive crystals with a feedback mirror*, JETP **86**, 614–627 (1998).
31. M. Schwab, C. Denz and M. Saffman, *Transverse modulational instability in counterpropagating two-wave mixing with frequency detuned pump beams*, J. Opt. Soc. Am. B **18**, 628–638 (2001).
32. M. Schwab. *Manipulation and Control of Self-Organized Transverse Optical Patterns*. Shaker, Aachen, 2001.
33. M. Schwab, C. Denz, A.V. Mamaev and M. Saffman, *Manipulation of optical patterns by a frequency detuning of the pump beams*, J. Opt. B: Quantum Semiclass. Opt. **3**, 318–327 (2001).
34. M. Saffman, A.A. Zozulya and D.Z. Anderson, *Transverse instability of energy-exchanging counterpropagating waves in photorefractive media*, J. Opt. Soc. Am. B **11**, 1409–1417 (1994).
35. A.I. Chernykh, B.I. Sturman, M. Aguilar and F. Agulló-López, *Threshold for pattern formation in a medium with local photorefractive response*, J. Opt. Soc. Am. B **14**, 1754–1760 (1997).
36. T. Honda, H. Matsumoto, M. Sedlatschek, C. Denz and T. Tschudi, *Spontaneous formation of hexagons, squares and squeezed hexagons in a photorefractive phase conjugator with virtually internal feedback mirror*, Opt. Commun. **133**, 293–299 (1997).
37. C. Denz, M. Schwab, M. Sedlatschek, T. Tschudi and T. Honda, *Pattern dynamics and competition in a photorefractive feedback system*, J. Opt. Soc. Am. B **15**, 2057–2064 (1998).
38. O. Sandfuchs, F. Kaiser and M.R. Belić, *Dynamics of transverse waves and zig-zag instabilities in photorefractive two-wave mixing with a feedback mirror*, J. Opt. Soc. Am. B **18**, 505–514 (2001).

39. O. Sandfuchs, F. Kaiser and M.R. Belić, *Spatiotemporal pattern formation in counterpropagating two-wave mixing with an externally applied field*, J. Opt. Soc. Am. B **15**, 2070–2078 (1998).
40. E.V. Degtiarev and M.A. Vorontsov, *Spatial filtering in nonlinear two-dimensional feedback systems: phase-distortion suppression*, J. Opt. Soc. Am. B **12**, 1238–1247 (1995).
41. M.A. Vorontsov and B.A. Samson, *Nonlinear dynamics in an optical system with controlled two-dimensional feedback: black-eye patterns and related phenomena*, Phys. Rev. A **57**, 3040–3049 (1998).
42. T. Ackemann, B. Schäpers, J.P. Seipenbusch, Yu.A. Logvin and W. Lange, *Drifting and locking behavior of optical patterns – an investigation using Fourier filtering*, Chaos Solitons Fractals **10**, 665–673 (1999).
43. A.V. Mamaev and M. Saffman, *Selection of unstable patterns and control of optical turbulence by Fourier space filtering*, Phys. Rev. Lett. **80**, 3499–3502 (1998).
44. M. Schwab, C. Denz and S. Juul Jensen, *Stabilization and manipulation of self-organized patterns in a photorefractive feedback system*, Asian J. Phys. **7**, 470–482 (1998).
45. T. Ackemann, B. Giese, B. Schäpers and W. Lange, *Investigations of pattern forming mechanisms by Fourier filtering: properties of hexagons and the transition to stripes in an anisotropic system*, J. Opt. B: Quantum Semiclass. Opt. **1**, 70–76 (1999).
46. G.K. Harkness, G.L. Oppo, E. Benkler, M. Kreuzer, R. Neubecker and T. Tschudi, *Fourier space control in an LCLV feedback system*, J. Opt. B: Quantum Semiclass. Opt. **1**, 177–182 (1999).
47. E. Benkler, M. Kreuzer, R. Neubecker and T. Tschudi, *Experimental control of unstable patterns and elimination of spatiotemporal disorder in nonlinear optics*, Phys. Rev. Lett. **84**, 879–882 (2000).
48. S. Juul Jensen, M. Schwab and C. Denz, *Manipulation, stabilization and control of pattern formation using Fourier space filtering*, Phys. Rev. Lett. **81**, 1614–1617 (1998).
49. B. Thüring, A. Schreiber, M. Kreuzer and T. Tschudi, *Spatio-temporal dynamics due to competing spatial instabilities in a coupled LCLV feedback system*, Physica D **96**, 282–290 (1996).
50. M.A. Vorontsov and A.Yu. Karpov, *Kerr-slice based nonlinear interferometer with two-dimensional feedback: control of roll and hexagon formation*, Opt. Lett. **20**, 2466–2468 (1995).
51. M.A. Vorontsov and A.Yu. Karpov, *Pattern formation due to interballoon spatial mode coupling*, J. Opt. Soc. Am. B **14**, 34–50 (1997).
52. G. D'Allesandro and W.J. Firth, *Spontaneous hexagon formation in a nonlinear optical medium with feedback mirror*, Phys. Rev. Lett. **66**, 2597–2600 (1991).

11 Transverse Patterns in Active Photorefractive Oscillators

We have seen in the preceding chapters that a transverse modulation instability of a single beam or counterpropagating beams is a general mechanism that leads to pattern formation in nonlinear optics. This instability occurs in media where the combination of nonlinear phase modulation and linear, i.e. diffractive, phase retardation leads to the creation and oscillation of spatial sidebands. The nonlinear stage of this instability leads to regular transverse structures, which can be solitary waves or symmetric patterns with a predominantly hexagonal symmetry.

An alternative route to optical pattern formation is to place a nonlinear medium with gain in an optical cavity that supports a spectrum of linear transverse-mode patterns. We then obtain a system that represents an oscillator or an amplifier, depending on the gain of the system. Above a certain threshold, the system operates as an oscillator, while for values below the threshold, we obtain an amplifier. In the oscillator mode, a superposition of the cavity modes is excited. The resulting pattern may be static or dynamic, depending on whether or not the modes are frequency degenerate. Such a device, where a nonlinear gain medium is placed in a resonant cavity, is nothing other than a laser. It is therefore no surprise that the resulting patterns are connected closely to the well-known Hermite–Gaussian and Laguerre–Gaussian transverse-mode patterns of any laser oscillator.

In principle, patterns such as those described in this chapter could have been observed in the early days of the laser. However, at that time optical cavities were mostly limited to the lowest-mode pattern or a few of the lowest-mode patterns because strongly limiting apertures, were used, which did not allow one to observe these transverse phenomena. Therefore, by allowing the transverse degrees of freedom to participate in the interaction, one can obtain striking, new phenomena.

There are some basic differences between pattern formation due to a transverse modulation instability and pattern formation in resonant cavities. In the in the case of modulation instability, a nonlinear phase shift is necessary for oscillation, whereas in the cavity case it is sufficient to have a medium with nonlinear gain. A nonlinear phase shift may, nevertheless, occur in the cavity case due to a frequency shift between the pump and the oscillating field. These two systems are also distinct with regard to the types of patterns they produce. Patterns due to modulation instabilities may be classified as "mixed" modes, whereas the patterns observed in resonant cavities are close to "pure" modes. This terminology may be understood in the following way. The paraxial wave equation admits Hermite–Gaussian and Laguerre–Gaussian functions as solutions. It is well known that any Hermite–Gaussian or Laguerre–Gaussian function of a given order is self-similar upon Fourier transformation. This implies that the intensity pattern due to a single Hermite–Gaussian or Laguerre–Gaussian mode is self-similar in the near and far fields, and at any plane in between. This also applies to superpositions of mode patterns. Likewise, patterns due to mode superpositions with widely spaced values of the mode index will be approximately self-similar in intensity. Therefore, we refer to these patterns as *pure modes*, since they are form-invariant under linear propagation. In contrast, the transverse modulation instabilities we investigated in the preceding chapters result in *mixed modes*, in the sense that they are not invariant under propagation.

In the case of optical cavities, the Fresnel number F characterizes the system size, just as the aspect ratio does in systems that show spontaneous pattern formation. The Fresnel number is defined as the radius of the smallest aperture in the cavity divided by the resonator length and by the wavelength of the pump beam. When F is small, a description on the basis of the empty-cavity modes is relevant, because the number of modes remains small, while in systems with large F, it is more suitable to adopt a global viewpoint in which optical turbulence or defects can be described.

For lasers, which can be described near threshold in a way that is equivalent to the complex Ginzburg–Landau equation [1], it is well known that in the low-F limit, even a few mode patterns can lead to rich spatio-temporal dynamics [2, 3, 4]. However, the dynamics of lasers are rather fast, making observations of the spatio-temporal dynamics of these patterns quite involved. In contrast, the dynamics of photorefractive oscillators are quite slow, allowing easy observation and analysis of pattern dynamics by conventional recording methods. In that regard it is interesting to consider the basic phenomena of the transverse dynamics of a photorefractive oscillator in the weakly multimode case, i.e. where F is of the order of several units to gain a more detailed insight into the dynamics of laser systems. We shall show in the following sections that the patterns arising in these oscillators can be well resolved, and show a rich phenomenology from simple mode patterns to complex spatio-temporal behavior.

Beyond these similarities to lasers, photorefractive oscillators represent attractive systems in their own right for studying pattern formation in oscillators in general, because they have several special features. Although the patterns that can be observed in such an oscillator depend in principle on the geometry of the optical oscillator, the oscillating modes are influenced by the properties of the amplifying medium. In comparison with passive resonators, the presence of the active medium imposes constraints that limit

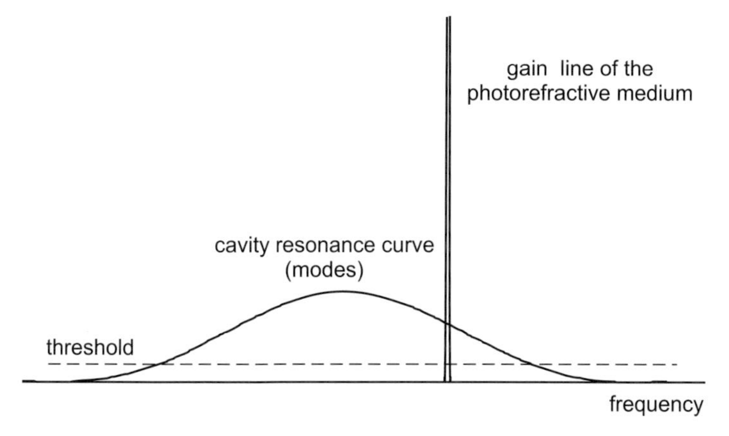

Fig. 11.1. The resonance line of the modes of the oscillator is much larger than the narrow line of photorefractive amplification, leading to strong frequency pulling

the number of oscillating modes in practice. This is true for both lasers and photorefractive oscillators.

In addition, photorefractive oscillators are expected to show distinct resonance properties in comparison with lasers. On the one hand, because the principles of light amplification by the photorefractive effect are different from those of the laser, the influence of losses and the behavior of the system for multimode operation should be different. On the other hand, owing to the large nonlinearity of photorefractive materials and the narrow gain line of photorefractive two-beam coupling, which is governed by the line width of the pump wave, and owing to diffusion and recombination process, which provide dissipative dynamics with a damping rate of the same order as the frequency detuning between the pump and oscillator fields, a strong frequency pulling is present in photorefractive oscillators. Therefore, transverse modes with relative frequency shifts of tens of Hz are observed, even though the empty-cavity modes have natural-frequency differences in the MHz range. This behavior is illustrated in Fig. 11.1. We shall see later on that frequency pulling also leads to frequency degeneracy among the modes and therefore to simultaneous excitation of modes. The resulting slowly moving patterns can be recorded directly and easily with video cameras, allowing easy access to the dynamical effects of transverse patterns.

When a unidirectional ring oscillator operating in the self-oscillation mode is considered, many interesting aspects of multiple-transverse-mode oscillations can be observed. In general, both frequency-degenerate families of modes and nondegenerated ones can be observed, depending on the geometry of the oscillator. The dynamics of the resulting transverse intensity distributions of photorefractive oscillators often have an odd, irregular structure, which depends strongly on the Fresnel number F of the system.

One of the most spectacular examples of these spatio-temporal effects was found in a unidirectional oscillator containing a BSO nonlinear crystal at the beginning of the 1990s by Arrechi and coworkers [5]. These researchers found three different characteristic effects of transverse coupling. For small F, when the signal field is coherent in space even though it is fluctuating in time, they found a single-mode regime where different transverse modes are excited one at a time in a regular sequence (*periodic alternation*). For intermediate Fresnel numbers, this scenario changes into a chaotic sequence (*chaotic alternation*). For larger F, the field has a multimode behavior, and *spatio-temporal chaos* can be found. In this regime, the local fluctuations display a non-Gaussian character, as expected from turbulence theories [6].

These effects can in fact be found in various different photorefractive oscillators with any photorefractive gain medium, and describe the behavior when the oscillator is running freely in a self-oscillation mode above threshold. However, the origin of the spatio-temporal dynamics in the various regimes of the Fresnel number is strongly influenced by the superposition of internal, nonlinearity-driven dynamics and external sources of dynamics such as temperature changes.

As a first step towards understanding these spatio-temporal phenomena in photorefractive ring oscillators, we shall give in the following sections a summary of the theoretical descriptions that have been developed to describe photorefractive oscillators. The aim of our experimental investigations described in the sections following those theoretical considerations was to quantify these complex pattern dynamics. By carefully stabilizing the experimental configuration and using specially adapted analysis tools, we were able to obtain deeper insight into the origins of pattern dynamics and competition. We shall show that by using actively stabilized systems, the effects of cavity length changes, of mode competition due to oscillator dynamics and of transverse dynamics due to nonlinear coupling can be distinguished. This is an essential step towards understanding the nonlinear effects of pattern formation that arise in these systems and avoiding the misunderstandings that may easily occur when one observes pattern dynamics in these systems without knowledge of the underlying cavity effects.

11.1 The Unidirectional Photorefractive Oscillator

Most of the theoretical models developed to describe spatio-temporal phenomena in unidirectional photorefractive oscillators are based on a multiple-transverse-mode analysis using mode decompositions.

A first, simplified description was formulated in 1987 by Anderson and Saxena [7]. Subsequently, this describtion was extended to the more realistic case of Laguerre–Gaussian transverse modes in 1992 by D'Allesandro [8] and later formulated in a more general way by Jost and Saleh [9, 10]. These general considerations showed that investigating unidirectional photorefractive

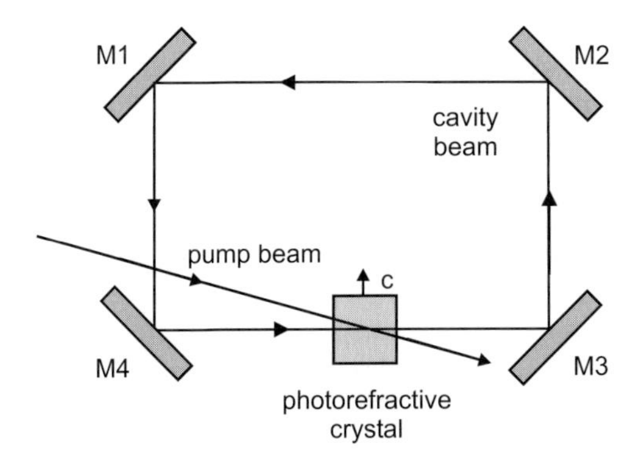

Fig. 11.2. Schematic of a unidirectional ring oscillator with photorefractive gain via two-beam coupling

systems might provide considerable insight into pattern formation dynamics in general, and not just the dynamics of this specific type of system.

In the following, we shall consider a simple unidirectional ring oscillator as depicted in Fig. 11.2. Let us assume that the photorefractive gain medium is pumped by a plane reference wave

$$\boldsymbol{A}_{\mathrm{p}}(\boldsymbol{r},t) = A_{\mathrm{p}}(\boldsymbol{r}) \exp\left[\mathrm{i}(\boldsymbol{k}_{\mathrm{p}} \cdot \boldsymbol{r} - \omega_{\mathrm{p}} t)\right] + \mathrm{c.c.}\ . \tag{11.1}$$

The notation is straightforward compared with the systems described up to now; $A_{\mathrm{p}}(\boldsymbol{r})$ is the complex pump amplitude, $\boldsymbol{k}_{\mathrm{p}}$ being the pump beam wave vector and ω_{p} is its frequency. The pump is assumed to propagate at a small angle relative to the z axis. In order to satisfy the condition of an undepleted pump, the pump is restricted to variations that satisfy the inequality $|A_{\mathrm{r}}(\boldsymbol{r},t)|^2/I_{\mathrm{p}} = |A_{\mathrm{r}}(\boldsymbol{r},t)|^2/|A_{\mathrm{p}}(\boldsymbol{r},t)|^2 \ll 1$, where A_{r} is the complex amplitude of the slowly varying oscillator beam.

The oscillating beam that establishes itself in the ring oscillator is also assumed to travel at a small angle to the z axis in the photorefractive crystal and to have an electric field

$$\boldsymbol{A}_{\mathrm{r}}(\boldsymbol{r},t) = A_{\mathrm{r}}(\boldsymbol{r},t) \exp\left[\mathrm{i}(\boldsymbol{k}_{\mathrm{r}} \cdot \boldsymbol{r} - \omega_{\mathrm{r}} t)\right] + \mathrm{c.c.}\ , \tag{11.2}$$

where $\boldsymbol{k}_{\mathrm{r}}$ is the oscillator beam wave vector for a reference mode in the cavity. The reference mode is assumed to have the wave number k_{r}. We can assume that the frequency $\omega_{N,0,0} = \omega_{\mathrm{c}}$ of the longitudinal mode N in the empty cavity that lies nearest to the pump frequency ω_{p} is only slightly detuned from the reference mode frequency $\omega_{\mathrm{r}} = k_{\mathrm{r}} c$, resulting in a frequency detuning δ_{cr}. The detuning between the reference wave ω_{r} and the pump frequency ω_{p} is given by $\delta_{\mathrm{pr}} = \omega_{\mathrm{p}} - \omega_{\mathrm{r}}$.

11.1.1 Photorefractive Nonlinearity Equations

Owing to the interference of the oscillator field with the pump field, the total optical field in the photorefractive medium is $\boldsymbol{A}(\boldsymbol{r},t) = \boldsymbol{A}_\mathrm{p}(\boldsymbol{r},t) + \boldsymbol{A}_\mathrm{r}(\boldsymbol{r},t)$ This creates an intensity pattern

$$I(\boldsymbol{r},t) = I_0(\boldsymbol{r},t)\left[1 + \left(\frac{A_\mathrm{p}(\boldsymbol{r})A_\mathrm{r}^*(\boldsymbol{r},t)}{I_0(\boldsymbol{r},t)}\exp\left[\mathrm{i}(\boldsymbol{K}\cdot\boldsymbol{r} - \delta_\mathrm{pr}t)\right] + \mathrm{c.c.}\right)\right] , \quad (11.3)$$

where $I_0(\boldsymbol{r},t) = I_\mathrm{p}(\boldsymbol{r}) + |A_\mathrm{r}(\boldsymbol{r},t)|^2$, and $\boldsymbol{K} = \boldsymbol{k}_\mathrm{p} - \boldsymbol{k}_\mathrm{r}$ is the grating wave vector. In this equation, we have neglected terms with frequencies equal to $2\omega_\mathrm{p}$ because they are not expected to be supported by the oscillator. Therefore, only a dominating transmission grating is active in the photorefractive crystal. The total intensity in the crystal consists of an unmodulated and a modulated part as given by $I(\boldsymbol{r},t) = I_0(\boldsymbol{r},t) + I_1(\boldsymbol{r},t)$, where the modulated part is $I_1(\boldsymbol{r},t) = A_\mathrm{p}(\boldsymbol{r})A_\mathrm{r}^*(\boldsymbol{r},t)\exp(\mathrm{i}\boldsymbol{K}\cdot\boldsymbol{r}) + \mathrm{c.c.}$.

As described in Chap. 3, the spatio-temporally varying intensity produces a spatio-temporally varying electric field within the crystal and leads to refractive index variations described by $n = n_0 + \Delta n_\mathrm{pr}(\boldsymbol{r},t)$, where n_0 is the unperturbed (static) refractive index, and $\Delta n_\mathrm{pr}(\boldsymbol{r},t)$ is the refractive index change induced by the photorefractive effect, which is typically small compared with the first, static term. This process was described in Chap. 3 by a reduced set of Kukhtarev equations and leads, by the normal procedure, to a differential equation relating the index change to the intensity distribution. For $I_1/I_0 \ll 1$ we obtain

$$\left(\frac{\partial}{\partial t} + \frac{1}{\tau(\boldsymbol{r},t)}\right)\Delta n_\mathrm{pr}(\boldsymbol{r},t) = \mathrm{i}\frac{\gamma' n_0^3 r_\mathrm{eff}}{2\tau_\mathrm{c} I_\mathrm{p}} I_1(\boldsymbol{r},t) , \quad (11.4)$$

where $\tau(\boldsymbol{r},t)$ is the intensity-dependent material time constant and γ' the complex coupling constant, respectively. Compared to the precedent chapters, γ' is defined in a slightly different way only including the different contributions of the space charge field, adapted to the usual formulation in oscillator systems.

There are several ways to write the index grating appearing in (11.4), which is in general given by (see also 3.7)

$$\begin{aligned}\Delta n_\mathrm{pr}(\boldsymbol{r},t) &= -\frac{1}{2}n_0^3 r_\mathrm{eff}\tilde{E}_\mathrm{sc}(\boldsymbol{r},t)\\ &= -\frac{1}{2}n_0^3 r_\mathrm{eff}\left(E_\mathrm{sc}(\boldsymbol{r},t)\exp\left(\mathrm{i}(\boldsymbol{K}\cdot\boldsymbol{r} - \delta_{pr}t)\right) + c.c.\right) . \quad (11.5)\end{aligned}$$

(see also 3.7)

If we desire an explicit solution for the refractive index change, and if we allow some conditions to be valid, a relatively simple form is possible. First, for $\partial\tau/\partial t \ll 1$, the solution of (11.4) is space–time separable. Second, when the external field is small or not present, and there is a relatively large probe-to-pump ratio ($0.01 \ll (I_0 - I_p)/I_p \ll 1$), the refractive index distribution within the photorefractive crystal is approximately proportional to the optical intensity distribution, as the simple phenomenological model suggests [11]. Note that when these conditions are violated, the refractive index distribution becomes spatially distorted relative to the intensity distribution and the index may contain longitudinal variations that invalidate the uniform-field limit. In this case, the refractive index can be expanded in the same subset of modes as used for the field distribution. Thus, we may write the refractive index in the same form as the intensity distribution [9]:

$$\begin{aligned}\Delta n_{\mathrm{pr}}(\boldsymbol{r},t) &= \sum_m (G_m(t)W_m(\boldsymbol{r})\exp(\mathrm{i}\boldsymbol{K}_m\cdot\boldsymbol{r}) + \mathrm{c.c.}) \\ &+ \sum_m \sum_{p\neq m} (G_{pm}(t)W_p(\boldsymbol{r})W_m^*(\boldsymbol{r})\exp(\mathrm{i}\boldsymbol{K}_{pm}\cdot\boldsymbol{r}) + \mathrm{c.c.}\ ,\end{aligned} \quad (11.6)$$

where $G_m(t)W_m(\boldsymbol{r})$ is the pump–mode coupling contribution, and $G_{pm}(t)W_p(\boldsymbol{r})W_m^*(\boldsymbol{r})$ is the mode-mode coupling contribution to slowly varying spatio-temporal amplitude of the refractive index change.

11.1.2 Stability Analysis

To analyze the question of whether our resonant system with an active photorefractive nonlinearity has a stable steady-state operation mode, we make use of the ray transfer matrix algorithm [12]. This algorithm is useful if the transformation of a beam through several different optical systems has to be evaluated in the paraxial-beam approximation. In that case, a beam is completely described by its transverse displacement x and its slope or angle α relative to the optical z axis. The output parameters (x_2, α_2) depend on the input values (x_1, α_1) through the transfer matrix $\mathbf{M}$:

$$\begin{pmatrix} x_2 \\ \alpha_2 \end{pmatrix} = \begin{pmatrix} A & B \\ C & D \end{pmatrix} \begin{pmatrix} x_1 \\ \alpha_1 \end{pmatrix} , \quad (11.7)$$

where A, B, C, D are the parameters characterizing the change of the beam. All periodic sequences can be analyzed with this formalism with regard to their stability. A sequence is called stable if the trace $(A+D)$ is able to fulfill $-1 < 0.5(A+D) < 1$ [12].

If we apply such an analysis to an oscillator that uses nonlinear two-beam coupling, one round-trip of the optical beam is described by the product of the propagation matrix for the development of the beam in a free-space translation over the resonator length L with the matrices for reflection by the conventional mirrors with curvature R. Because we may assume to a first approximation that the photorefractive crystal is small compared with the length of the oscillator, the passage through the crystal can be neglected.

Therefore, we obtain the same beam transfer matrix as for the case of a passive resonator:

$$\begin{aligned}\mathbf{M} &= \mathbf{M}_{\text{prop}}\mathbf{M}_{\text{mir}}\mathbf{M}_{\text{prop}}\mathbf{M}'_{\text{mir}} \\ &= \begin{pmatrix} 1 & L \\ 0 & 1 \end{pmatrix}\begin{pmatrix} 1 & 0 \\ 2/R & 1 \end{pmatrix}\begin{pmatrix} 1 & L \\ 0 & 1 \end{pmatrix}\begin{pmatrix} 1 & 0 \\ -2/R & 1 \end{pmatrix} . \end{aligned} \tag{11.8}$$

The stability condition requires, in that case, that $|L/R| < 1$. Thus, not all sorts and shapes of resonators are suitable for feedback with amplification via two-beam coupling. In experimental realizations, unidirectional photorefractive oscillators are therefore constructed using spherical mirrors or a stabilizing lens that has precisely the purpose of fulfilling the stability criterion derived above.

11.1.3 Longitudinal Oscillator Modes

To analyze the longitudinal modes of an oscillator, the spatial extend of the active medium can be considered to be small relative to the length of the cavity. Thus, the conventional analysis of Fresnel and Kirchhoff, as used for passive resonators can be applied. The modes in the resonator are defined by their Fresnel number. If $F \ll 1$ (far-field diffraction), the diffraction effect for modes of higher order is very strong, leading to an attenuation of these modes. Here, only those modes of low order are able to "survive". The behavior changes if we allow higher Fresnel numbers. The coexistence of many modes of higher order with similar frequencies is possible. Therefore, mode competition may take place.

In our initial, simple description of the longitudinal-mode interaction in a ring oscillator with photorefractive gain, we shall follow mainly the derivations of Kogelnik [12] and Anderson and Saxena [7].

In addition to the equations that describe the nonlinearity in the photorefractive crystal, as formulated in Chap. 3, the field in the cavity has to be taken into account. When the effect of the photorefractive medium is assumed to be very slow compared with the cavity round-trip time and when the material is idealized as thin, the field dynamics can be formulated in the framework of the empty-cavity modes described above. We shall retain only the gratings formed by each mode with the pump beam and ignore the intermode gratings that describe the interaction of signals with each other, in the weak-field limit used in this first step of the analysis (see (11.6)). Moreover, we assume negligible pump beam depletion and a small modulation index of the intensity pattern in the photorefractive medium. The nonlinear polarization induced in the photorefractive medium is obtained by considering the Bragg scattering of the pump field by the index gratings in the direction of the resonator field.

The complex interaction of the modes with the external fields and with their feedback into the resonator is dominated by the nonlinear interaction of the pump wave and the resonator field in the photorefractive medium. This leads to an interference field and subsequently a refractive index grating, which reacts on the pump wave again, which in turn will modify the grating. The evolution of the field in the resonator is governed by Maxwell's equations, which define the new resonator field that interacts with the pump wave. Therefore, we may expand the resonator field into the complete set of the normal eigenmodes of the cavity:

$$\boldsymbol{A}_{\mathrm{r}}(\boldsymbol{r},t)=\sum_n S_n(t)\mathrm{e}^{-\mathrm{i}[\omega_n t+\phi_n(t)]}\boldsymbol{U}_n(\boldsymbol{r})+\mathrm{c.c.} \tag{11.9}$$

The spatial eigenmodes are taken to be plane waves in the case of a unidirectional ring resonator.

If we neglect the effect of transverse distributions for the moment, the longitudinal part can be written as

$$\boldsymbol{U}_n(\boldsymbol{r})=\hat{e}_n\mathrm{e}^{\mathrm{i}(\boldsymbol{k}_n\cdot\boldsymbol{r})}=\hat{e}_n\mathrm{e}^{\mathrm{i}k_n z}\,, \tag{11.10}$$

where $\boldsymbol{k}_n$ is the wave vector for the nth mode. Note that the magnitudes of the wave vectors for the pump and resonator fields differ from the free-space propagation values in a photorefractive medium. The orthogonality of the eigenmodes is expressed by

$$\int_0^L \boldsymbol{U}_m^*(z)\cdot\boldsymbol{U}_n(z)\mathrm{d}z=\delta_{mn}L\,. \tag{11.11}$$

Because the electric-field amplitude $S_n(t)$ and the phase $\phi_n(t)$ vary little in an optical-frequency period, they are regarded as real quantities. The optical fields that develop in the resonator and the comparatively static field in the photorefractive medium are connected by an induced optical polarization.

Now, $A_{\mathrm{r}}(\boldsymbol{r},t)$ as given by (11.9) can be substituted into the photorefractive nonlinear equations, leading in the case of negligible intermode interaction, to

$$I(\boldsymbol{r},t)=I_0\left(1+\sum_n M_n\exp\{\mathrm{i}[\boldsymbol{K}_n\cdot\boldsymbol{r}-\delta_{\mathrm{cr}}t-\phi_n(t)]\}+\mathrm{c.c.}\right)\,, \tag{11.12}$$

where the incoming intensity is $I_0(\boldsymbol{r},t)=I_{\mathrm{p}}(\boldsymbol{r})+I_{\mathrm{r}}(\boldsymbol{r},t)=|A_{\mathrm{p}}|^2+\sum_n|S_n(t)|^2$, the modulation terms are $M_n(t)=(\hat{e}_n\cdot\hat{e}_{\mathrm{p}}^*)A_{\mathrm{p}}S_n(t)/I_0(t)$ and the grating vectors for the various index modulations are $\boldsymbol{K}_n=\boldsymbol{k}_{\mathrm{p}}-\boldsymbol{k}_n$.

Using Kukhtarev's equations ((3.14)–(3.22)), the space charge field results in a similar decomposition

$$E_{\mathrm{sc}}(\boldsymbol{r},t)=\sum_n E_{\mathrm{sc,n}}(t)\exp\left(\mathrm{i}\boldsymbol{K}_n\cdot\boldsymbol{r}\right)+\mathrm{c.c.} \tag{11.13}$$

and the refractive index modulation is given by

$$\Delta n_{\mathrm{pr}}(\boldsymbol{r},t) = \sum_n \Delta n_{\mathrm{n,pr}}(t) \exp\{\mathrm{i}[\boldsymbol{K}_n \cdot \boldsymbol{r} + \nu_n t + \Theta_n(t)]\} + \mathrm{c.c.}\,, \tag{11.14}$$

where $\nu_n t + \Theta_n(t)$ is the frequency of the nth Fourier component of the refractive index grating. The equations of motion of both terms are then given by the appropriate extensions of (11.4). The most important term in these new equations is the instantaneous phase difference between the interference and the refractive index grating, which may be formulated in terms of the mode decomposition:

$$\Psi_n(t) = \omega_{\mathrm{p}} t - [\omega_n t + \phi_n(t)] - [\nu_n(t) + \Theta_n(t)]\,. \tag{11.15}$$

Bragg scattering of the optical fields from the index grating creates a nonlinear polarization in the photorefractive medium. We shall neglect effects that arise because of scattering of the resonator field back into the pump field in the weak-field limit. The pump field incident upon the index grating scatters only in the direction of the resonator field. Here, the Bragg condition is automatically satisfied, and this acts as a source of polarization for the resonator field, given by

$$\begin{aligned}\boldsymbol{P}_{\mathrm{nl}}(\boldsymbol{r},t) &= 2\epsilon_0\, \Delta n_{\mathrm{pr}}(\boldsymbol{r},t)\boldsymbol{P}(\boldsymbol{r},t)\\ &= \epsilon_0 \hat{e}_{\mathrm{p}} P \sum_n \Delta n_{\mathrm{n,pr}}(t) \exp[-\mathrm{i}\Psi_n(t)]\\ &\quad\times \exp\{\mathrm{i}[\omega_n t + \phi_n(t) - \boldsymbol{K}_n \cdot \boldsymbol{r}]\} + \mathrm{c.c.}\,,\end{aligned}$$

where we have dropped all phase mismatch terms that will average to zero over the crystal volume. Having now obtained the nonlinear polarization induced in the photorefractive medium for a multimode resonator field, we are able to derive the equations of motion for the amplitudes and frequencies of oscillation from the self-consistency equations.

If we apply the above equations for only *one oscillating mode*, we find an instantaneous, typical beat frequency phenomenon: the frequency of the oscillation is shifted relative to the pump frequency. The direction of the shift depends on the relative change of the cavity length – whether there has been an increase or decrease in the cavity length relative to the length at the resonator frequency. This effect is well known as *mode pulling* and leads to a variety of competition effects in the cavity.

For *two modes*, there is no longer the possibility of mode locking [7]. For the stationary equilibrium state of the system, which is given by the phase angle

$$\Psi_{i,\mathrm{s}} = \frac{\Omega_i - \omega_{\mathrm{p}}}{\gamma} = \Delta_i \tag{11.16}$$

and is equal to the normalized cavity detuning (Ω_i is the frequency of the passive resonator), we find the two possible mode intensities in the steady state given by [7]

$$I_{1,\mathrm{s}} + I_{2,\mathrm{s}} = \frac{\beta_1}{1+\Delta_1^2} - 1 \,, \tag{11.17}$$

$$I_{1,\mathrm{s}} + I_{2,\mathrm{s}} = \frac{\beta_2}{1+\Delta_2^2} - 1 \,. \tag{11.18}$$

Here, β_i are the normalized amplification factors. These equations show that only in the case of identical gain factors, losses and detunings of the cavity is simultaneous oscillation of both modes possible in the steady state. In all other cases, the dynamics of the system lead to a steady-state situation where one mode completely inhibits the oscillation of the other, regardless of the initial conditions; this mode is the mode with the larger gain or the mode with the smaller magnitude of cavity detuning. The steady-state intensity of the favored mode is that which would have been obtained if only one mode had been oscillating in the cavity.

The analysis for *three or more modes* leads to the same result. The mode with the highest amplification will win the competition and its intensity value in the stationary state is equal to that for single-mode operation. In conclusion, there is no possibility for different longitudinal modes to coexist in a resonator with photorefractive two-beam-coupling gain, except for those modes that have exactly the same gain factors in the photorefractive amplifier.

11.2 Transverse-Mode Patterns

Up to now, we have neglected the transverse extension of the oscillator signal. If this extension is taken into account, the oscillator equations become more difficult, because the transverse structures need to be decomposed into the appropriate oscillator modes.

To obtain a simple insight into the behavior in the transverse case, we may extend the description of the eigenmodes (11.9) and include the transverse distribution of the modes:

$$\boldsymbol{U}_n(\boldsymbol{r}) = \hat{e}_n A_n(x,y) \cdot \mathrm{e}^{\mathrm{i}k_n z} \,. \tag{11.19}$$

As before, these eigenmodes are assumed to be orthogonal, with a normalization factor given by

$$\int_{V_{\mathrm{res}}} |U_n(\boldsymbol{r})|^2 \,\mathrm{d}^3 r = L \int_S \int |A_n(x,y)|^2 \,\mathrm{d}x \,\mathrm{d}y = L N_n \,, \tag{11.20}$$

where the double integral is taken over the transverse dimensions of the resonator. Thus, both the modulation of the refractive index and the polarization become dependent on the position in space. If we proceed with the same calculation as in the longitudinal case, we can compare the intensity in the quasi-stationary case with the intensity in the case of plane waves [7]:

$$I_{1,\mathrm{s}} = \frac{\beta_1 - \Delta_1^2 - 1}{\beta_1} \frac{\displaystyle\iint_S |A_1(x,y)|^2 \, \mathrm{d}x \, \mathrm{d}y}{\displaystyle\iint_S |A_1(x,y)|^4 \, \mathrm{d}x \, \mathrm{d}y} . \tag{11.21}$$

Thus the transverse distribution changes the stationary intensity of the modes. In the case of two modes, we obtain

$$I_{i,\mathrm{s}} = \frac{\alpha_i}{B_i} - \frac{T_{ij}}{B_i} \frac{\alpha_j}{B_j} (1 - C)^{-1} , \tag{11.22}$$

where $i, j = 1, 2$ and the coupling constant C between the two modes

$$C = \frac{T_{12} T_{21}}{B_1 B_2} = \frac{\displaystyle\iint_S |A_1(x,y)|^2 |A_2(x,y)|^2 \, \mathrm{d}x \, \mathrm{d}y}{\displaystyle\iint_S |A_1(x,y)|^4 \, \mathrm{d}x \, \mathrm{d}y \iint_S |A_2(x,y)|^4 \, \mathrm{d}x \, \mathrm{d}y} . \tag{11.23}$$

In the case $T_{12} = T_{21} = 0$, when there is no overlap between the modes, the two modes are no longer coupled to each other and both modes acquire the same intensity as in the single-mode case. If $C \gg 1$, the competition is weak, and the modes do not influence each other. If, in contrast, we have $C \ll 1$, the competition is so strong that only one mode is able to oscillate. Finally, the case of $C \approx 1$ describes the case that is most sensitive to disturbances. This is the case of neutral coupling, where competition is strong, but both modes are still able to oscillate simultaneously as long as their cavity detuning, gain and loss are equal. Any difference in cavity detuning or gain factors leads to complete inhibition of the oscillation of one mode by the favored mode. Therefore, the competition processes described above result in a dynamic state, in which the modes may change arbitrarily owing to external influences.

11.2.1 Decomposition into Laguerre–Gaussian Modes

For a more detailed insight into the transverse competition processes, let us consider first the modes of an empty cavity with round apertures, the Laguerre–Gaussian modes. For each longitudinal mode of the cavity, there is a set of transverse modes, which are Laguerre–Gaussian modes. These are stationary solutions of Maxwell's equations in a cavity in the absence of any active, (e.g. photorefractive) medium. The Laguerre–Gaussian modes are functions of the transverse coordinates r and φ and of the longitudinal coordinate z. They are identified by three integers: $p \geq 0$, the radial index (i.e. the number of zeros along the radial direction); $m \geq 0$, the angular index (the number of zeros in the angular direction); and i, an index that takes only two values, 1 and 2. Thus, the shape of the Laguerre–Gaussian modes in the plane $z = 0$ is

$$A_{(p,m,i)}(r, \varphi, z = 0) = A_{pm}(r) B_m^{(i)}(\varphi)$$

$$= 2(2r^2/\eta_1)^{m/2}\left(\frac{p!}{(p+m)!}\right)^{1/2} \times L_p^m(2r^2/\eta_1)\exp(-r^2/\eta_1)B_m^{(i)}(\varphi) , \quad (11.24)$$

where η_1 is the square of the minimum beam waist. $L_p^m(x)$ is the associated Laguerre polynomial of order p and index m,

$$L_p^m(x) = (-1)^p \sum_{n=0}^{p} \frac{(-1)^n}{n!}\binom{p+m}{p-n} x^n , \quad (11.25)$$

and

$$B_m^{(i)}(\varphi) = \begin{cases} \frac{1}{\sqrt{2\pi}} & \text{if } m = 0 , \\ \frac{1}{\sqrt{\pi}}\sin(m\varphi) & \text{if } m > 0 \text{ and } i = 1 , \\ \frac{1}{\sqrt{\pi}}\cos(m\varphi) & \text{if } m > 0 \text{ and } i = 2 . \end{cases} \quad (11.26)$$

Examples of Laguerre–Gaussian modes are classified into different orders of families, which are given by

$$q = 2p + m . \quad (11.27)$$

Examples of Laguerre–Gaussian modes belonging to different (*mode families*) are shown in Fig. 11.3.

Each mode has an oscillation frequency

$$\omega_{(n,p,m)} = \frac{2\pi c}{\Lambda}n + \frac{c}{\Lambda}(2p + m + 1)a , \quad (11.28)$$

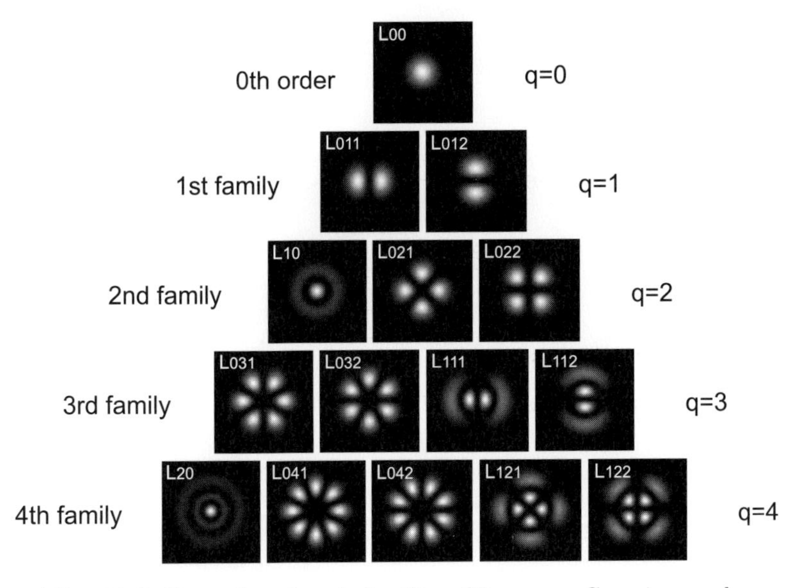

Fig. 11.3. Examples of mode families of Laguerre–Gaussian modes

where a is a parameter that depends on the cavity geometry. Therefore, the cavity geometry is an essential factor in determining the mode spectrum that can be achieved in any given oscillator. In the following, we shall consider only those geometries that allow self-imaging inside the resonator, and thus do not display any forced transverse interaction due to image rotation, transverse shifts or scale changes.

For an empty cavity with rectangular apertures, Hermite–Gaussian modes are excited; these are given by

$$A_{(m,n,i)}(x, y, z = 0) = A_0 H_m(\sqrt{2}x/\eta_i) H_n(\sqrt{2}y/\eta_i) \times \exp\left[-(x^2 + y^2)/\eta_i\right] ,$$

where

$$H_m(x) = (-1)^m \exp(x^2) \frac{\mathrm{d}^m \exp(-x^2)}{\mathrm{d}u^m} . \tag{11.29}$$

The mode families of Hermite–Gaussian modes are given by

$$q = m + n , \tag{11.30}$$

Examples of Hermite–Gaussian modes belonging to various mode families are shown in Fig. 11.4.

All pattern formation processes in ring oscillators depend on the linear spacing of the cavity modes and the relations between them. Therefore, one has to distinguish clearly the different types of cavities in order to understand the succession of longitudinal and transverse modes that they produce.

Several typical mode spectra for a cavity are shown in Fig. 11.5. In Fig. 11.5a the spectrum is shown for the case of a cavity with nearly plane mirrors, a *quasi-plane mirror cavity*. The spacing between longitudinal modes

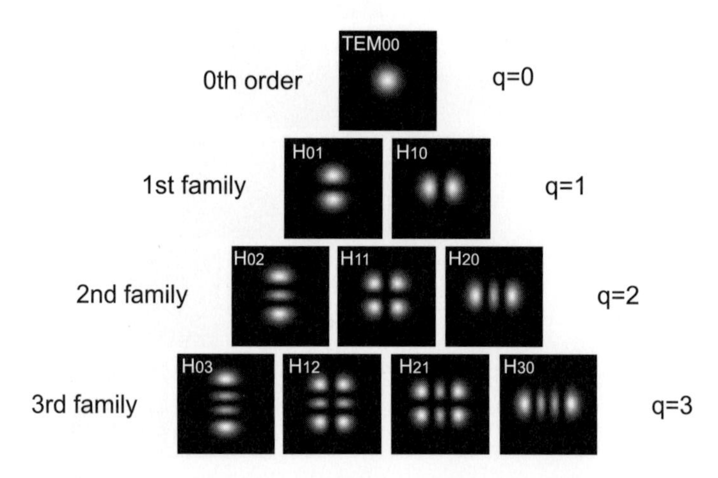

Fig. 11.4. Examples of mode families of Hermite–Gaussian modes

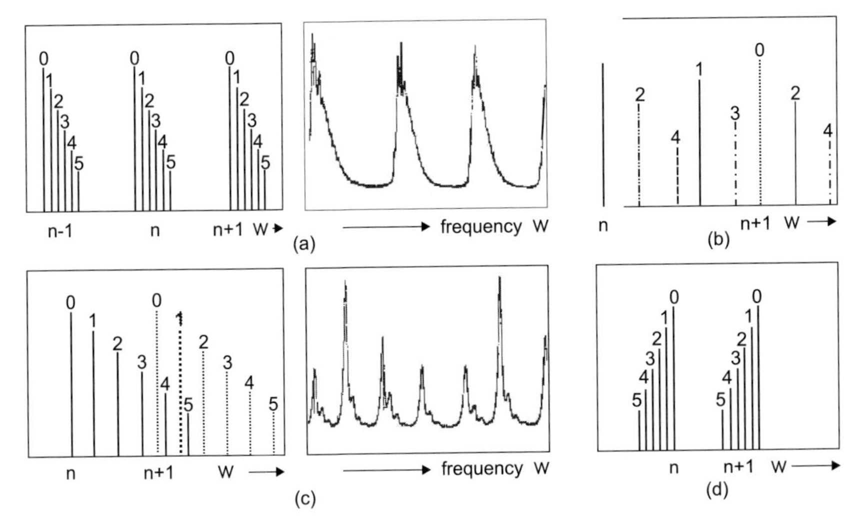

Fig. 11.5. Mode spectra of a laser cavity. (**a**) Quasi-plane mirror case (two planar mirrors): the transverse-mode spacing is much smaller than the free spectral range, and the transverse modes (*shorter lines*) are near their corresponding longitudinal modes (*long lines* labeled "0"); (**b**) Nearly confocal cavity (two spherical mirrors): the transverse modes in the region shown correspond to several different longitudinal modes. (**c**) Semi-confocal cavity with a single lens or one spherical mirror: the spacing of the longitudinal modes gives rise to overlap of different mode families. (**d**) Semiplanar cavity with two lenses: the distance between the two lenses can be adjusted to reverse the sequence of the transverse modes. Here, the transverse modes have smaller frequencies than that of the corresponding longitudinal mode

here is much larger than the spacing between transverse modes. Each set of transverse modes is near the corresponding longitudinal mode. In the case of an ideal, infinitely extended plane mirror cavity, all transverse modes would be degenerate; they would all be located at the same frequency as the corresponding longitudinal mode. In that case, the number of transverse modes per free spectral range is infinite. In reality, the degeneracy is lifted as soon as we consider real mirrors with a limited size. An experimental measurement of the modes of a quasi-planar resonator, which is the type of resonator used with a the photorefractive cavity in the experiments described later in this chapter, is shown in Fig. 11.5a. In Fig. 11.5b, the transverse-mode spacing is shown for the case of a nearly confocal cavity, which can be realized using curved mirrors. This spacing is a large fraction of the longitudinal-mode spacing, and the transverse modes close to each longitudinal mode have different longitudinal indices. Therefore, longitudinal modes of different order may coincide.

Fig. 11.5c,d show two configurations that are of experimental relevance. By replacing one of the mirrors by a spherical mirror or by introducing a lens into the oscillator (Fig. 11.5c), the spacing of the transverse modes can be controlled via the detuning through the focal length f of the lens and the cavity length L. In this case, the frequencies of the transverse modes have a larger spacing than in the case of the quasi-planar oscillator, but are not as separated as in the case of a nearly confocal system. Thus, higher-order transverse modes of a longitudinal mode enter the region of the next mode family. The corresponding experimental measurement of the modes shows the same behavior, and it is difficult to associate a single longitudinal mode with its corresponding mode family.

Figure 11.5d shows a mode spectrum of a cavity with two lenses. This configuration has the advantage that the free spectral range is strongly dependent on the distance between the two lenses. This is essential for experimental realizations, where it is often difficult to change the whole oscillator length, but much easier to change the distance between two lenses inside the cavity. If the lenses are properly adjusted, the transverse modes can be made to occur at lower frequencies than the corresponding longitudinal modes.

11.2.2 Field Equations of Transverse-Mode Patterns

Let us return to the description of the transverse modes by a resonator mode decomposition. The evolution of the field A in the resonator is controlled by Maxwell's equation

$$\nabla_{\perp}^2 \boldsymbol{A}(\boldsymbol{r},t) - \frac{1}{c^2}\frac{\partial^2 \boldsymbol{A}}{\partial t^2} = \frac{2n_0 \Delta n_{\mathrm{pr}}(\boldsymbol{r},t)}{c^2}\frac{\partial^2 \boldsymbol{A}}{\partial t^2} \,. \tag{11.31}$$

Applying the slowly-varying-amplitude approximation to (11.31) and decomposing $\boldsymbol{A}$ into A_{p} and A_{r} gives the following equation for the resonator field amplitude [8]:

$$\begin{aligned} &\left(2\mathrm{i}k_{\mathrm{r}}\frac{\partial}{\partial z} + \frac{2\mathrm{i}\omega_{\mathrm{c}}}{c^2}\frac{\partial}{\partial t} + \nabla_{\perp}^2\right) A_{\mathrm{r}} \\ &= \frac{n_0^4 \omega_{\mathrm{p}}^2}{c^2} r_{\mathrm{eff}} E_{\mathrm{sc}}(\boldsymbol{r},t) A_{\mathrm{r}}(\boldsymbol{r}) + \frac{2\omega_{\mathrm{c}}\delta_{\mathrm{cr}}}{c^2} A_{\mathrm{r}} \,. \end{aligned} \tag{11.32}$$

At this stage, the similarity to the laser case is obvious: there is a cavity frequency ω_{c}, a pump frequency that is equivalent to the atomic frequency ω_{p} and the resonator field which is equivalent to the laser field oscillating at a reference frequency ω_{r}. Therefore, it is possible to decompose the resonator field into the empty cavity modes $\boldsymbol{U}_n(\boldsymbol{r})$ in a manner equivalent to the laser case. Here, we shall use the Laguerre–Gaussian basis,

$$S_{\mathrm{r}}(r,\varphi,z,t) = \sum_{(p,m,i)} f_{(p,m,i)}(z,t) A_{(p,m,i)}(r,\varphi,z) \,. \tag{11.33}$$

This allows to reduce the coupled partial differential equations describing the photorefractive oscillator field (11.32) into a set of linear partial differential equations for the mode amplitudes [8]

$$\frac{\partial}{\partial z} f_{(p,m,i)} + \frac{1}{c}\frac{\partial}{\partial t} f_{(p,m,i)} = -\frac{\mathrm{i}}{c}\delta_{cr} f_{(p,m,i)} \\ - \mathrm{i}\alpha \int dr d\varphi A^*_{(p,m,i)}(r,\varphi,z) \cdot E^*_{\mathrm{sc}}(r,\varphi,z,t) A_p(r,\varphi,z) , \tag{11.34}$$

where α is given by $\alpha = n_0^4 \omega_p^2 r_{\mathrm{eff}}/(2k_{\mathrm{r}}c^2)$. In the discussion of this approach, I will mainly follow the thorough analysis given by D'Allessandro [8].

The advantage of decomposing the field along the empty-cavity modes is that the behavior of each mode amplitude during the propagation outside the photorefractive medium is very simple. Each mode is just phase-shifted with respect to the reference frequency ω_r and is attenuated due to the losses at the mirrors. The amplitude of a certain (p, m, i) mode will be transformed in the cavity during propagation so that the boundary conditions at the medium entrance faces can be formulated from this condition. They can be used to write the field equation in a reduced way. Moreover, we can assume that we operate the system in the mean-field limit, supposing that the losses in the cavity are small (the photorefractive medium has no influence and is small) so that only the modes that are "near" the pump frequency are excited. As the output of the resonator is a finite quantity, we can also suppose that the gain per single pass is a small quantity, of the order of the losses. This means in a more formal way that on the one hand we take the intensity transmittivity of the resonator mirrors $T_M = 1_R$, where R is the intensity reflectivity of the mirrors to be a small parameter. Then the gain, which is given by the right-hand side of equation 11.35, needs to be of the same order of magnitude as T_M. On the other hand, the requirement that only the modes having a frequency near the pump frequency are active, means, that the phase shift during propagation caused by the difference between the active mode frequencies and the reference frequency ω_r is small. In other words, the resonator field is the sum of the transverse modes whose frequency difference with respect to the pump frequency is of the order of the cavity line width, giving the following conditions: $T_M \ll 1$ (smallness parameter) , $\mid \omega_{\mathrm{n,p,m}} - \omega_r \mid \cdot \Lambda/c \simeq O(T_M) \ll 1$. The last condition is the description for the fact that the detuning is much smaller than the free spectral range. The mean field limit finally can be expressed as $\alpha \mid E_{\mathrm{sc}} \mid L \simeq O(T_M) \ll 1$.

Assuming that also the cavity length is large enough so that the Laguerre-Gaussian modes do not change appreciable during propagation, it is possible to introduce these assumptions into the resonator field equation by expanding (11.35) to the first order of this small frequency difference, which can be formulated as an expression to the first order in T_M. This results in the equation [8]

$$\frac{d}{dt'} \hat{f}_{(n,p,m,i)} = -\kappa \left\{ \left(1 + \mathrm{i}\Delta_{\mathrm{r}}^{(n)} + \mathrm{i}\tilde{a}(2p+m)\right) \hat{f}_{(n,p,m,i)}\right.$$

$$+ 2\mathrm{i}C \int drd\varphi r E_{\mathrm{sc}}(r,\varphi,t) \cdot A_p(r,\varphi) A^*_{(p,m,i)}(r,\varphi) \Big\} \,, \qquad (11.35)$$

where $\kappa = cT_M/\Lambda$, $\Delta_{\mathrm{r}}^{(n)} \equiv \delta_{\mathrm{r}}^{(n)} \Lambda/(cT_M)$, $\delta_{\mathrm{r}}^{(n)} \equiv w_{\mathrm{N},0,0} - \omega_r$, $\tilde{a} \equiv a/T_M$, and $2C \equiv \alpha L/T_M$. For the case of all transverse modes belonging to the same longitudinal mode ($a \ll \pi$), we can write the equation for the electric field as a single partial differential equation rather than a set of ordinary differential equations for the mode amplitudes. In this case, we can write the field as the sum over all tranverse modes of a chosen longitudinal mode N [8]

$$A_r(r,\varphi,z',t') = \sum_{(p,m,i)} \hat{f}_{(p,m,i)}(t') A_{(p,m,i)}(r,\varphi) \qquad (11.36)$$

and use the property of the Laguerre-Gaussian modes

$$\left(\frac{\eta_1}{4} \nabla^2 + 1 + \frac{r}{\eta_1} \right) A_{(p,m,i)} = -(2p+m) A_{(p,m,i)} \qquad (11.37)$$

in order to sum the entire set of (11.35) and obtain

$$\begin{aligned} \frac{\partial A_r}{\partial t'} &= -\kappa \left\{ \left[1 + \mathrm{i}\Delta_{cr} - \mathrm{i}\tilde{a} \left(\eta_1 \frac{\delta_\perp^2}{4} + 1 - \frac{r^2}{\eta_1} \right) \right] \right\} A_r \\ &\quad + \kappa \left\{ 2\mathrm{i}C A_p(r,\varphi) E^*_{\mathrm{sc}}(r,\varphi,t') \right\} \,, \end{aligned} \qquad (11.38)$$

where $\Delta_{\mathrm{cr}} \equiv \Delta_{\mathrm{r}}^{(N)} = \delta_{\mathrm{cr}} \Lambda/(cT_M)$.

11.2.3 Complete Model Equations

Now the model of the unidirectional photorefractive cavity can be obtained by combining the field equations with the material equations in order to get a complete description of the transverse effects in the system. First, let us assume that the pump and the resonator field do not change considerably in the material, so that we can neglect any z dependence in the resonator and in the material equations, and assuming that for a thin material slice $t' \approx t$, we can obtain the final equation [8]

$$\begin{aligned} \hat{f}_{(n,p,m,i)} &= -\kappa \Big\{ \left(1 + \mathrm{i}\Delta_{\mathrm{cr}}^{(n)} + \mathrm{i}\tilde{a}(2p+m) \right) \hat{f}_{(n,p,m,i)} \\ &\quad - C_{\mathrm{pr}} \int drd\varphi r E_{\mathrm{sc}}(r,\varphi) \cdot (n_D - \tilde{n}_e) \cdot A^*_{(p,m,i)}(r,\varphi) \Big\} \,, \end{aligned} \qquad (11.39)$$

where the material equations have now been introduced into the field equation using the normalized charge densities $N_D = (N_D^+ - N_D)/n_0$ and $\tilde{n}_e = (n_e - n_0)/n_0$, $n_0 = sN_D I/\zeta N_A$. Moreover,

$$C_{\mathrm{pr}} \equiv 2C \frac{e n_0}{\epsilon K_{\mathrm{pr}}} \qquad (11.40)$$

is a dimensionless parameter which measures the coupling strength between the pump and the resonator fields.

Now, we can use for the final step of simplification the fact that the time scale τ_{dr} is much longer than the other two τ_r and $1/\kappa$. This assumption is equivalent to considering a class A laser, where the time scales of the population inversion and the polarization can be neglected compared to the one of the laser field. Exploiting these relationships, we can use the results of the material equations to include them into the field equation. Due to the smallness of the expressions $\delta_{\mathrm{pr}} \cdot \tau_r \ll 1$ and $\delta\mathrm{pr} \cdot \tau_{\mathrm{dr}} \simeq O(1)$, we can project the resonator mode onto the Gauss-Laguerre mode of the empty resonator using the projection operator $\mathcal{P}_{(\surd,\Updownarrow,\rangle)}$. This finally gives the field equation in normalized field amplitudes [8]

$$\frac{d}{dt'} f_{(n,p,m,i)}(\bar{t}) = - \Big[(1 + i\Delta_{\mathrm{pr}}) f_{(n,p,m,i)}(\bar{t}) - \frac{B}{1 + i\Delta_r^{(n)} + i\tilde{a}(2p+m)} \cdot \mathcal{P}_{(p,m,i)} \left(\frac{|A_{\mathcal{P}}|^2 A_{\mathcal{R}}}{|A_{\mathcal{P}}|^2 + |A_{\mathcal{R}}|^2} \right) \Big] . \tag{11.41}$$

where $\bar{t} = t/\tau_{dr}$, $\Delta_{pr} \equiv \delta_{pr}\tau_{dr}$. The coefficient B measures the efficiency of the energy transfer between pump and resonator field and is proportional to C_{pr}. In this model, it plays the same role as the pump intensity in an equation for a homogeneously broadened laser. Moreover, B can be shown to be the bifurcation parameter for the resonator field threshold [8].

These equations, one for each active Laguerre–Gaussian mode, are the final result. Let us summarize here the main assumptions that have led to this result in order that we can remember the features of the model:

- The equations for the photorefractive material are linearized. This means that only a small modulation should appear between the pump and the resonator fields. However, we have shown in Chap. 8, where a similar approximation was made for the analysis of the system considered there, that this assumption is valid at least for a modulation of the order of 50%.
- A more stringent assumption is that the slice of the photorefractive medium is very thin. As a consequence, neither the pump beam nor the resonator field changes significantly in the slice and absorption can be neglected. This assumption is necessary in the theory to apply the mean-field hypothesis, which can only be done if we assume that the losses in the cavity are small and that the field does not change significantly in the photorefractive medium. However, an extended treatment that includes longitudinal diffusion and diffraction effects in the medium in the equations [13] gives the same qualitative results as does the analysis presented here, justifying this assumption for our purposes.
- The resonator field is decomposed into Laguerre–Gaussian modes, which are a good basis onto which a laser field in a resonator with spherical mirrors can be projected in general, provided that the pump beam is not

narrower than the beam waist. Therefore, this assumption is also included in the analysis of the photorefractive system.
- The two fast equations are adiabatically eliminated, which shows that the system has some equivalence to class A lasers, where the same procedure is applied.

Now that these equations have been derived, it is now necessary to find the stationary states and their stability in order to be able to compare the results with the experimentally found patterns. This is not a trivial task, and will not be performed in detail here. We refer to the work of D'Allesandro [8] for details. The problem can be made somewhat simpler if we suppose that the equilibrium configuration of the field is a pure Gaussian mode, with the pump field also having a Gaussian shape. The most important feature that can be deduced from the transverse modes that can be supported by a photorefractive ring oscillator is that the dynamics are very similar to those of a homogeneously broadened laser. Simple patterns can be formed by a single Laguerre–Gaussian mode. More complicated stationary patterns that result from superpositions of these modes, and periodic patterns of rotating vortices can also be found. These results exhibit a close analogy to the patterns observed in laser resonators.

There is, however, a difference of great importance: the timescale. The timescale of the dynamics in the transverse plane of a laser field is of the order of the light frequency. The timescale of the resonator field in the case of a photorefractive nonlinearity is the dielectric relaxation time of the crystal. Because photorefractive materials are insulators, this time constant is of the order of seconds. In a very formal interpretation, it is therefore useful to describe the crystal as a "time expander". The photorefractive unidirectional ring resonator is a convenient tool for studying simple laser patterns of class A lasers on a timescale that is much more easily accessible by standard experimental observation techniques.

As in the laser case, the final state depends on the initial conditions: for the same parameter values, it is possible to obtain different final states by starting from different initial conditions. As in laser dynamics, the dynamical effects depend strongly on the number of active modes involved in the pattern formation process. For small values of B and large values of the transverse mode spacing $\tilde{a}$, the pump is resonant with just one family of transverse modes at a time. By changing the value of the detuning it is then possible to obtain a vortex of charge ± 1, or some combination of those vortices. If the tranverse-mode spacing is decreased, more families of modes become resonant with the pump beam and the pattern becomes more complicated. In this case, rotating patterns may occur; these are superpositions of different Laguerre–Gaussian modes, oscillating at different frequencies. The reason for this rotation is beating phenomena between the modes. In our experimental realizations, the above scenarios could be realized exactly.

11.2.4 Analogy Between Photorefractive Oscillators and Class A Lasers

Another, more general approach is to consider the similarities between the equations describing a photorefractive oscillator in the uniform-field limit with an undepleted pump and the complex Ginzburg–Landau equation with space-dependent coefficients. We shall follow here the derivation of Jost and Saleh [10] and of Weiss and coworkers [14] to demonstrate the usefulness of such an approach.

One of the disadvantages of the theoretical evaluations mentioned above is that although they are able to demonstrate some of the spatio-temporal effects observed in the experiments, the general spatio-temporal complexity present in the experiments cannot be described. This can be attributed to approximations that do not allow one to investigate the parameter regions where this complexity is observed. This, in turn, is mainly due to the fact that the modal expansion requires a truncation to a limited number of modes for computational studies to be performed. Even if a very large number of modes is considered, it is difficult to estimate accurately the parameters associated with each of the modes for a given experimental configuration (e.g. the mode gain and loss, and the mode frequency.), and therefore the accuracy of the modal decomposition is compromised.

In the following, we shall concentrate on the similarity of the equations to more general forms and therefore shall refer to the slowly varying complex amplitude of the refractive index grating described in (11.6) in the form

$$\Delta n_{\mathrm{pr}} = G(\boldsymbol{r},t)\exp(\mathrm{i}\boldsymbol{K}\cdot\boldsymbol{r}) + \mathrm{c.c.}\ . \tag{11.42}$$

Introducing the normalized variables $g = -\mathrm{i}G/n_{\mathrm{pr}}|A_{\mathrm{p}}|^2_{\max}$ and $v = A_{\mathrm{r}}/|A_{\mathrm{p}}|_{\max}$, where $|A_{\mathrm{p}}|_{\max}$ is the maximum value of the magnitude of the pump field, and substituting the expressions for Δn_{pr} and I_1 into (11.4), we obtains the index grating equation [10]

$$\frac{\partial g}{\partial t} = \frac{1}{\tau_{\mathrm{c}}}\left[\frac{\gamma}{A^*_{\mathrm{p}}|A_{\mathrm{p}}|_{\max}}v^* - \left(1+\frac{|v|^2}{i_{\mathrm{p}}}\right)g\right] , \tag{11.43}$$

where $i_{\mathrm{p}} = I_{\mathrm{p}}/|A_{\mathrm{p}}|^2_{\max}$. In the next step, we have to derive again the wave equation that governs the resonator beam. The resonator cavity is assumed to have weak longitudinal losses (due to cavity imperfections, etc.) that are described by the distributed loss coefficient α_{L}. Hence, the resonator field $A_{\mathrm{r}}(\boldsymbol{r},t)$ must satisfy the equation [15]

$$\left(\nabla^2_{\perp} - \frac{\alpha_{\mathrm{L}}}{2\pi}\sqrt{\mu_0\epsilon}\frac{\partial}{\partial t} - \mu_0\epsilon\frac{\partial^2}{\partial t^2}\right)\boldsymbol{A}_{\mathrm{r}} = -\frac{1}{\epsilon}\boldsymbol{\nabla}(\boldsymbol{\nabla}\cdot\boldsymbol{P}_{\mathrm{nl}}) + \mu_0\frac{\partial^2}{\partial t^2}\boldsymbol{P}_{\mathrm{nl}}\ , \tag{11.44}$$

where $\boldsymbol{P}_{\mathrm{nl}}$ is the nonlinear polarization of the cavity medium. The nonlinear polarization is related to the refractive index change by $\boldsymbol{P}_{\mathrm{nl}}(\boldsymbol{r},t) =$

$2\epsilon \boldsymbol{A}_{\rm p}(\boldsymbol{r},t)\,\Delta n_{\rm pr}(\boldsymbol{r},t)$, which can be substituted into (11.44). We shall consider a crystal with a single axis and a large electro-optic coefficient, which was the medium of choice in our experiments. We consider the uniform-field limit as described above, including the assumption that the gain is weak (just above threshold), that cavity losses are small and that the slowly varying longitudinal variations can be neglected. Consequently, we consider the case $v(\boldsymbol{r},t) \to v(x,y,t)$, $g(\boldsymbol{r},t) \to g(x,y,t)$ and $A_{\rm p}(\boldsymbol{r}) \to A_{\rm p}(x,y)$. In this case, the right-hand side of (11.44) is considerably simplified. When eh introduce the normalized field and index variables, the wave equation reduces to

$$\frac{\partial v}{\partial t} = -\frac{1}{\tau_\alpha}\left(v - \tau_\alpha A_{\rm p}|A_{\rm p}|_{\rm max}\omega_{\rm p}\frac{n_1}{\epsilon_{\rm r}}g^* - \mathrm{i}\frac{\tau_\alpha c^2}{2\omega_{\rm p}}\nabla_\perp^2\right), \tag{11.45}$$

where the cavity lifetime is described by the time constant $\tau_\alpha = [c\alpha_l/4\pi + \mathrm{i}(\Omega-\omega_{\rm p})]^{-1}$. Eq. (11.43) and (11.45) and the transverse boundary conditions together describe the evolution of the complex optical field in a unidirectional photorefractive ring oscillator. In the limit where $|\tau_\alpha| \ll |\tau_{\rm c}|$, which is always obtained for the available photorefractive crystals with high gain, we can eliminate the field variable in (11.45) adiabatically. If we set the time derivative in (11.45) equal to zero, solve for g and substitute into (11.43), we obtain a reduced equation for v,

$$\frac{\partial v}{\partial t} = d(a - |v|^2)v + \mathrm{i}b\,\nabla_\perp^2 v\,, \tag{11.46}$$

where

$$d = \frac{1}{i_{\rm p}(x,y)\tau_{\rm c}^*(x,y)}\,, \tag{11.47}$$

$$a(x,y) = i_{\rm p}(x,y)\left(\frac{\tau_\alpha\omega_{\rm p}\gamma^* n_1}{\epsilon_r} - 1\right), \tag{11.48}$$

$$b(x,y) = \frac{\tau_\alpha c^2}{2\omega_{\rm p}\tau_{\rm c}^*(x,y)}\,. \tag{11.49}$$

Equation (11.46) has exactly the form of a complex Ginzburg-Landau equation. Since v is a slowly varying small quantity, the higher-order terms proportional to $\partial_t\,\nabla_\perp^2 v$ and $|v|^2\;\nabla_t^2 v$ have been neglected. The inclusion of these higher-order terms would yield a new, modified Ginzburg-Landau equation.

Because the Ginzburg-Landau equation is well-known as the equation-commonly used to describe lasers, we shall add here a brief comparison with the analysis of a laser to emphasize the similarities to and differences from the photorefractive-oscillator case. From a mathematical point of view, the following points are important:

- The derivation of the complex Ginzburg–Landau equation for lasers depends on a near-threshold condition for the reduction of the three Maxwell–Bloch equations [16]. No analogous approximation has to be made for the reduction of the two equations of the unidirectional photorefractive system.

- However, the near-threshold condition of the analysis of the laser is analogous to the condition of an undepleted pump for the photorefractive ring. The photorefractive complex Ginzburg–Landau equation is valid beyond the threshold as long as the undepleted-pump approximation is satisfied.
- In the case of the laser, different length scales may have different timescales, leading to a substantial complication in the analysis. The situation is different for the photorefractive system, where both long and short spatial scales are associated with a fast timescale and can therefore be eliminated simultaneously [8]. Therefore, the adiabatic elimination of the fast equations is straightforward, and there is no need for complex mathematical techniques such as the central manifold technique [17] or the special scalings [16] used in laser analysis. As a result, the square of the transverse operator does not appear and we do not need to impose a constraint on the transverse spatial scale.

From a physical point of view, the following correspondences are found between the parameters of a class A laser and those of a photorefractive oscillator [14]:

- The pump parameter of a laser corresponds to the refractive index modulation in a photorefractive oscillator. In a drift-type oscillator, the electric field applied to the crystal corresponds to the pump in a real laser since Δn_{pr} is directly proportional to the electric field.
- The detuning in a laser corresponds to the detuning between the pump and oscillator fields in a photorefractive oscillator. This means that a selection of transverse-mode families can occur through resonator detuning, as in lasers. The difference from the laser is that the detuning required in a photorefractive oscillator is smaller than that required in a laser to excite the same transverse-mode family. Moreover, the photorefractive oscillator exhibits strong frequency pulling owing to the narrow pump beam line (see Fig. 11.1).
- Mode beats appear on a very slow timescale in photorefractive oscillators. The beat frequency is small owing to the small separation of different transverse modes caused by the pump beam profile (see Fig. 11.19).
- There is a transverse-mode selection in photorefractive oscillators analogous to that in lasers. This selection occurs for a laser when the spectral width of the gain line is narrower than the free spectral range of the resonator. In the photorefractive oscillator, the selection does not occur because of the finite gain linewidth, but because of the finite width of the empty-resonator modes. For simultaneous oscillation of several transverse modes, it is necessary, but not sufficient, that the linewidths of the resonator modes overlap.
- In photorefractive oscillators, the diffusion part of the nonlinearity causes self-defocusing. This can influence the dynamics when at least a few transverse-mode families are active. In this case the self-defocusing can cause a transfer of energy to the lower transverse-mode families. In the

case of emission within only one transverse-mode family, the defocusing plays no role.

To summarize, a photorefractive oscillator behaves like a "very slow" class A laser with additional self-defocusing and a frequency shift. In the case of only one active transverse-mode family, the photorefractive system behaves completely identically to a class A laser [14].

In the experimental realization that we shall present in the next section, a lens is placed inside the resonator for stability and symmetry reasons. However, it has a sufficiently large focal length that we can expect the photorefractive complex Ginzburg–Landau equation to provide at least a qualitative description of the experimentally observed spatio-temporal dynamics, from which the dependences on the parameters can be determined. Let us consider here two cases that are important in the experiment:

– In the case of a uniform pump beam, (11.46) can be written in the standard form

$$\frac{\partial v'}{\partial t'} = v' + (1 + \mathrm{i}c_1)\nabla_\perp^2 v' - (1 - c_2)v'|v'|^2, \tag{11.50}$$

where c_1, c_2 and the primed quantities have been scaled in an appropriate manner. The relative values of c_1 and c_2 influence the system dynamics and, for the photorefractive system, these are mutually dependent quantities. They determine whether the operation is far below the modulational instability (see Chap. 8 for details), providing a stable system, or far above the modulational instability, giving rise to spatio-temporal chaotic states. The transition region near the modulational instability curve gives rise to a variety of dynamic phenomena, including phase turbulence or vortices, and spatio-temporal intermittency.

When we consider the necessary condition for chaotic dynamics [10], a moving grating, which can be obtained with an applied field or a photovoltaic crystal field, combined with low losses and a large cavity detuning, seems to be necessary to observe spatio-temporal dynamics. However, in all experimental realizations, photorefractive diffusion-dominated crystals such as $BaTiO_3$ or $KNbO_3$ have been used which exhibit only a small contribution from the photovoltaic field. It therefore seems sufficient to a have very small contribution of this kind to observe considerable spatio-temporal dynamics.

The most interesting point is that by changing the appropriate control parameters, it should be possible to begin the operation of the photorefractive oscillator below the modulational instability at a stable point. Subsequently, by changing the electric field across the photorefractive crystal, decreasing the resonator losses or increasing the Fresnel number, one should be able to move to a larger c_1 value with an unchanged c_3. Hence, for some initial parameter values, it should be possible to begin at a stable operating point and then, by reducing the resonator losses or increasing the Fresnel number of the cavity, to obtain a chaotic operation point. We have shown in our experiments that this change can in fact be observed. Moreover, we could show, for both the unidirectional photorefractive oscillator and for a slightly different phase-conjugate photorefractive oscillator, that the mean number density of optical defects increases with increasing Fresnel number of the cavity.

- Gil et al. [18] have obtained numerical solutionx of the complex Ginzburg–Landau equation using space-dependent coefficients to simulate the pump profile in a laser. They showed that stable, rotating spatial patterns may arise in association with a given topological charge. In contrast, uniform coefficients gave a stable pattern that did not rotate. We shall show in the experimental results described below that a corresponding long-term stability of rotating spatial patterns with the same topological charge, called *circling optical vortices*, can be found in the photorefractive system. The derivations made here indicate that these rotations may be attributed to the spatial variations of the pump.

11.3 Transverse-Mode Pattern Dynamics, Competition and Selection

In our experimental setup, which is depicted in Fig. 11.6, we used a ring resonator with an even number, four, of highly reflecting mirrors, which allows one to excite all modes, in contrast to a setup with an odd number of mirrors, where only symmetric modes can be observed. Mirror M_3 was mounted on a piezoelectric-actuator, which enabled us to change the resonator length in a defined way within a range of about 5 μm. By this means, different transverse modes could be adjusted in a defined way.

A small amount of the cavity intensity was coupled out of the cavity at mirror M1, allowing us to measure the integral temporal evolution of the cavity signal with a photodiode, as well as observe the transverse patterns with a CCD camera.

The $BaTiO_3$ gain medium was pumped by an argon ion laser operating at 514 nm in the extraordinary polarization in order to exploit the largest electro-optic coefficient of $BaTiO_3$, which is $r_{42} = 1640$ pm/V. A lens inside the cavity as used to increase the optical stability of the system. This lens also allowed us to adjust the number of transverse modes that could exist in the free spectral range of the cavity, over a broad range. To change the Fresnel number of the resonator, an iris aperture was inserted into the cavity. This aperature allowed us to adjust the losses of different modes by adjusting its diameter, giving higher-order modes higher losses owing to their larger diameter. Thus, the number of transverse modes in self-oscillation could be varied by changing the diameter of the iris.

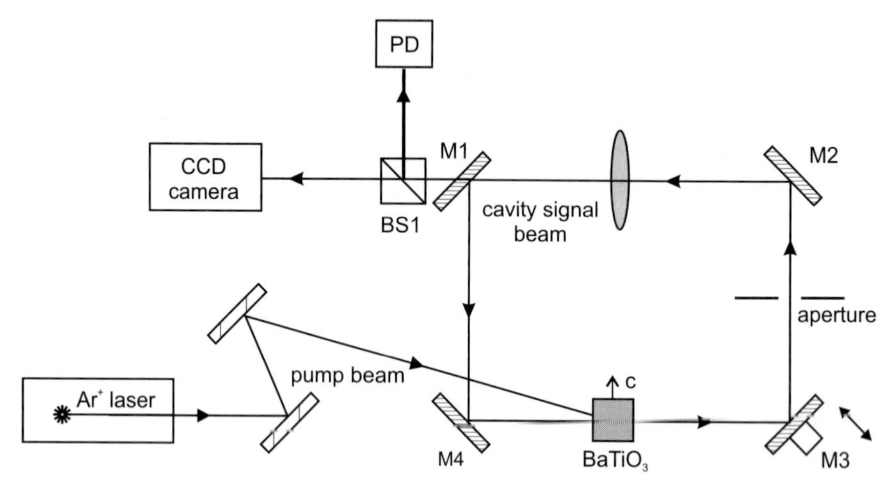

Fig. 11.6. Experimental configuration of a unidirectional photorefractive oscillator. M_i are highly reflecting mirrors, and PD is a photodiode. Mirror M_3 is mounted on a piezoelectric actuator in order to scan the free spectral range of the cavity. The active gain medium is a diffusion-dominated $BaTiO_3$ crystal

11.3.1 Active Stabilization

To excite the transverse-mode families in the resonator in a reproducible and defined way, independently of external effects such as thermal fluctuations and mechanical vibrations, both of which may have a severe influence on the cavity length, the cavity length was actively stabilized [19]. Our measurements showed that even changes of only a few kelvin in the room temperature could change the cavity length by several micrometers, and therefore we required a means to stabilize the oscillator length at the submicrometer level. For that purpose, a reference beam with a perpendicular polarization was fed into the cavity, counterpropagating relative to the cavity signal. The complete experimental configuration, including this active stabilization, is depicted in Fig. 11.7.

In this setup, the two beams, the reference beam and the resonator signal, are separated before the cavity signal passes through the crystal. They are given orthogonal polarizations in order to pass through the cavity along the same path without interacting. The reference beam counterpropagates relative to the signal beam, but circumvents the photorefractive crystal. Although this arrangement does not take account for changes in the optical path length in the crystal, it allows us to realize a high-finesse resonator in the reference path. In contrast, the path containing the crystal is a very low-finesse resonator owing to absorption in the photorefractive crystal. Thus, by stabilizing the reference beam cavity, we are able to stabilize at the same time the cavity containing the photorefractive nonlinearity as long as the difference

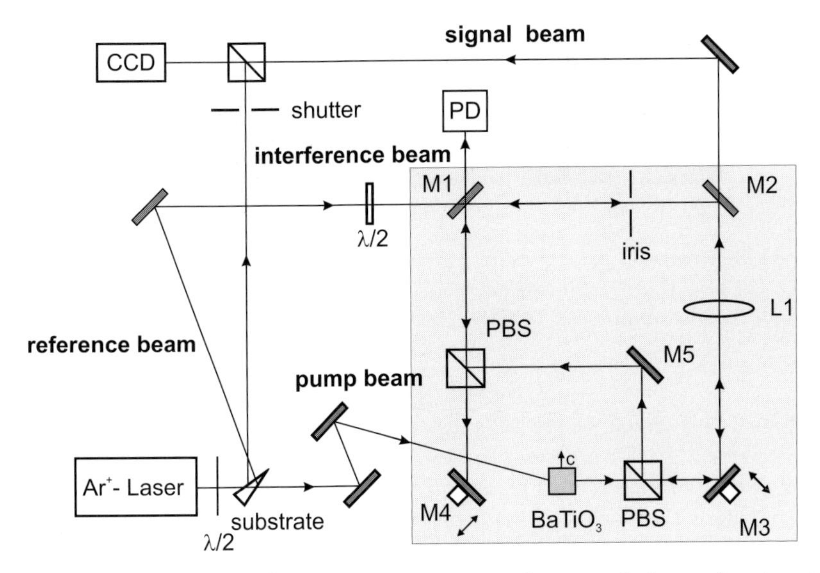

Fig. 11.7. Complete experimental setup of the unidirectional photorefractive ring oscillator, including the active-stabilization unit

between the two paths can be neglected compared with the resonator length. In order to fulfill this criterion, we included a very short unstabilized part (this part is enlarged in Fig. 11.7 in order to show its function), constructed symmetrically on a stabilizing, mechanically damped aluminum base plate. In this configuration, changes in the path and thermal effects will influence both arms in the same manner, and only changes of the path in the crystal itself can contribute to any difference between the paths of the reference and cavity signals.

The active stabilization is interferometric, combined with an electronic feedback compensation by adjustment of a piezoelectrically driven mirror (M4). The principle of the stabilization is based on a control signal, provided by lock-in amplifier, that adjusts the reference signal to achieve maximum transmission, and then maintains it by active adjustment of the mirrors. Because the response time of the crystal is much larger than the adjustment time, high-frequency mirror-positioning movements do not affect the resonator. With this stabilization, changes both in the cavity and in the pump wave frequency are compensated.

11.3.2 Pure Cavity Modes for Low Fresnel Numbers

The influence of the Fresnel number $F \equiv a^2/L\lambda$ becomes significant for transverse modes in a cavity when comparing it with the outer radius of a Hermite–Gaussian or Laguerre–Gaussian mode of order n ($n > 1$). It is

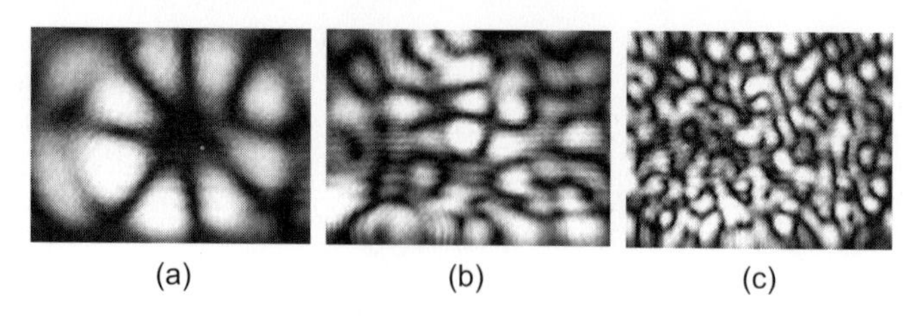

Fig. 11.8. Examples of transverse patterns for different Fresnel numbers F in the cavity. (**a**) Low Fresnel number ($F \approx 3$), (**b**) medium Fresnel number ($F \approx 10$) and (**c**) high Fresnel number ($F \approx 1000$)

approximately given by $s_n \approx \sqrt{nL\lambda/\pi}$. Therefore, the highest mode that may still oscillate for a given Fresnel number is defined by $n_{\max} = \pi F$. Thus, the Fresnel number is also a measure of the order of those modes that can pass through the oscillator without any disturbance. Although the Fresnel number in a photorefractive unidirectional oscillator is still modulated by the intensity-dependent nonlinear response function of the photorefractive medium, the Fresnel number of the empty cavity remains an important order parameter for describing the system.

In our experimental realization, the Fresnel number was varied by varying the aperture of an iris in the cavity [19]. When the aperture is almost closed, no self-oscillation pattern can be formed. At a certain threshold value of the diameter, the TEM_{00} mode appears. For a small Fresnel number of the oscillator, the spatial coherence is large and the transverse structure of the oscillator signal is dominated by pure basic, dynamically stable modes. When the diameter of the aperture is slowly increased, higher-order modes are excited. Figure 11.8a shows, as an example, a pure Laguerre–Gaussian mode of order four (L_{042}).

In our system, it is possible to excite pure modes up to an order of about 10. For larger apertures, a region of medium Fresnel numbers ($F \approx 10 - 100$) is reached where pure modes disappear. Instead, the structures become more complex. Patterns that are similar to pure modes appear only in the region near the optical axis. An example of this area is shown in Fig. 11.8b.

An autocorrelation analysis of these patterns showed that the transverse spatial correlation decreases with increasing Fresnel number. For higher F ($F > 10^3$), a completely uncorrelated, speckle-like pattern appears that is no longer stationary in time.

Using our active-feedback stabilization, we varied the oscillator length in a quasi-stationary way within the free spectral range. Because the gain line of the photorefractive medium is much smaller than that of the oscillator, the crystal defines the mode which will be excited in the oscillator. A change in the oscillator length thus results in a shift of the oscillator modes relative to the gain line of the photorefractive medium. Thus it is possible to scan

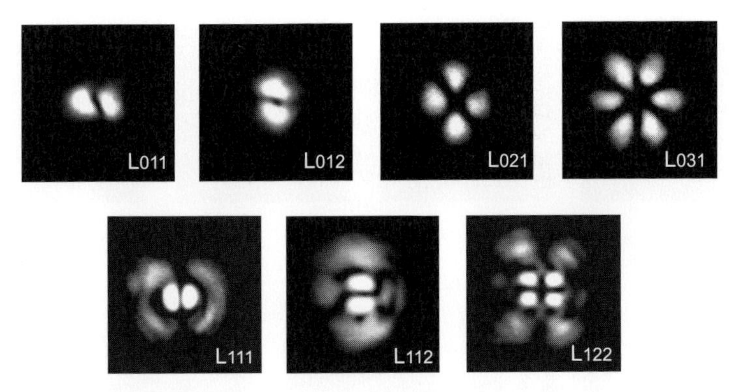

Fig. 11.9. Examples of transverse modes of different mode families. First mode family, L_{011} and L_{012}, second mode family: L_{021}, third mode family: L_{031}, L_{111}, L_{112}, fourth mode family: L_{122}

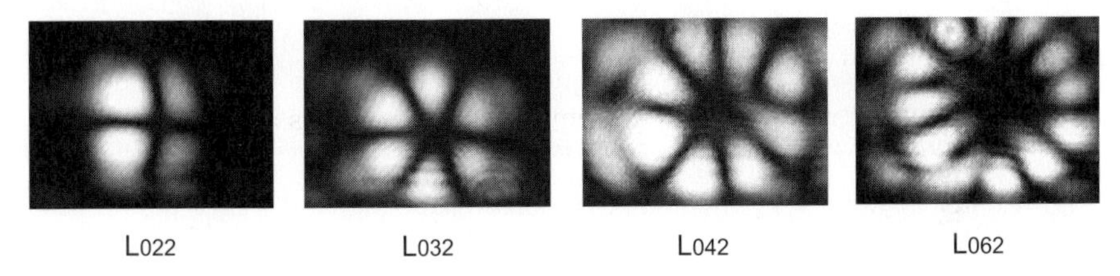

Fig. 11.10. Examples of transverse Laguerre–Gaussian "flower" modes of different mode families

through all the modes that are accessible in the free spectral range. If the change is quasi-stationary, allowing the system to build up an oscillation at each value of the length, one can obtain all modes that are able to oscillate in that configuration. Figure 11.9 shows some examples of different longitudinal-mode families, and Fig. 11.10 shows examples of Laguerre–Gaussian "flower" modes of various mode families.

We were also able to excite various doughnut-like patterns. These are essentially bright fields with a dark center at the optical axis. The optical fields of those patterns contain a phase singularity (a vortex). The doughnut order, which equals the number of vortices, is the number of their transverse-mode family. We could show that the doughnuts excited in our experiment indeed had the number of vortices defined by this rule.

In general, pure doughnut-like patterns of order n appear if a resonator is able to support $n + 1$ excitable transverse-mode families per free spectral range. In that case an order-n doughnut can usually be obtained by exciting the corresponding transverse-mode family without simultaneous excitation of the other transverse-order families. Owing to the resonator length and the lens that we used to stabilize the system, about five transverse-mode

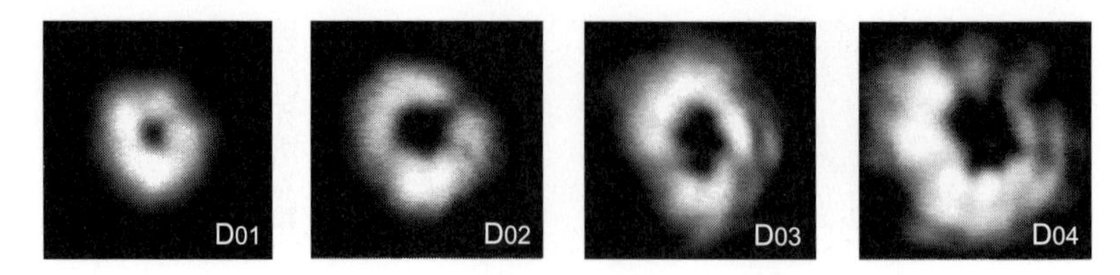

Fig. 11.11. Examples of transverse doughnut modes of different mode families

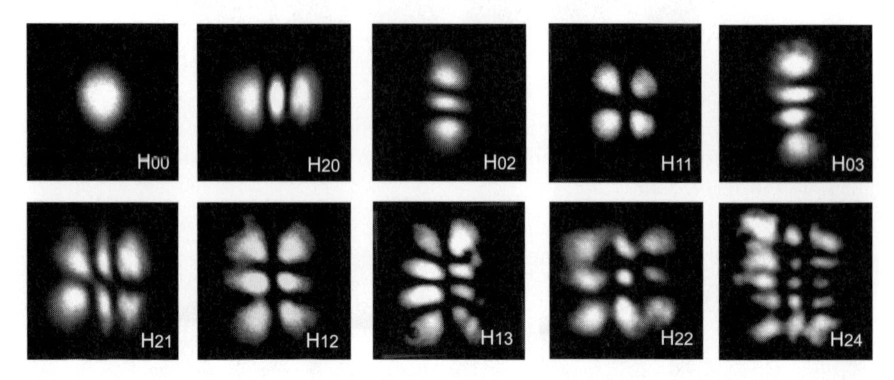

Fig. 11.12. Examples of transverse Hermite–Gaussian modes of different mode families. In addition to the TEM_{00} basic mode, modes of the second mode family (H_{20}, H_{02}, H_{11}), the third mode family (H_{03}, H_{21}, H_{12}) and the fourth family (H_{13}, H_{22}), and the H_{24} mode of the sixth family are shown

families could be excited per free spectral range. Figure 11.11 shows some of the doughnut patterns excited in our system. These patterns are essentially bright ring-like structures with a dark center, although in some cases some residues of the Laguerre flower patterns are still apparent.

In the case of an even slightly rectangular aperture, Hermite–Gaussian modes could also be excited in our system. Figure 11.12 shows examples of experimentally excited patterns with this symmetry.

11.3.3 Linear Superposition of Cavity Modes

If a transverse mode of higher order lies in the region of the next mode family, linear superpositions of modes of different families can also be observed, according to the linear laser mode theory.

Several examples of superpositions of modes of the same symmetry group and the same order are depicted in Fig. 11.13. Figure 11.14 shows superpositions of different symmetry groups, which may appear because of asymmetries in the system. The left side shows numerical simulations of pure modes, and the right side shows the corresponding patterns in our photorefractive oscillator. Figure 11.14a shows a superposition of a Laguerre–Gaussian (031) mode

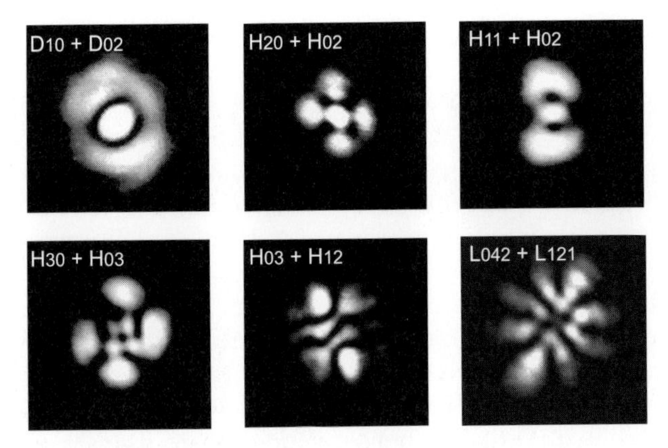

Fig. 11.13. Examples of superpositions of modes of the same symmetry, but of different mode families

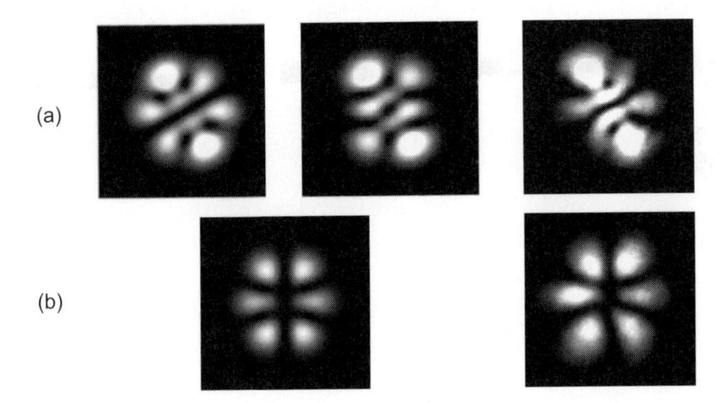

Fig. 11.14. Superpositions of modes of different symmetry groups excited because of slight disturbances in the polar symmetry. The images on the *left* show calculations of laser modes, and the images on the *right* show experimental patterns in the unidirectional photorefractive cavity. (**a**) Superpositions of an L_{031} mode with an H_{03} mode (*left*) and of an H_{12} mode with an H_{03} mode, rotated by 30° with respect to the axis of the previous pattern (*middle*). (**b**) Superposition of an L_{031} mode with an H_{12} mode

and a Hermite–Gaussian (03) mode at the left. Beside, a superposition of a Hermite–Gaussian (12) mode with a Hermite–Gaussian (03) mode, rotated by 30° with respect to the axis of the previous pattern, shown in the middle. Both of these patterns could be the origin of the pattern observed, shown at the right. Figure 11.14b shows a superposition of a Laguerre–Gaussian (031) mode with a Hermite–Gaussian (12) mode, which is caused by disturbances in the cylindrical symmetry.

11.3.4 Mode Pattern Dynamics

Dynamics of transverse patterns may occur for several reasons. For example, the starting dynamics of the system may cause changes in the resonator length due to heating effects. Thermal drift in the oscillator may cause changes in the resonator length that are sufficient to tune the cavity from one mode to the other, causing mode alternation. Chen and Abraham [20] could show, in the case of a bidirectional photorefractive oscillator, that intensity fluctuations in the oscillator were due to temperature drift.

In our unidirectional system, a similar behavior was observed [19] (see Fig. 11.15). To quantify this behavior, we restricted our system by means of the aperture to a Fresnel number that allowed only the basic Gaussian mode to be excited. Owing to temperature changes, the signal disappeared periodically, as the system was pushed periodically below the self-oscillation threshold. In this configuration, the temperature was increased continuously by about 2 K. This led to a linear change in the resonator length. In this case even the change in the refractive index of the air resulted in a change of the cavity length of about $\Delta L \approx 2400$ nm $\approx 4.5\ \lambda$. As a consequence, the oscillator emits after each change in λ, which corresponds to the longitudinal mode distance, a new, next transverse mode.

This example shows clearly that with an aperture that allows a higher number of modes to be excited and a nonmonotonic temperature variation, complex dynamics, including periodic or "chaotic" mode alternations, can be expected to appear. Several sequences of such mode changes, which can be

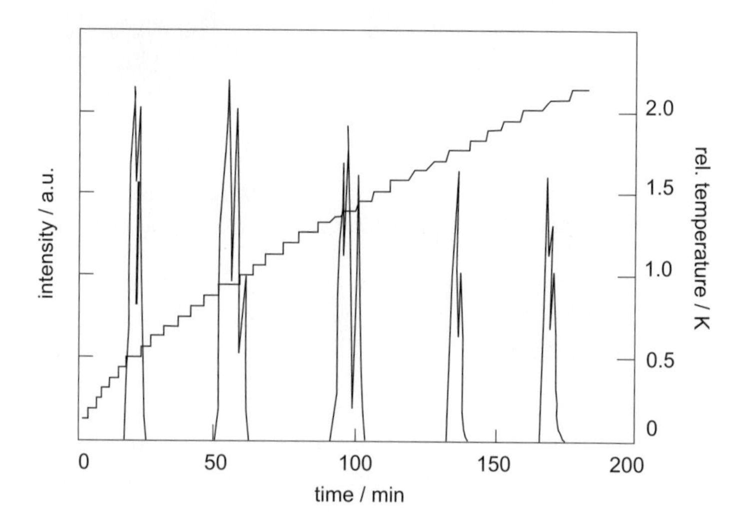

Fig. 11.15. Cavity signal compared with the variation of the room temperature for a cavity that allow only the Gaussian ground mode to oscillate

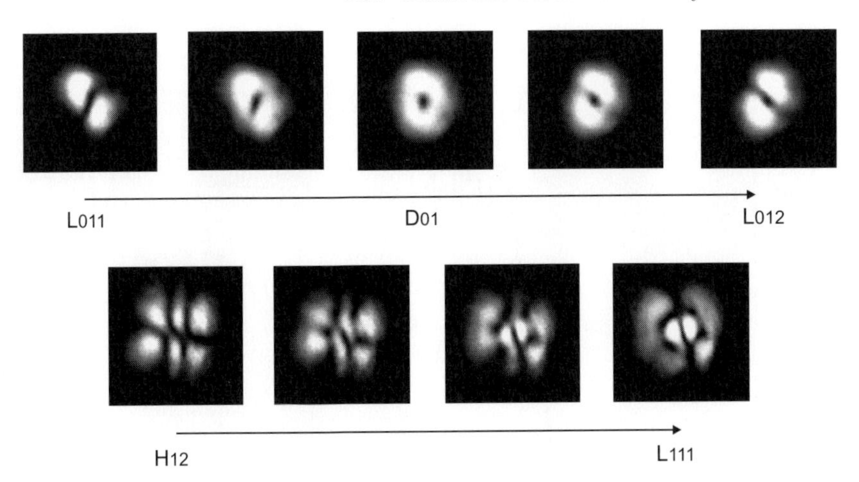

Fig. 11.16. Mode alternations for two different Fresnel numbers of the cavity. Top: mode alternation between low-order Laguerre and doughnut modes. Bottom: Alternation between Hermite and Laguerre modes of the third mode family

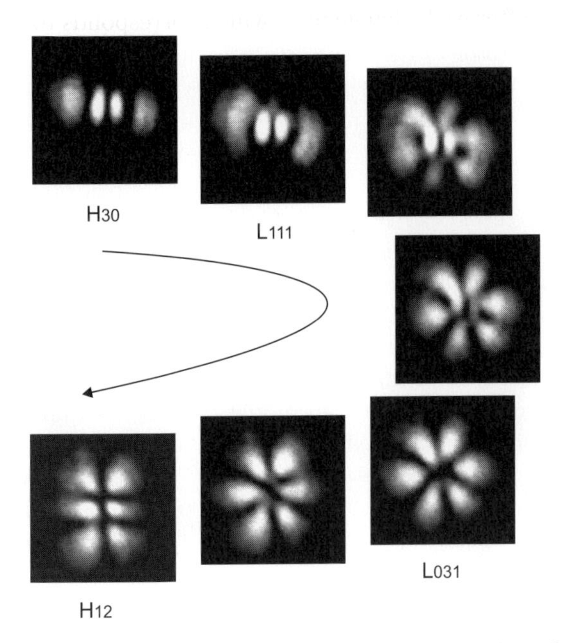

Fig. 11.17. Complex alternation of modes of the third mode family

observed easily in the photorefractive system owing to its slow timescale in the region of seconds, are depicted in Figs. 11.16 and 11.17. These sequences

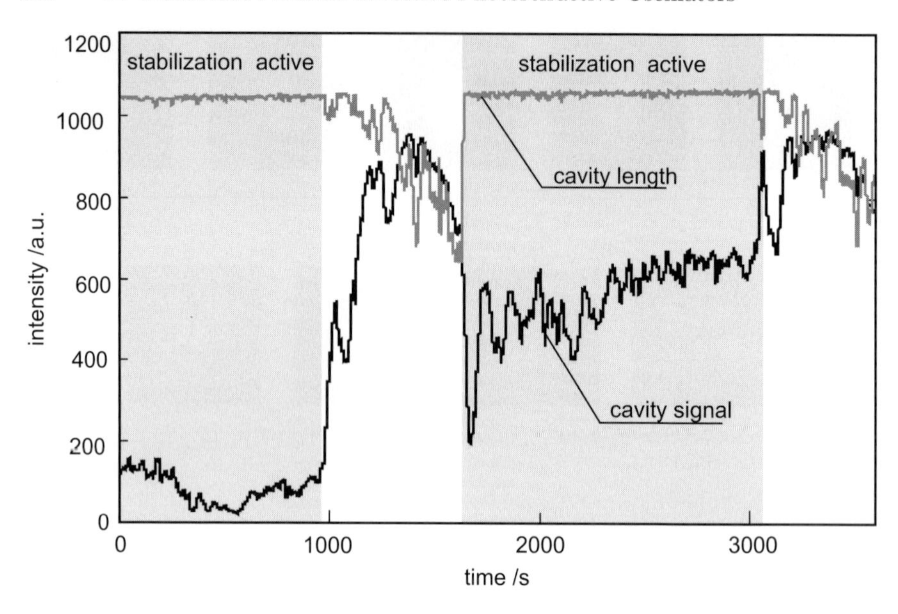

Fig. 11.18. Comparison of the change of the cavity length and the integral cavity signal when the active stabilization is switched on and off

can be interpreted as periodic or chaotic alternations of modes, although changes in the external cavity length are the origin of these dynamics.

Our active stabilization is able to compensate for these length changes. When the active stabilization is switched off, modes begin to alternate in a nonperiodic manner owing to instabilities in the cavity length (Fig. 11.18). In contrast, when the stabilization is active, the intensity remains stable as well as the resonator length. In such a configuration, the only changes in the mode pattern are due to the mode competition effects that can be found in a perfectly stabilized system and that are due to pattern competition effects. Therefore, the active stabilization enables us to separate external sources of periodic and chaotic mode alternation from internal mode competition processes, thereby allowing us to investigate these phenomena in isolation and in detail.

11.3.5 Pattern Competition and Circling Optical Vortices

A second source of dynamics in the photorefractive oscillator is *mode beats*. These beats appear when two modes of different frequencies can oscillate simultaneously in the cavity. The superposition of these modes leads to a periodic temporal part in the intensity of the cavity signal, with the frequency of this period being the difference frequency of the competing patterns [19].

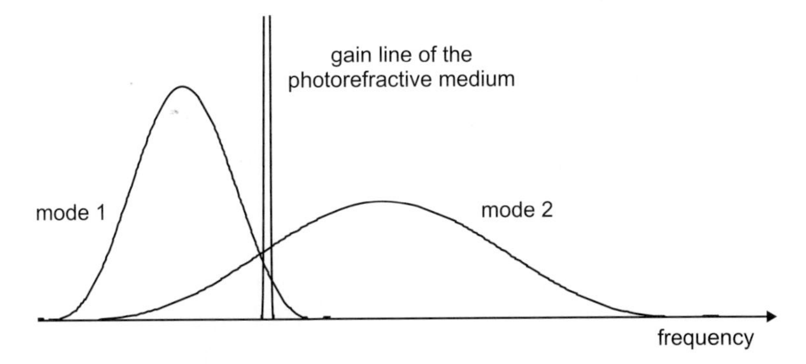

Fig. 11.19. Schematic explanation of the origin of mode beats: two modes of different resonant frequency can be excited simultaneously, owing to the strong mode pulling of the gain line

The origin of these mode beats is the significant difference between the linewidth of the emission of the resonator and the linewidth of the pump wave (see Fig. 11.1). Because the response time of the photorefractive medium is much larger than the decay time constant of the energy of the empty cavity, the resonances of the modes are much larger than the gain line of the photorefractive medium. The bandwidth of the modes is in the MHz range, whereas the amplification line of the photorefractive medium has a linewidth of less than 1 Hz. As a consequence, the frequencies of the cavity field are pulled in the direction of the pump frequency, an effect that is called *frequency pulling*.

When the photorefractive medium supplies sufficient gain, the threshold for oscillations can be very small. In this case it is possible to excite modes far away from the central, resonant frequency of the emission line. This allows two modes to oscillate simultaneously in a degenerate way. Figure 11.19 shows a case where mode 1 is excited at the frequency of the left, rising edge of the gain line, whereas mode 2 oscillates ar the frequency of the right, falling edge of the gain line. The resulting difference frequency corresponds to the width of the gain line.

Owing to our active stabilization, it was possible to induce defined changes in the length of the resonator that enabled us to change the relative positions of the cavity modes with respect to the gain line. By this means, it was possible to change the relative amplitudes of the two modes in such a way that several different mode beat scenarios could be obtained. An example of such a competition behavior of Laguerre and Hermite modes is shown in Fig. 11.20.

The patterns observed in lasers or active oscillators that support several transverse modes are characterized by the appearance of wavefront dislocations, or optical vortices. These are points where the intensity goes to zero,

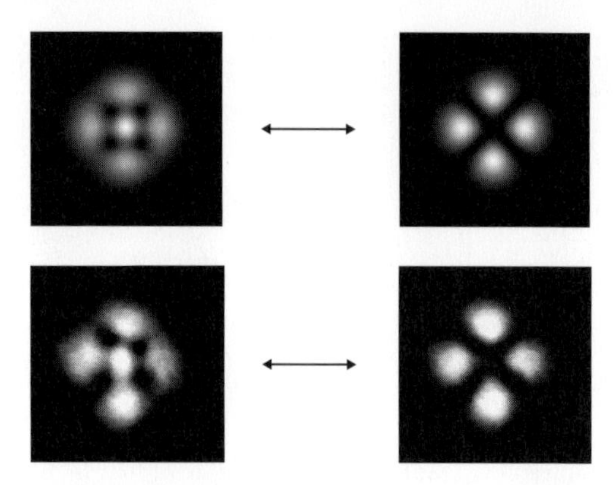

Fig. 11.20. Competition due to frequency pulling of two or more different modes into the gain line of the photorefractive medium. An L_{021} mode (*right*) competes with an L_{10} mode or a superposition of an H_{20} and an H_{01} mode (*left*). *Top row*, theory; *bottom row*, experiments

while the phase advance in a closed loop around the zero is a multiple of 2π. The multiple n is often called the charge of the vortex. Its sign indicates the helicity of the phase variation. Optical vortices are well known in linear optics and are also seen as the TEM_{01} or doughnut mode of a laser. Solutions of the Maxwell–Bloch equations have demonstrated, in a general context, the existence of dislocations in nonlinear optical media with gain also [21]. These dislocations have been observed in a number of experiments with lasers and photorefractive oscillators [5, 22, 23, 24] and have therefore been accepted as a useful means for gaining insight into the formation of spatio-temporal patterns in oscillating systems. We have used this technique in the work described below to analyze mode beat effects when two modes of different charge are excited at the same time.

The simultaneous excitation of a doughnut mode of topological charge m and a Gaussian mode or a singly charged doughnut mode reveals one of the most interesting competition effects for small Fresnel numbers. This competition results in a circular motion of vortices in the pattern. The angular speed of the rotation of the pattern is determined by the beat frequency of the interacting modes. The number of vortices is defined by the interacting modes. If the Gaussian TEM_{00} mode is involved, the central area on the optical axis has a high intensity. The number of circling vortices is then equal to the charge m of the doughnut mode. If, in contrast, the structure of the pattern is determined by a superposition of two doughnut modes of different charge, the number of vortices created depends on the sign of the charges. If a doughnut of charge 1 and another of charge m interact, charges of the same sign result in $|m|-1$ vortices, whereas charges of opposite sign result in $|m|+1$ vortices. The sense of rotation of a circling vortex depends on

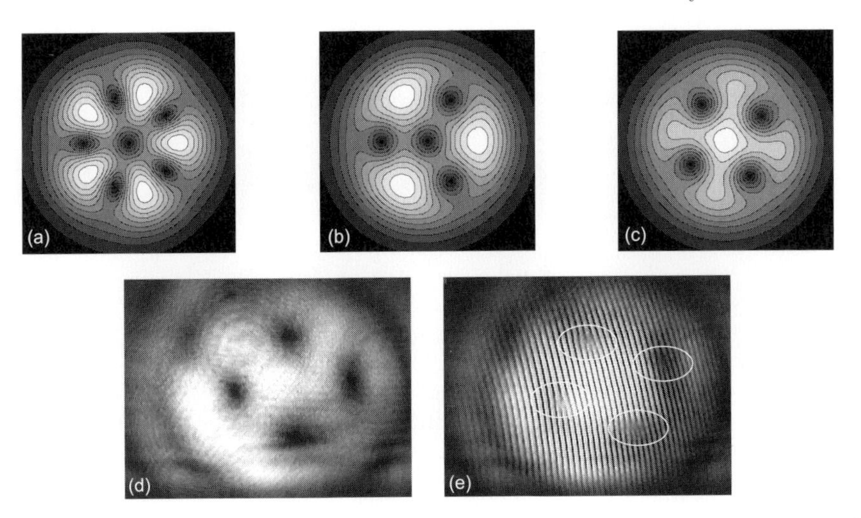

Fig. 11.21. Superposition of a doughnut mode of charge 4 with a doughnut mode of charge 1 (**a**), (**b**) or with the Gaussian TEM_{00} mode (**c**)–(**e**). (**a**)–(**c**) show theoretical calculations; in (**a**) the two doughnuts have charges of the same sign, whereas in (**b**) the charges are of opposite sign. (**d**) shows the experimental realization of this competition scenario. (**e**) shows an interferogram with four vortices. Because (**d**) and (**e**) were recorded at slightly different times, the vortices have rotated slighty

its helicity, or the sign of its charge. When two doughnut modes with $m = 1$ and $m = -3$ are superposed, the same pattern appears as in the case of two doughnut modes with $m = -1$ and $m = 3$. However, the sense of rotation of the vortices changes. In the first case, the sense of rotation is clockwise, whereas the second case results in a counterclockwise motion.

Therefore, optical vortices provide an ideal means to observe and analyze mode beats experimentally. In Fig. 11.21 an example of a superposition of a doughnut of charge 4 and the TEM_{00} mode is illustrated by theoretical calculations Figs. 11.21a–c and by experimental results from the photorefractive cavity Fig. 11.21d,e. The fact that the charges have the same sign can be seen clearly in the interferogram of the experimental pattern Fig. 11.21e. The superposition gives four circling vortices, in agreement with the theoretical calculations.

The interaction scenarios may become much more complex when modes of higher order contribute to the formation of circling vortices. Figure 11.22a shows nine vortices circling around the center of the mode. If two doughnut modes of higher order can be excited simultaneously, as shown in Fig. 11.22b, vortices may circle around two other stationary vortices. In the figure, seven vortices are circling around the center. Finally, the excitation of three modes at the same time can produce rotating vortices at different radii, leading to structures that become hard to investigate in a mode representation.

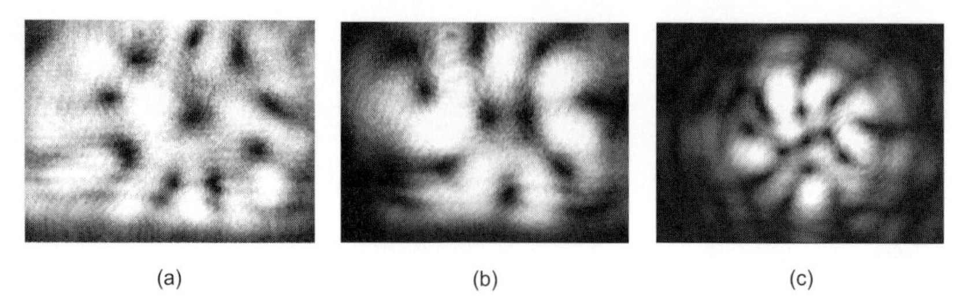

Fig. 11.22. Circling optical vortices due to the interaction of modes of higher order. (**a**) Circling of nine vortices around the optical axis due to the superposition of a doughnut mode with a higher-order mode; (**b**) excitation of two doughnut modes of higher order, resulting in seven circling vortices that circle around two stationary vortices; (**c**) excitation of three modes, leading to circling vortices at different radii. The windows of the transverse laser field in which these images were recorded have different sizes in each of the three images

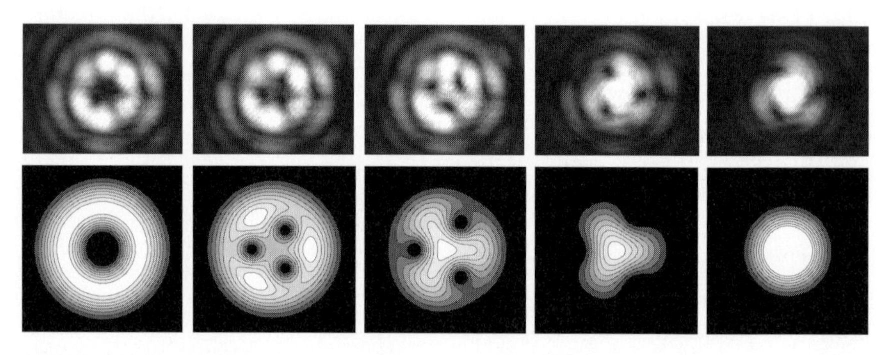

Fig. 11.23. Changes in the relative intensity distribution of a doughnut of charge 3 and a TEM_{00} mode due to changes in the oscillator length: experimental observations (*top*) and numerical calculations (*bottom*)

In the actively stabilized setup, these competition scenarios can be adjusted in a well-defined way. The radius of the circle containing the optical vortices depends on the relative amplitudes of the two modes. This ratio can be changed by changing the resonator length. As an example, Fig. 11.23 shows how a doughnut of charge $m = 3$ can be shifted towards the Gaussian mode in our photorefractive cavity and in theoretical calculations. Starting with the pure doughnut mode, all vortices are in the center and the radius of the vortices is zero. When the intensity of the TEM_{00} mode increases, the vortices are pushed outwards, until only the ground mode is left in the end. In this case, the radius of the vortices becomes infinite.

These results show that by use of an active stabilization, the pattern dynamics of a unidirectional photorefractive oscillator can be examined in a much more detailed and clear way. Characteristic mode dynamics that are due to cavity length changes can be clearly separated from mode-beating and interaction effects. By means of the active stabilization, the spontaneous pattern dynamics that result from the interaction of the nonlinearity with the propagation in the material and give rise to mode alternation, rotation or circling optical vortices can be varied and controlled in a definited way by varying the cavity length. Therefore, this method allows us to distinguish between the classically causal reasons for these effects and the effects of nonlinear pattern formation. A carefully aligned and stabilized oscillator shows no spontaneous pattern alternations, but exhibits the more interesting effects of mode beats and optical vortices. Our system is therefore an ideal means to seperate the nonlinear effects of mode pattern formation from other, classical effects with simple explanations.

11.3.6 Control of Competing Mode Patterns

If several different transverse modes can be excited within the gain line of the photorefractive oscillator, degenerate situations may appear. In this region, different tools for selecting one mode in preference to another are available. The first, straightforward tool is to exploit the oscillator geometry in order to select modes with a certain geometrical preference. This can be accomplished by introducing losses in a particular direction, thereby breaking the radial symmetry of the configuration. A similar effect can be induced by using nonradially symmetric mirrors or by introducing nonsymmetric optical elements into the oscillator. If lenses are used to produce this effect, the induced astigmatism is the origin of the selection of certain modes in the cavity. However, these techniques also imply a severe change in the cavity oscillation conditions, mainly by breaking the symmetry or inducing novel oscillation routes. Therefore, techniques that do not affect the cavity but still allow one to stabilize a desired mode pattern with respect to a second, degenerate mode pattern are required. Here, we shall focus on phase-only selection techniques based on the same idea as the minimally invasive method of pattern control in the single-feedback configuration (see Sect. 10.5).

The basis of this technique is to extract a small part of the degenerate signal from the oscillator and feed it back into the cavity after one or several symmetry operations , combined with phase shifts, have been applied to the pattern. One of the essential points of the technique is the exact reproduction of the spatial pattern into the oscillator, which includes the exact spatial extent as well as all spatial frequencies of the pattern. Therefore, we used a 1:1 imaging system in order to create this supplementary feedback loop.

To show the potential of the method, we have demonstrated the selection of symmetric and antisymmetric modes of the cavity. The symmetry classification of the modes is determined by the self-imaging features of the system when the mode is reflected in the central vertical axis. Symmetric modes do not change their transverse appearance under this transformation, whereas antisymmetric modes undergo an inversion of their field amplitude.

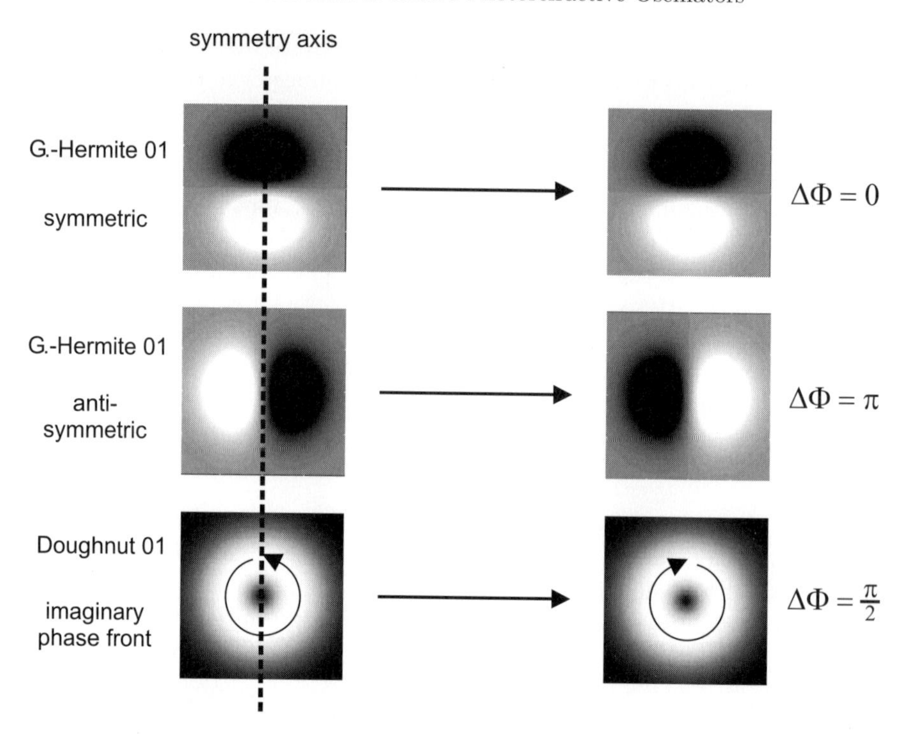

Fig. 11.24. A reflection in the vertical symmetry axis is equivalent to a phase shift that depends on the symmetry characteristic of the mode. This example shows modes of the first mode family. (Figure contributed by R. Nicolaus, Institute of Applied Physics, Darmstadt University of Technology)

This operation is equivalent to a phase shift by π of the whole transverse field, thereby allowing us to use phase shifts as a means to change the behavior of antisymmetric modes. An example is shown in Fig. 11.24, where three modes of the first mode family are subjected to a reflection in the vertical axis. The symmetric H_{01} mode does not change its field distribution, i.e. it undergoes a phase shift of 0, whereas the antisymmetric H_{01} mode is subject to an inversion of its field distribution because it has undergone a phase shift of π. The circularly symmetric doughnut mode has a spiraling phase front. Here, the mirror reflection induces a change in the rotation sense of the phase distribution, thereby changing the charge of the mode from $m = +1$ to $m = -1$. This operation is equivalent to a phase shift of $\pi/2$.

The experimental realization of such a control method is relatively simple. We just add an odd number of mirrors to the supplementary feedback arm, combined with a piezoelectric transducer. In this way, both the symmetry breaking and the adjustment of the transverse phase can be accomplished in the feedback arm. As in the case of the minimally invasive techniques

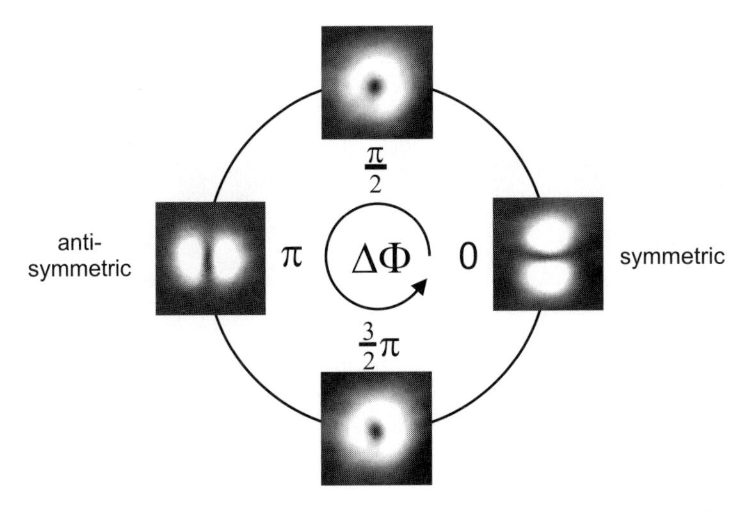

Fig. 11.25. Phase selection in the first-order mode family. Using a supplementary feedback loop that extracts only 2.6% of the intensity of the central cavity, the application of different phase shifts can cause a switch between different modes. (Figure contributed by R. Nicolaus, Institute of Applied Physics, Darmstadt University of Technology)

for single-feedback systems (see Sect. 10.5), we can tune the phase in such a way as to realize constructive interference in the feedback signal, thereby amplifying the field distribution of a mode, or to produce destructive interference, thereby damping the mode considerably. Because the oscillator cavity is not altered by this technique and only about 2.6% of the cavity intensity is coupled into the supplementary feedback arm, this technique can also be considered as minimally invasive.

Because the zeroth-order mode is the ground state, which is only one state, the first mode family that has degenerate patterns is the first mode family ($q = 1$); this familyy has a symmetric and an antisymmetric mode, as well as a doughnut mode of charge 1. If the oscillator is adjusted to a reasonably good rotational symmetry, all three modes can be excited and compete dynamically. If the supplementary control feedback loop is active, the degeneracy will be lifted, allowing one to enhance one mode with respect to the others. Therefore, changing the phase in the feedback arm allows one to stabilize and select each of the three modes (Fig. 11.25).

This technique can also be applied to higher mode families. However, because more modes become degenerate with increasing order, even when a symmetry-breaking operation is performed, only modes that have a unique symmetry characteristic can be effectively controlled. In this case, a combination of phase-dependent and geometry-dependent techniques can again be

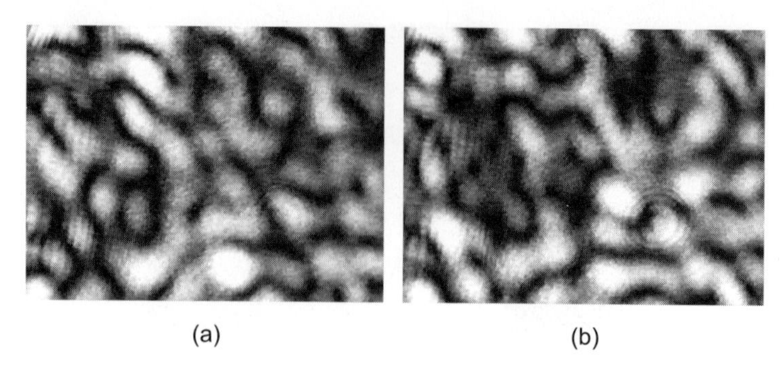

Fig. 11.26. Part of a transverse structure for a high Fresnel number around 1000. Images (**a**) and (**b**) have a temporal separation of 45 s, indicating the timescale of the dynamics in such a complex structure

used to obtain a defined selection of desired modes in a dynamic competition scenario.

11.3.7 Structures and Dynamics for High Fresnel Numbers

In all these investigations, the correspondence of the photorefractive oscillator to a class A laser was valid. Like a narrow-gain-line laser, a photorefractive oscillator can be tuned from one transverse-mode family to another by changing the resonator length. Patterns such as rotating vortices, astigmatism-induced pattern alternation (which is similar to induced drift instability in the case of spontaneous pattern formation due to an induced asymmetry in the system) and vortex patterns, which are typical of class A lasers, have also all been observed in our photorefractive oscillator.

What happens when the Fresnel number is increased to a region where many transverse modes can be active? For an intermediate Fresnel number, the influence of the boundaries decreases quickly, leading to the onset of several different regimes. First, temporal coherence is lost, but spatial coherence is still possible. This results in the formation of pure modes in the vicinity of the optical axis, but also in pronounced dynamic effects. For larger F, the spatial correlation is also lost. This situation is shown in Fig. 11.26. In this region, the pattern acquires a speckle-like nature, and spatial and temporal coherence are lost. Therefore, no spatial or temporal correlation can be found in the pattern any longer [25]. The correlation length decreases to the size of the small, grainy structures.

Moreover, complexity of the dynamics of the pattern increases dramatically. Although the dynamics occur on a timescale that is easy to observe owing to the slow response time of the photorefractive medium when the active stabilization is applied, the tracking of the various characteristic spatial structures becomes difficult. Figure 11.26 shows two different scenes that

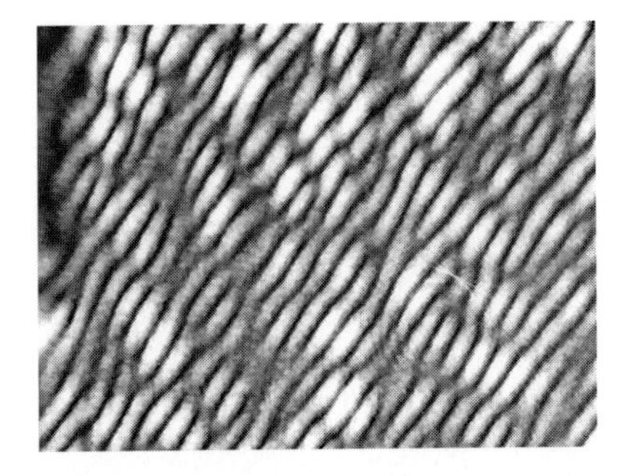

Fig. 11.27. Interferogram of the structure of Fig. 11.26 (for a high Fresnel number). Numerous vortices with different charges and signs can be identified, as can their birth and annihilation

are separated by a temporal spacing of 45 s. Referring to our considerations for the case of low Fresnel numbers, one can expect dynamics to be induced by several different mode beats. The number of mode beats is dramatically increased compared with the case of lower Fresnel numbers, because a much higher number of modes can now be excited simultaneously. However, the association of structures with particular modes becomes difficult, suggesting that we should use tools other than mode analysis for the analysis of these patterns.

A simple tool for the analysis of these patterns can be found in the behavior of the optical singularities, which are numerous in these structures. Figure 11.27 shows a structure similar to the one depicted in Fig. 11.26, this time with a plane reference wave superimposed on it in order to identify optical vortices. The number of vortices in this system is apparent. The sites of vortex formation and destruction are randomly distributed over the field. Vortices with higher charges can also easily be observed. The reduction of the influence of the boundary is also signaled by the reduction of the imbalance in topological charge. Since a regular field should have a balance between topological charges of opposite sign [26], an imbalance means that two phase singularities of opposite sign have been created close to the boundary and only one has remained within the boundary. This fact results in a scaling law of the number $U = \sqrt{N_1}/N$, the square root of the number N_1 of unbalanced charges divided by the number N of balanced charges, which scales as $F^{-1.5}$ below and $F^{-0.75}$ above the critical Fresnel number F_c. This relation was proved by Arecchi et al. [27]. Because the motion of the vortices and their mutual influence are characteristic for any given field, the classification of a resonator field by vortex statistics is possible. A general statistical approach using vortex dynamics can be found in [28].

In the framework of such a statistical approach, the transition of the structure that occurs when the aperature is varied from low to higher Fresnel numbers can be understood in the following way. It is possible to consider the

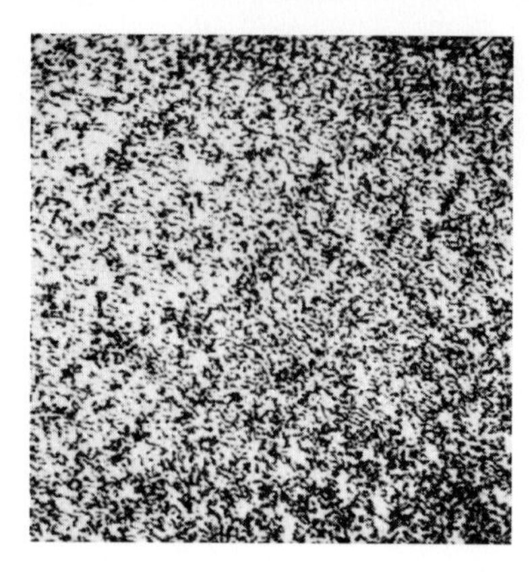

Fig. 11.28. Irregular, turbulent-like state of the resonator signal for a Fresnel number $F \approx 10^4$

photorefractive crystal as a collection of uncorrelated optical domains, each with a transverse size limited by the correlation length l_c intrinsic to the crystal excitations. The decorrelation mechanism in the photorefractive crystal is due to the amplification process based on diffusion of the space charge wave which provides the refractive index grating. The decorrelation length that we measured coincides with the diffusion length found by other authors [29]. The medium gain has an upper cutoff at a transverse wave number $1/l_c$, so that spatial details are amplified only up to that frequency; that is , these details are amplified provided that they are bigger than l_c. Thus, for a critical Fresnel number F_c that has a value such that the smallest scale is equal to the inverse cutoff wave number, we expect a transition from a boundary-dominated to a bulk-dominated regime, where the separation of the phase singularities becomes independent of F. In our analysis, we found this transition to occur at about $F \approx 50$. A completely uncorrelated self-oscillation pattern for an oscillator that is not bound by any artificial aperture, and has a Fresnel number of several thousand is shown in Fig. 11.28. In analogy to the uncorrelated-optical-domain picture of the transverse oscillator field, Hohenberg and Shraiman [6] showed that the spatio-temporal chaotic state can be seen as a coarse-grained set of uncorrelated "pixels", each of a size corresponding to the correlation length. Therefore, many authors interpret the patterns in an optical oscillator for high Fresnel number as turbulent patterns, in imitation of the classification in hydrodynamics [30]. This conclusion about the equivalence of this system to a hydrodynamic system is based mainly of the derivations on the scaling laws cited above. In fact, Arecchi et al. [27] showed that the local fluctuation field displays non-Gaussian statistics imposed by the complex nonlinear dynamics, whereas its spatial Fourier transform has a Gaussian distribution – a clear indication of spatio-temporal chaos.

For both cases of statistic distributions, the laser Ginzburg-Landau equation transforms into the complex Ginzburg-Landau equation. This is a completely diffractive equation, i.e. the coefficient of the Laplace operator is purely imaginary. An analysis of the complex Ginzburg–Landau equation allows one to calculate the radius of the cores of the vortices and thus the characteristic scale of the spatial structure in the case of turbulence. This radius is given by $x_0^2 \cong \lambda L/T$, independent of the aperture radius.

This result is qualitatively comparable to our experimental observations and those of other authors [30]: the spatial scale in the turbulent regime ceases to depend on the aperture radius when the aperture radius is larger than a certain critical value. This is equivalent to the statement that the pattern is no longer boundary-dependent [27], but bulk-dominated.

From another viewpoint, this effect can be explained equally well by diffusion of charges, which adds a real part to the coefficient of the Laplace operator in the corresponding complex Ginzburg-Landau equation. But the argument based on an infinite spatial scale may itself be sufficient to explain this behavior, without invoking particular material properties.

Diffusion tends to stabilize the solutions, and defect-mediated turbulence does not occur in a complex Ginzburg–Landau equation where diffusion prevails. The solutions of the complex Ginzburg–Landau equation have been shown to be modulationally stable [31] in this case. On the other hand, for a diffractive complex Ginzburg–Landau equation, instabilities occur at a lower pump intensity.

Thus, the common observation of turbulence-like structures in photorefractive oscillators would seem to indicate that these systems are more diffractive than diffusive and that material properties are not so important. This is supported by the fact that in crystals exhibiting diffusion in addition to drift phenomena, the same effects can be observed, so that we should attribute only a minor contribution to the effect of diffusive dynamics, and emphasize diffractive dynamics.

Summary

When the ability of nonlinear photorefractive materials to amplify light is exploited, a laser beam can be fed back into the material repeatedly, thus creating an active oscillation system. In this chapter, a unidirectional photorefractive feedback system has been investigated. This system can be shown to be formally equivalent to a certain class of lasers, the class A lasers, but operates with a considerably slower time constant. We have implemented an active stabilization technique that allows us to separate nonlinear transverse effects in the oscillator from external effects. The emphasis of this chapter was on the dependence of the patterns and their dynamics on the Fresnel number of the system, studied with the help of the active stabilization technique. For low Fresnel numbers, the patterns are dominated by the boundaries and are formed by simple cavity modes. Owing to the strong frequency pulling of photorefractive oscillators, mode competition becomes possible, resulting in circling optical vortices. This area allows us to implement novel control techniques for manipulating the transverse pattern by breaking the degeneracy using symmetry considerations and phase-sensitive control. For higher Fresnel numbers, the influence of the boundaries decrease and dynamic states dominate that are less correlated dominate. As a consequence, several tens of thousands of uncorrelated channels can be created by in self-oscillation. If this field can be separated by masks or by a pixelated saturable absorber into information channels, each representing the information in a single bit, large digital storage systems can be constructed in the resonator. Each channel can then be programmed from outside as described above, and will oscillate independently of the others.

References

1. R. Graham and H. Haken, *Laserlight – first example of a second-order phase transition far from thermal equilibrium*, Z. Phys. **237**, 31 (1970).
2. M. Brambilla, M. Cattaneo, L.A. Lugiato, R. Piravona, F. Parti, A.J. Kent, G.L. Oppo, A.B. Coates, C.O. Weiss, C. Green, E.J. D'Angelo and J.R. Tredicce, *Dynamical transverse laser patterns I*, Phys. Rev. A **49**, 1427 (1994).
3. A.B. Coates, C.O. Weiss, C. Green, E.J. D'Angelo, J.R. Tredicce, M. Brambilla, M. Cataneo, L.A. Lugiato, R. Pirovano, F. Prati, A.J. Kent and G.L. Oppo, *Dynamical transverse laser patterns II*, Phys. Rev. A **49**, 1452 (1994).
4. D. Hennequin, C. Lepers, E. Louvergneaux, D. Dangoisse and P. Glorieux, *Spatio-temporal dynamics of a weakly multimode CO_2 laser*, Opt. Comm. **93**, 318 (1992).
5. F.T. Arecchi, G. Giacomelli, P. Ramazza and S. Residori, *Experimental evidence of chaotic itinerancy and spatiotemporal chaos in optics*, Phys. Rev. Lett. **65**, 2531 (1990).
6. P.C. Hohenberg and B.I. Shraiman, *Chaotic behaviour of an extended system*, Physica D **37**, 109 (1989).
7. D.Z. Anderson and R. Saxena, *Theory of multimode operation of a unidirectional ring oscillator having photorefractive gain: weak-field limit*, J. Opt. Soc. Am. B **4**, 164 (1987).
8. G. D'Allesandro, *Spatiotemporal dynamics of a unidirectional ring oscillator with photorefractive gain*, Phys. Rev. A **46**, 2791 (1992).
9. B.M. Jost and B.E.A. Saleh, *Complex Ginzburg–Landau and extended Kuromoto–Sivashinksy equations for unidirectional photorefractive ring resonators*, Opt. Commun. **205**, 44 (1995).
10. B.M. Jost and B.E.A. Saleh, *Spatiotemporal dynamics of coupled-transverse-mode oscillations in unidirectional photorefractive ring resonators*, Phys. Rev. A **51**, 1539 (1995).
11. P. Günther and J.-P. Huignard. *Photorefractive Materials and their Applications II.* Springer, Berlin, 1989.

12. H. Kogelnik and T. Li, *Laser beams and resonators*, Appl. Opt. **5**, 1550 (1966).
13. F.T. Arecchi, S. Boccaletti, G.P. Puccioni, P. Ramazza and S. Residori, *Pattern formation and competition in photorefractive oscillators*, Chaos **4**, 491 (1994).
14. K. Staliunas, M.F.H. Tarroja, G. Slekyas, C.O. Weiss and L. Dambly, *Analogy between photorefractive oscillators and class A lasers*, Phys. Rev. A **51**, 4140 (1995).
15. B.M. Jost and B.E.A. Saleh, *Effect of using a $B_{12}SiO_{20}$ light amplifier on the formation and competition of modes in optical resonator*, J. Opt. Soc. Am. B **11**, 1864 (1994).
16. P. Mandel, M. Georgiou and T. Erneux, *Determination of the lifetime width of the argon l1-hole state*, Phys. Rev. A **47**, 1539 (1995).
17. G.-L. Oppo, G. D'Allesandro and W.J. Firth, *Spatiotemporal instabilities of lasers in models reduced via center manifold techniques*, Phys. Rev. A **44**, 4712 (1991).
18. L. Gil, K. Emilsson and G.L. Oppo, *Dynamics of spiral waves in a spatially inhomogeneous Hopf bifurcation*, Phys. Rev. A **45**, R567 (1992).
19. G. Balzer, C. Denz, O. Knaup and T. Tschudi, *Circling vortices and pattern dynamics in a unidirectional photorefractive ring oscillator*, Chaos, Solitons and Fractals **10**, 725 (1999).
20. Z. Chen and N. B. Abraham, *Pattern dynamics in a bidirectional photorefractive ring resonator*, Appl. Phys. B **60**, 183 (1995).
21. P. Coullet, L. Gil and F. Rocca, *Defect-mediated turbulence*, Opt. Commun. **73**, 403 (1989).
22. C.O. Weiss, *Spatio-temporal structures. Part II*, Phys. Rep. **219**, 311 (1992).
23. S.R. Liu and G. Indebetouw, *Periodic and chaotic spatiotemporal states in a phase-conjugate resonator using a photorefractive $BaTiO_3$ phase-conjugate mirror*, J. Opt. Soc. Am. B **8**, 1507 (1992).
24. D. Hennequin, L. Ambly, D. Dangoisse and P. Glorieux, *Basic transverse dynamics of a photorefractive oscillator*, J. Opt. Soc. Am. B **11**, 676 (1994).
25. R. Blumrich, T. Kobialka and T. Tschudi, *Behavior of the self-oscillation pattern in phase-conjugate ring resonator*, J. Opt. Soc. Am. B **7**, 2299–2305 (1990).
26. M.V. Berry, *Quantal phase factors accompanying adiabatic changes*, Proc. R. Soc. London A **392**, 45 (1984).
27. F.T. Arecchi, S. Boccaletti, G. Giacomelli, G.P. Puccioni, P. Ramazza and S. Residori, *Boundary-dominated versus bulk dominated regime in optical space–time complexity*, Int. J. Bif. Chaos **4**, 1281 (1994).
28. K. Staliunas, A. Berzanskis and V. Jarutis, *Vortex statistics in optical speckle fields*, Opt. Commun. **120**, 23 (1995).
29. F.T. Arecchi, S. Boccaletti, G. Giacomelli, G.P. Puccioni, P. Ramazza and S. Residori, *Patterns, space–time chaos and topological defencs in nonlinear optics*, Physica D **61**, 55 (1992).
30. F.T. Arecchi, S. Boccalett, P. Ramazza and S. Residori, *Transition from boundary- to bulk-controlled regimes in optical pattern formation*, Phys. Rev. Lett. **70**, 2277 (1993).
31. L. Gil P. Coullet and J. Lega, *Defect-mediated turbulence*, Phys. Rev. Lett. **62**, 1619 (1989).

12 Conclusion and Outlook

Having analyzed the complex mode structure in a unidirectional photorefractive oscillator, we have finished our passage through the various light–matter interactions in photorefractive media. We have started from the simplest interaction – the propagation of a single beam in a nonlinear medium – and proceeded to complex oscillator systems. At all stages, above a certain threshold the interaction results spontaneously in the formation of novel structures, which may be single entities such as solitons, or may be symmetric patterns or uncorrelated turbulent-like structures.

Throughout the book, we have focused our attention on the photorefractive nonlinearity, which has several attractive advantages that enable us to extract much more general information from the experiments than simple results on the pure photorefractive light–matter interaction. First, the photorefractive nonlinearity is saturable. Therefore, the nonlinear effects do not depend on the intensity itself, but on the ratio of the beam intensity to a background illumination or a pumping beam. This feature allows to realize all possible transverse-structure phenomena at very low laser intensities – generally a few microwatts – making them easily accessible in experiments and attractive candidates for applications. Second, the saturation behavior includes two stages. The first stage is a quasi-linear dependence of the refractive index on the intensity for low intensities of the signal relative to the pump beam (the undepleted-pump approximation). In this region, the photorefractive nonlinearity approaches the ideal Kerr nonlinearity. Therefore, may simple theoretical models based on a Kerr nonlinearity can be examined by using photorefractive systems operating in this regime. The second stage occurs at higher intensity ratios, Δn_{pr} becomes almost independent of the intensity ratio, and instead is quasi-constant. Therefore, the nonlinearity can reach significantly high values. For complete saturation, strongly uncorrelated, turbulent states that are often not accessible with nonlinearities that are intensity-dependent become accessible. Therefore, the work described here has used the photorefractive nonlinearity to access the whole range of transverse-pattern formation scenarios far from equilibrium described in Chap. 1, beginning from solitary-beam formation and symmetric spontaneous pattern formation and proceeding to mode competition and highly uncorrelated states in active unidirectional cavities. We think that the complexity of pattern formation, competition and transformation into uncorrelated structures shows that these systems are much more than pure examples of pattern formation – they are, at the same time ideal models for investigating the whole complexity of pattern formation on a timescale that is easily accessible.

Thus, on the one hand, our investigations have allowed us to obtain detailed insight into transversaly resolved light–matter interactions at various stages of complexity. The photorefractive nonlinearity allowed us to realize, for the first time, configurations that enabled us to investigate and verify various theoretical predictions from general soliton theory and from models of spontaneous pattern formation. On the other hand, we have also been able to find new, attractive fields of application such as soliton lattices and adaptive waveguides that may allow one to realize new concepts in nonlinear photonics. Both directions – the investigation of possible applications of photorefractive nonlinear systems and the investigation of general system features using the photorefractive nonlinearity as a model – are essential for developing the field of transverse-optical-pattern formation out of the stage of pure analysis into the stage of useful applications in the new area of optical information processing. Our investigations described throughout this book form building blocks that will allow us to implement new concepts in nonlinear photonics, which will probably change our view of optical information-processing systems considerably in the future.

Index

Springer Tracts in Modern Physics

148 **X-Ray Scattering from Soft-Matter Thin Films**
Materials Science and Basic Research
By M. Tolan 1999. 98 figs. IX, 197 pages

149 **High-Resolution X-Ray Scattering from Thin Films and Multilayers**
By V. Holý, U. Pietsch, and T. Baumbach 1999. 148 figs. XI, 256 pages

150 **QCD at HERA**
The Hadronic Final State in Deep Inelastic Scattering
By M. Kuhlen 1999. 99 figs. X, 172 pages

151 **Atomic Simulation of Electrooptic and Magnetooptic Oxide Materials**
By H. Donnerberg 1999. 45 figs. VIII, 205 pages

152 **Thermocapillary Convection in Models of Crystal Growth**
By H. Kuhlmann 1999. 101 figs. XVIII, 224 pages

153 **Neutral Kaons**
By R. Beluševi 1999. 67 figs. XII, 183 pages

154 **Applied RHEED**
Reflection High-Energy Electron Diffraction During Crystal Growth
By W. Braun 1999. 150 figs. IX, 222 pages

155 **High-Temperature-Superconductor Thin Films at Microwave Frequencies**
By M. Hein 1999. 134 figs. XIV, 395 pages

156 **Growth Processes and Surface Phase Equilibria in Molecular Beam Epitaxy**
By N.N. Ledentsov 1999. 17 figs. VIII, 84 pages

157 **Deposition of Diamond-Like Superhard Materials**
By W. Kulisch 1999. 60 figs. X, 191 pages

158 **Nonlinear Optics of Random Media**
Fractal Composites and Metal-Dielectric Films
By V.M. Shalaev 2000. 51 figs. XII, 158 pages

159 **Magnetic Dichroism in Core-Level Photoemission**
By K. Starke 2000. 64 figs. X, 136 pages

160 **Physics with Tau Leptons**
By A. Stahl 2000. 236 figs. VIII, 315 pages

161 **Semiclassical Theory of Mesoscopic Quantum Systems**
By K. Richter 2000. 50 figs. IX, 221 pages

162 **Electroweak Precision Tests at LEP**
By W. Hollik and G. Duckeck 2000. 60 figs. VIII, 161 pages

163 **Symmetries in Intermediate and High Energy Physics**
Ed. by A. Faessler, T.S. Kosmas, and G.K. Leontaris 2000. 96 figs. XVI, 316 pages

164 **Pattern Formation in Granular Materials**
By G.H. Ristow 2000. 83 figs. XIII, 161 pages

165 **Path Integral Quantization and Stochastic Quantization**
By M. Masujima 2000. 0 figs. XII, 282 pages

166 **Probing the Quantum Vacuum**
Pertubative Effective Action Approach in Quantum Electrodynamics and its Application
By W. Dittrich and H. Gies 2000. 16 figs. XI, 241 pages

167 **Photoelectric Properties and Applications of Low-Mobility Semiconductors**
By R. Könenkamp 2000. 57 figs. VIII, 100 pages

168 **Deep Inelastic Positron-Proton Scattering in the High-Momentum-Transfer Regime of HERA**
By U.F. Katz 2000. 96 figs. VIII, 237 pages

169 **Semiconductor Cavity Quantum Electrodynamics**
By Y. Yamamoto, T. Tassone, H. Cao 2000. 67 figs. VIII, 154 pages

Springer Tracts in Modern Physics